INDUSTRIAL ECOLOGY

T. E. Graedel
Distinguished Member of Technical Staff
AT&T Bell Laboratories

B. R. Allenby
Research Vice President, Technology and Environment
AT&T

Prentice Hall, Englewood Cliffs, New Jersey 07632

Graedel, T. E.
    Industrial ecology / T.E. Graedel, B.R. Allenby.
       p.   cm.
    Includes index.
    ISBN 0-13-125238-0
    1. Industry—Environmental aspects. 2. Product life cycle—
Environmental aspects. 3. Commercial products—Environmental
aspects.   I. Allenby, Braden R. II. Title.
TS149.G625   1995

94-17479
   CIP

Publisher: Marcia Horton
Project Manager: Mona Pompili
Cover Designer: DeLuca Design
Copy Editor: Peter Zurita
Manufacturing Buyer: Bill Scazzero
Supplements Editor: Alice Dworkin
Editorial Assistant: Delores Mars

© 1995 by AT&T
Published by Prentice Hall
A Simon & Schuster Company
Englewood Cliffs, New Jersey 07632

Printed in the United States of America

10 9 8 7 6 5 4 3

ISBN   0-13-125238-0

Prentice-Hall International (UK) Limited, *London*
Prentice-Hall of Australia Pty. Limited, *Sydney*
Prentice-Hall Canada, Inc., *Toronto*
Prentice-Hall Hispanoamericana, S.A., *Mexico*
Prentice-Hall of India Private Limited, *New Delhi*
Prentice-Hall of Japan, Inc., *Tokyo*
Simon & Schuster Asia Pte. Ltd, *Singapore*
Editora Prentice-Hall do Brasil, Ltda., *Rio de Janeiro*

To our children
Laura, Martha, Richard, and Kendra
in anticipation of a sustainable world

Contents

Foreword

We live in an increasingly complex age. We confront its quagmires daily, from shrinking ozone layers to the genetic engineers who would have us build not a better mousetrap, but a better mouse.

When it comes to settling such issues, to putting human values to science and technology, we may not be up to the task. We certainly won't be if we're depending on young minds masterfully schooled in a specialty, but blind to the ethical, environmental, and political implications of what they do.

It is with interest, then, that I have watched the development of the field of industrial ecology, a field that would nurture a new generation of engineers, scientists, business people, and public policy experts. With them we may enter a new century with an environmental ethic guiding each business decision.

The book is a pioneering effort in the development of this field. It acknowledges the interdisciplinary nature of the environmental issues we must confront and the link between technology and the environment. It speaks to the family ties we share as a world and the obligations that come with them. And it underscores not only the planet's frailty but also the urgency of our task.

Robert E. Allen
Chairman and Chief Executive Officer
AT&T Corporation

Preface

It has been a maxim of many years standing that the goals of industry are incompatible with the preservation and enhancement of the environment. It is unclear whether that maxim was ever true in the past, but there is certainly no question that it is untrue today. The more forward-looking corporations and the more forward-looking nations recognize that providing a suitable quality of life for Earth's citizens will involve not less industrial activity but more, not less reliance on new technologies but more, and that providing a sustainable world will require close attention to industry–environment interactions. This awareness of corporations, of citizens, and of governments promises to ensure that corporations that adopt responsible approaches to industrial activities will not only avoid problems but will benefit from their foresight.

Indeed, the involvement of industry is crucial if the world is to achieve sustainable development. Robert Sievers of the University of Colorado points out that governments have many critical short-term issues demanding their attention: achieving economic stability, feeding growing populations, establishing politically viable nations, dismantling nuclear arsenals, making the transition from centrally controlled to free-market economies, and so forth. The activities of many governments may thus be rather insular for the foreseeable future. At the same time, corporations are increasingly multinational, have longer time horizons, and depend for their survival and prosperity on relatively stable global business conditions and on responding to the desires and concerns of many different cultures and populations. Private firms, not governments, choose, develop, implement, and understand technology. Hence, responsible corporations may turn out to be among the global leaders in the transition between nonsustainable and sustainable development.

There are three time scales of significance in examining the interactions of industry and environment. The first is that of the past, and concerns itself almost entirely with remedies for dealing with inappropriate disposal of industrial wastes. The second time scale is that of the present, and deals largely with complying with regulations and with preventing the obvious mistakes of the past. Hence, it emphasizes waste minimization, avoidance of known toxic chemicals, and "end-of-pipe" control of emissions to air, water, and soil. Corporate environment and safety personnel are often involved, as are manufacturing personnel, in making small to modest changes to processes that have proved their worth over the years. In neither of these time scales do today's industrial process designers and engineers play a significant role.

The third time scale is that of the future. The industrial products and processes that are being designed and developed today will dictate a large fraction of the industry–environment interactions over the next decade. Thus, process and product design engineers hold much of the future of industry–environment interactions in their hands, and nearly all are favorably disposed toward designing with the environment in mind. Their problem is that doing so requires knowledge and perspective never given to them during their college or professional educations and not readily available in their current positions. Remedying this situation for the product and process design engineers of today and tomorrow is the primary focus of this book. We have also tried, however, to craft the book so that it will be a useful addition to curricula in policy, business, environmental science, law, and other related specialties, because it is equally important that students in those disciplines understand more about technology's role in mitigating or contributing to environmental problems.

Industrial ecology, which we define in the first chapter, is a very new way of thinking about economy–environment interactions. As applied to manufacturing, it requires familiarity with industrial activities, environmental processes, and societal interactions, a combination of specialties that is rare. Accordingly, we have purposely attempted to make this volume useful for those whose primary background is either industrial or environmental, or for those who interact with specialists in those disciplines. Many chapters include discussions of common industrial approaches to the issue being discussed as well as perspective on the environmental impacts produced. (Although we have tried to cite examples from across spectra of industries and countries, our familiarity with the electronics industry and with U.S. firms inevitably puts a bias in these examples; the reader should keep that limitation in mind.) For the latter, we emphasize impacts on long time and space scales, as shorter horizons are often better-recognized and better-regulated either internally or externally.

The book is divided into five parts. Part I, "Introductory Topics", provides a brief definition of the topic, outlines the approach of the book, and presents a short history of industry–environment interaction. Part II, "Relevant External Factors", describes the societal interactions that influence industrial technology. In Part III, "Evaluating Life Cycles", budgets are used to introduce life-cycle tools in industrial ecology. Part IV, "Design for Environment", treats specific topics related to the environmentally responsible design of products and processes. The final part, Part V, "Forward-Looking Topics", deals with techniques for incorporating industrial ecology into manufacturing activities and with visions for industry–environment interactions in the more distant future.

We have purposely mixed the philosophical with the practical. Avoiding that mix creates a textual product that is easier to write and easier to comprehend, but one that is much less relevant. The essence of industrial ecology is that it is the combination of technology with society, and that combination has many facets and many implications. The industrial ecologist needs to appreciate the societal interactions and to understand something of the interactions of industrial activity with the environment. Only at that point is there a logical framework in which to place goals and techniques.

We hope this book will find use as a text in university engineering curricula as well as in other academic programs dealing with environment–technology issues and policies. Industrial ecology as a discipline requires that all who are connected with it have some knowledge of each of its components, so we are reluctant to suggest that some parts of this book are for one curriculum and other parts for another. Indeed, we hope that business students will explore engineering checklists and that engineering students will study the implications of governmental and corporate policies. In addition to its usefulness to students, our experience suggests that practicing industrial engineers may find the volume a useful reference, and some of its features (such as the detailed appendices) were included partly with that audience in mind.

Industrial ecology is a subject that comes alive not only in the classroom, but especially on the factory floor. Thus, the student will benefit from immersing himself or herself in activities such as industrial ecology assessments of common products, manufacturing facility tours (as many and as diverse as possible), and industrial collaborations. Many opportunities will present themselves to innovative faculty members having good contacts with the local industrial community.

We are grateful to many people for their help during the preparation of this book. Portions of it are drawn from the Ph.D. dissertation of B. R. A., and we thank K. Keating, C. Uchrin, and J. Waldman of Rutgers University for advice and assistance. The following AT&T colleagues have provided information in many helpful discussions: W. Boyhan, B. Dambach, W. Glantschnig, R. A. Laudise, J. M. Morabito, G. C. Munie, C. K. N. Patel, N. L. Sbar, J. C. Sekutowski, J. C. Tully, and Y. Zaks. External reviews were performed by D. Allen, J. W. Bulkley, K. Callahan, R. Holt, I. Horkeby, D. R. Lynch, D. H. Marks, S. Morgan, B. Piasecki, R. E. Schuler, R. H. Socolow, D. Stein, and V. Thomas, and the book is much the better for their comments. For the use of illustrations and examples, we thank D. Benshoof (Best Lock Corporation), D. H. Moody (University of Michigan), I. Horkeby (Volvo Car Corporation), and T. McCarthy (Eli Lilly and Company). The support of R. White, President of the National Academy of Engineering, and B. Guile and D. Richards of the Academy staff, has been invaluable. We have appreciated our interactions with the staff at Prentice Hall, especially Marcia Horton, Delores Mars, and Mona Pompili, who have helped us turn a rough manuscript into what we feel is a most attractive book. Finally, we thank our managers at AT&T and at Bell Laboratories, particularly W. O. Baker, W. F. Brinkman, D. Chittick, R. S. Freund, R. A. Laudise, K. B. McAfee, E. McKeever, C. K. N. Patel, A. A. Penzias, L. Seifert, W. P. Slichter, and J. C. Tully, who have provided enthusiastic support and assistance over many years.

T. E. Graedel
B. R. Allenby

PART I: INTRODUCTORY TOPICS

<table>
<tr><td>CHAPTER
1</td><td># Humanity and Environment</td></tr>
</table>

1.1 OPENING COMMENTS

In 1968, Garrett Hardin of the University of California, Santa Barbara, published an article in *Science* magazine that has become more famous with each passing year. Hardin titled his article "The Tragedy of the Commons"; its principal argument was that a society that permitted perfect freedom of action in activities that adversely influenced common properties was eventually doomed to failure. Hardin cited as an example a community pasture area, used by any local herdsman who chooses to do so. Each herdsman, seeking to maximize his financial well-being, concludes independently that he should add animals to his herd. In doing so, he derives additional income from his larger herd but is only weakly influenced by the effects of overgrazing, at least in the short term. At some point, however, depending on the size and lushness of the common pasture and the increasing population of animals, the overgrazing destroys the pasture and disaster strikes all.

A modern version of the tragedy of the commons has been discussed by Harvey Brooks of Harvard University. Brooks points out that the convenience, privacy, and safety of travel by private automobile encourages each individual to drive to work, school, or stores. At low levels of traffic density, this is a perfectly logical approach to the demands of modern life. At some critical density, however, the road network commons is incapable of dealing with the traffic, and the smallest disruption (a stalled vehicle, a delivery truck, a minor accident) dooms drivers to minutes or hours of idleness, the exact opposite of what they had in mind. Examples of frequent collapse of road network commons systems are now legendary: Los Angeles, Tokyo, Naples, and on and on.

The common pasture and the common road network are examples of societal systems that are basically local in extent, and can be addressed by local societal action. In some cases, the same is true of portions of the environmental commons. Improper trash disposal or soot emissions from a combustion process are basically local problems, for example. Perturbations to water and air do not follow this pattern, however. The hydrosphere and the atmosphere are examples not of a "local commons", but of a "global commons", a system that can be altered by individuals the world over for their own gain, but, if abused, can injure all. Much of society's activities are embodied in industrial activity, and it is the relationship between industry and the environment, especially the global commons, that is the topic of this book.

It is undeniable that modern technology has provided enormous benefits to the world's peoples: a longer life span, increased mobility, decreased manual labor, and widespread literacy, to name a few. Nonetheless, there is growing concern about the relationships between industrial activity and Earth's environment. These concerns gather credence as we place some of the impacts in perspective. Since 1700, the volume of goods traded internationally has increased some 800 times. In the last 100 years, the world's industrial production has increased more than 100–fold. In the early 1900s, production of synthetic organic chemicals was minimal; today, it is over 225 billion pounds per year in the United States alone. Since 1900, the rate of global consumption of fossil fuel has increased by a factor of 50. What is important is not just the numbers themselves, but their magnitude and the relatively short historical time they represent.

Together with these obvious forcing functions, several underlying trends deserve attention. The first is the diminution of regional and global capacities to deal with anthropogenic emissions. For example, carbon dioxide production associated with human economic activity has grown dramatically (Fig. 1.1), largely because of extremely rapid growth in energy consumption. This pattern is in keeping with the evolution of the human economy to a more complex state, increasing growth in materials use and consumption, and an increased use of capital. The societal evolution has been accompanied by a shift in the form of energy consumed, which is increasingly electrical (secondary) as opposed to biomass or direct fossil fuel use (primary), the result being the now familiar exponential increase in atmospheric carbon dioxide since the beginning of the Industrial Revolution (Fig. 1.2). Thus, human activities appear to be rapidly consuming the ability of the atmosphere to act as a sink for the by-products of our economic practices.

Human population growth is, of course, a major factor fueling this explosive industrial growth and expanded use and consumption of materials. Since 1970, human population has grown 8-fold: it is now approximately 5.3 billion and is anticipated to peak at between 10 and 15 billion late in the twenty-first century. It is generally recognized that human population growth has exhibited this exponential growth since the Industrial Revolution; what is not frequently realized, but is critical, is how closely human population growth patterns are tied to technological and cultural evolution. As Fig. 1.3 shows, the three great jumps in human population have accompanied the initial development of tool use, the agricultural revolution, and the Industrial Revolution. The Industrial Revolution actually consisted of both a technological revolution and a "neo-

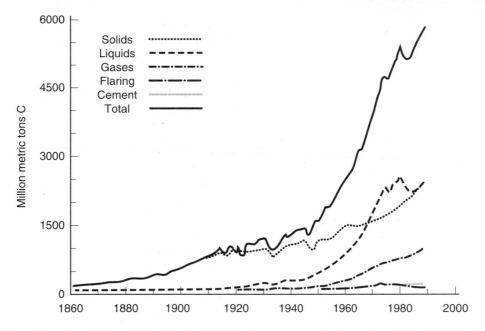

Figure 1.1   Global CO_2 emissions from fossil fuel burning, cement production, and gas flaring, 1860–1988. (T. A. Boden, P. Kanciruk, and M. P. Farrell, *Trends '90, A Compendium of Data on Global Change*, Report ORNL/CDIAC-36, p. 89, Oak Ridge, TN: Oak Ridge National Laboratory, 1990.)

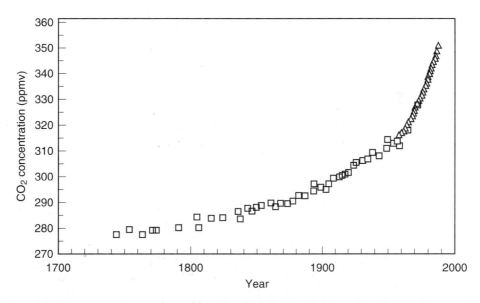

Figure 1.2   The increase in atmospheric carbon dioxide since 1700. (J. T. Houghton, G. J. Jenkins, and J. J. Ephraums, eds., *Climate Change: The IPCC Scientific Assessment.* Cambridge, UK: Cambridge University Press, 1990.)

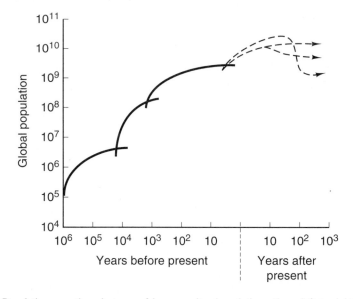

Figure 1.3   Population growth and stages of human cultural evolution. From left to right, the historical stages are tool use, the agricultural revolution, and the Industrial Revolution. The fourth (present and future) stage, shown in dashed lines, is that of ubiquitous technology and substantial environmental impact. Three possible scenarios for the fourth stage are pictured: one in which population stability is achieved by gradual, ordered approaches; one in which a reduced population stability occurs through a directed program of decreased use of technology; and one in which it is achieved by unmanaged growth followed by an unmanaged crash. (Based on E. S. Deevey, Jr., The human population. *Scientific American, 203*(3) (1960): 194–206; and M. G. Wolman, The impact of man. *EOS- Trans. AGU, 71* (1990): 1884–1886.)

agricultural" revolution (the advent of modern agricultural practices), which created what appeared to be unlimited resources for population growth. Our current population levels, economies, and cultures are inextricably linked to how we use, process, dispose of, and recover or recycle natural and synthetic materials and energy, and the innumerable products made from them.

The foregoing discussion demonstrates that the planet and its population are far from a steady state and clearly on an unsustainable path. Two possible routes toward long-term stability can be postulated: (1) a managed reduction of growth until a long-term sustainable population/technology/cultural steady state (which we will call "carrying capacity") is achieved, or (2) an unmanaged crash of one or more of the parameters (population, culture, technology) until stability at some undesirable low level is approached. Figure 1.3 suggests such possibilities.

This perspective has significant implications. When we objectively view the recent past—and 200 years is recent even in terms of human cultural evolution, and certainly in terms of our biological evolution—one fact becomes clear: The Industrial Revolution as we now know it is not sustainable. We cannot keep using materials and resources the way we are now. But what are we to do?

1.2 THE MASTER EQUATION

A useful way to focus thinking on the most efficient response that society can make to environmental stresses is to examine the predominant factors involved in generating those stresses. As is obvious, the stresses on many aspects of the Earth system are strongly influenced by the needs of the population that must be provided for, and by the standard of living that population desires. One of the more famous expressions of these driving forces is provided by the "master equation":

$$\text{Environmental impact} = \text{population} \times \frac{\text{GDP}}{\text{person}} \times \frac{\text{environmental impact}}{\text{unit of per capita GDP}} \quad (1.1)$$

where GDP is a country's gross domestic product, a measure of industrial and economic activity. Let us examine the three terms in this equation and their probable change with time.

Both Earth's population and the rate of population change are increasing rapidly, as seen in Fig. 1.4 (an expansion of the most recent part of Fig. 1.3). For a specific geographical region (city, country, continent), the rate of population change is given by

$$R = (R_b - R_d) + (R_i - R_e) \quad (1.2)$$

where the subscripts refer to birth, death, immigration, and emigration, respectively. Different factors can dominate the equation during periods of high birth rates, war, enhanced migration, plague, and the like. For the world as a whole, of course, $R_i = R_e = 0$. Given the rate of change, the population at a future time can be predicted by

$$P = P_0 e^{Rt} \quad (1.3)$$

where P_0 is the present population, t is the number of years in the projection, and R is expressed as a fraction. If R remains constant, the equation predicts an infinite population if one looks far enough into the future. Such a scenario is obviously impossible; at some point in the future, R will have to approach zero or go negative and the population growth will thus be adjusted accordingly.

The time and height of Earth's eventual human population peak are both quite uncertain and are strongly dependent on whether or not governments choose to encourage or discourage fertility as well as the (unknown) carrying capacity of the planet. Even in the mildest reasonable scenario, however, a global population of twice the present level is anticipated, and higher levels yet are not thought outside the realm of responsible prediction.

The second term in Eq. (1.1), the per capita gross domestic product, varies substantially among different countries and regions, responding to the forces of local and global economic conditions, the stage of historical and technological development, governmental factors, weather, and so forth. The general trend, however, is positive, as seen in Table 1.1. This table is related to the aspiration of humans for a better life. Although

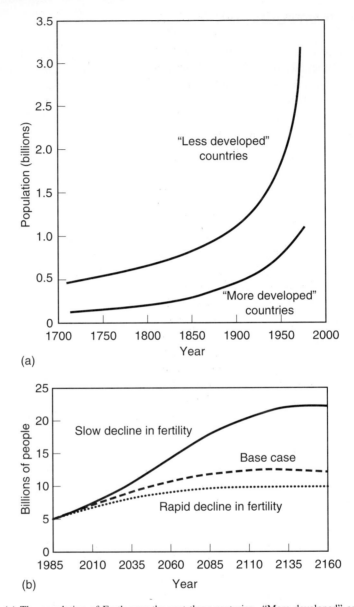

Figure 1.4 (a) The population of Earth over the past three centuries. "More developed" countries are the United States, Canada, Europe (west and east), the former USSR, Japan, Australia, and New Zealand. (T. E. Graedel and P. J. Crutzen, *Atmospheric Change: An Earth System Perspective*. New York: W. H. Freeman, 1992.) (b) Scenarios for world population development under different fertility scenarios. The base case supposes that countries with high and nondeclining fertility levels will begin the transition toward lower fertility by the year 2005 and undergo a substantial decline over the following 40 years. All countries reach replacement fertility levels by 2060. In the rapid decline case, countries not yet in transition toward lower fertility begin the transition immediately. For countries already in transition, total fertility declines at twice the rate of the base case. In the slow decline case, transition toward lower fertility begins after 2020 in most low-income countries. For countries in transition, declines are half the rate for the base case. (Reproduced from *World Development Report 1992*, Copyright 1992 by The International Bank for Reconstruction and Development/ The World Bank. Reprinted by permission of Oxford University Press, Inc.)

Table 1.1   Growth of Real per Capita Income in More Developed and Less Developed Countries, 1960–2000 *

Country Group	1960–1970	1970–1980	1980–1990	1990–2000
More Developed Countries	4.1	2.4	2.4	2.1
Sub-Saharan Africa	0.6	0.9	−0.9	0.3
East Asia	3.6	4.6	6.3	5.7
Latin America	2.5	3.1	−0.5	2.2
Eastern Europe	5.2	5.4	0.9	1.6
Less Developed Countries	3.9	3.7	2.2	3.6

* The figures are average annual percentage changes; and for "Less Developed Countries," entries are weighted by population. The 1990–2000 figures are estimated.
Source: Data from The World Bank, *World Development Report 1992.* Oxford, UK: Oxford University Press, 1992.

GDP and quality of life may not be directly related, we can expect GDP growth to continue, particularly in developing countries.

The third term in the master equation, the degree of environmental impact per unit of per capita gross domestic product, is an expression of the degree to which technology is available to permit development without serious environmental consequences and the degree to which that available technology is deployed. The typical pattern followed by nations participating in the Industrial Revolution of the eighteenth and nineteenth centuries is shown in Fig. 1.5. The abscissa can be divided into three segments: the

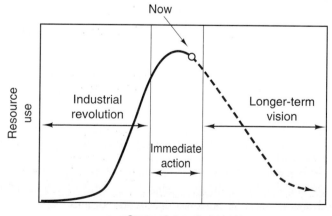

Figure 1.5   The typical life cycle of the relationship between the state of technological development of society and its resulting environmental impact.

unconstrained Industrial Revolution, during which the levels of resource use and waste increased very rapidly; the period of immediate remedial action, in which the most egregious examples of excess were addressed; and the period of the longer-term vision (not yet implemented) in which one can postulate that environmental impacts will be reduced to small or even negligible proportions while a reasonably high quality of life is maintained.

Of the trends for the three terms of the master equation, the one that perhaps has the greatest degree of support for its continuation is the second, the gradual improvement of the human standard of living, defined in the broadest of terms. The first term, population growth, is not primarily a technological issue but a social issue. Although different countries approach the issue differently, the upward trend is clearly strong. The third term, the amount of environmental impact per unit of output, is primarily a technological term, though societal and economic issues provide strong constraints to changing it rapidly and dramatically. It is this third term in the equation that offers the greatest hope for a transition to sustainable development, and it is modifying this term that is the central tenet of industrial ecology.

1.3   INDUSTRIAL ECOLOGY: THE CONCEPT

No firm exists in a vacuum. Every industrial activity is linked to thousands of other transactions and activities and to their environmental impacts. A large firm manufacturing high-technology/low-material products will have tens of thousands of suppliers located all around the world and changing on a daily basis. It may manufacture and offer for sale hundreds of thousands of individual products to myriads of customers, each with her or his own needs and cultural characteristics. Each customer, in turn, may treat the product very differently, a consideration when use and maintenance of the product may be a source of potential environmental impact (e.g., used oil from automobiles). When finally disposed of, the product may end up in almost any country, in a high-technology landfill, an incinerator, beside a road, or in a river that supplies drinking water to local populations.

In such a complex circumstance, how has industry approached its relationships with the outside world? Satisfying the needs of its customers has always been well done. Industry has, however, been less adept at identifying some of the long-term consequences of the ways in which it goes about satisfying needs. Examples of a few of these interactions have been collected by James Wei of Princeton University; we adapt and display that information here as Table 1.2. The table indicates the difficulties created for society in a world in which industrial operations are perceived as essentially unrelated to the wider world.

It is important to note that the relationships in Table 1.2 were not the result of disdain for the external world by industry. Several of the solutions were, in fact, great improvements over the practices they replaced, and their eventual consequences could not have been forecast with any precision. What was missing, however, was any attempt to relate the techniques for satisfying customer needs to any possible environmental con-

Table 1.2   Relating Current Environmental Problems to Industrial Responses
to Yesterday's Need

Yesterday's Need	Yesterday's Solution	Today's Problem
Nontoxic, non-flammable refrigerants	Chlorofluoro-carbons	Ozone hole
Automobile engine knock	Tetraethyl lead	Lead in air and soil
Locusts, malaria	DDT	Adverse effects on birds, mammals
Fertilizer to aid food production	Nitrogen and phosphorus fertilizer	Lake and estuary eutrophication

sequences. Although making such attempts does not ensure that no deleterious impacts will result from industrial activity, these actions have the potential to avoid the most egregious of the impacts and to contribute toward incremental changes in the impacts that are now occurring or can be well forecast. How are such attempts best made?

The approach to industry-environment interactions that is described in this book to aid in evaluating and minimizing impacts is called "industrial ecology" (IE). Industrial ecology as applied in manufacturing involves the design of industrial processes and products from the dual perspectives of product competitiveness and environmental interactions. Our view is that such an approach to manufacturing is sustainable over the long term if properly defined and executed, continually updated in light of new data and understanding, and properly supported by enlightened government policies. This systems-oriented vision accepts the premise that industrial design and manufacturing processes are not performed in isolation from their surroundings, but rather are influenced by them and, in turn, have influence on them.

In Chapter 7, we will present an extensive definition of industrial ecology that uses a biological analogy to describe the perspective from which IE views an industrial system. The essence of IE can, however, be briefly stated:

> Industrial ecology is the means by which humanity can deliberately and rationally approach and maintain a desirable carrying capacity, given continued economic, cultural, and technological evolution. The concept requires that an industrial system be viewed not in isolation from its surrounding systems, but in concert with them. It is a systems view in which one seeks to optimize the total materials cycle from virgin material, to finished material, to component, to product, to obsolete product, and to ultimate disposal. Factors to be optimized include resources, energy, and capital.

In this definition, the emphasis on *deliberate* and *rational* differentiates the industrial-ecology path from unplanned, precipitous, and perhaps quite costly and disruptive alternatives. By the same token, *desirable* indicates the goal that industrial ecology practices will support a sustainable world with a high quality of life for all, as opposed to, for example, an alternative where population levels are controlled by famine.

One of the most important concepts of industrial ecology is that, like the biological system, it rejects the concept of waste. Dictionaries define waste as useless or worthless material. In nature, however, nothing is eternally discarded; in various ways, all materials are reused, generally with great efficiency. Nature has adopted this approach because acquiring these materials from their reservoirs is costly in terms of energy and resources, and thus something to be avoided whenever possible. In our industrial world, discarding materials wrested from the Earth System at great cost is also generally unwise. Hence, materials and products that are obsolete should be termed *residues* rather than *wastes*, and it should be recognized that wastes are merely residues that our economy has not yet learned to use efficiently. We will sometimes use the term *wastes* in this book where the context refers to material that is or has been discarded, but we encourage the use instead of the term *residues*, thereby calling attention to the societal value contained in obsolete products of all sizes and types.

A full consideration of industrial ecology would include the entire scope of economic activity, such as mining, agriculture, forestry, manufacturing, and consumer behavior. However, it is obviously impossible to cover the full scope of industrial ecology in one volume, especially an introductory text. Accordingly, we limit the discussion in most of this book to manufacturing activities, although some of the final chapters explore the subject in more general terms. That industrial ecology encompasses all human activity should not be forgotten, however, and most of the principles we discuss are equally applicable to other aspects of the subject.

1.4 LINKING INDUSTRIAL ECOLOGY AND ENVIRONMENTAL SCIENCE

Global budgets, natural and perturbed, demonstrate in a graphic way the connection between biological ecology and industrial ecology. The contrast between traditional environmental approaches to industrial activity and those suggested by industrial ecology can be demonstrated by considering several time scales and types of activity, as shown in Table 1.3. The first topic, remediation, deals with such things as removing toxic chemicals from soil. It concerns past mistakes and is very costly. The second topic, treatment, storage, and disposal, deals with the proper handling of residual streams from today's industrial operations. The costs are embedded in the price of doing busi-

Table 1.3 Aspects of Industry-Environment Interactions

Activity	Time Focus
Remediation	Past
Treatment, storage, disposal	Present
Industrial ecology	Future

ness, but contribute little or nothing to corporate success except to prevent criminal actions and tort suits. Neither of these activities is industrial ecology. In contrast, industrial ecology deals with practices of the future, and seeks to guide industry to cost-effective methods of operation that will render benign its interactions with the environment and will optimize the entire manufacturing process for the general good, and, we believe, for the financial good of the corporation. Corporate executives are familiar with the liabilities of past and present industry–environment interactions. A challenge to the industrial ecologist is to demonstrate that viewing this interaction from the perspective of the future is an industrial asset, not a liability.

Industrial ecology implies more than just optimizing its own materials budgets, because its interactions with the environment require input and guidance from the environmental science community. We might picture the interaction as shown in Fig. 1.6. On the left side of the diagram are the linkages of *industrial metabolism*: those involving suppliers and supplied, manufacturers and customers, and so forth. To the right are the linkages of *environmental metabolism*: relationships between biota and the atmosphere, interactions between estuarine waters and sea water, and so forth. An elementary form of industrial ecology occurs when engineers strive for efficiency and for minimal external interactions within the industrial system. An advanced form occurs when their

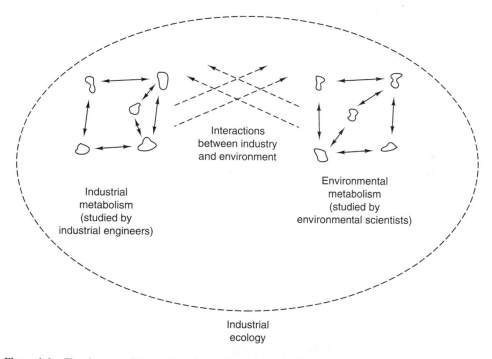

Figure 1.6   The elements of industrial ecology. The irregular shapes denote reservoirs for materials and the solid lines flow among them. The dashed lines suggest the sharing of concepts and information. (Adapted from a diagram devised by R. H. Socolow, Princeton University.)

efforts are guided by environmental scientists, thus integrating considerations from both realms. Industrial ecology thus embodies the linkages of industrial activities with environmental science. The goal is to make industrial decisions today that will be viewed with favor 20 or 30 years from now.

We began this chapter by discussing the "tragedy of the commons", in which a large number of individual actions, clearly beneficial over the short term to those making the decisions, eventually overwhelm a common resource and produce tragedy for all. Originally formulated to describe such local commons resources as public grazing lands, the concept was extended to global resources with discoveries such as the Antarctic ozone hole.

Industrial processes and products interact with many different commons regimes, and design engineers should interpret the concept of the commons in a very broad way. The concept certainly includes local venues such as city air and natural habitats. Regional resources are also included, groundwater and precipitation (and their possible chemical alteration) being examples. The global commons call other regimes to mind: the deep oceans (can oil spills or ocean dumping significantly degrade this resource, or is humanity's influence modest?), the Antarctic continent (thirty years of international scientific activity without environmental controls have left a dubious legacy), and, of course, the atmosphere (and its ozone and climate). Industrial ecologists should think even one step further—to the near-space environment, where decades of debris disposal threatens to result in collisions costly in both spacecraft performance and investment funds, and where clever design can minimize debris generation and perhaps employ recycling as well.

In practice, every product and process interacts with at least one commons regime and probably with several, fragile and robust, monitored and ignored, close and distant. Analyses of industrial designs may produce very different results if the same product or process is to be used under the sea, in the Arctic, or on a spacecraft rather than in a factory. Just as industrial ecology extends to all parts of the life cycle, it extends as well to all commons regimes it might affect. In short, we define *the commons* as everywhere that the human influence might be felt.

1.5 AN OVERVIEW OF THIS BOOK

This book is directed toward codifying and explicating ways in which to transform industrial activity from what is largely a nonsustainable system to resemble more and more closely a sustainable system. This is done by dividing the information into sections, as shown in Fig. 1.7. The first box, introductory topics, is composed of the information in this and several following chapters. The descriptions are intended to set the stage by examining trends and patterns of industrial development and environmental impact, particularly those places where the links between industry and environment are readily apparent.

Once the concept of industrial ecology is in hand, we move on to its evaluation for specific products, a family of methods generally called "life-cycle assessment" (LCA).

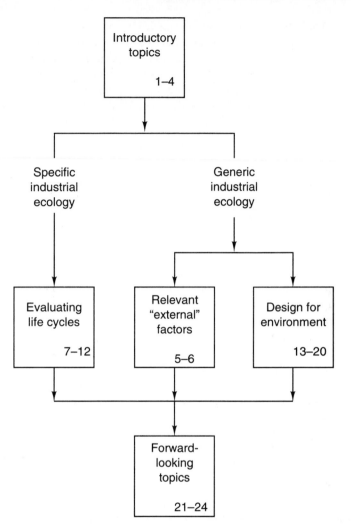

Figure 1.7   The structural outline of this book. Each box refers to the book chapters indicated.

LCA can be thought of as a *specific protocol*, in that in its optimum embodiment, it determines where products, processes, or facilities are less than environmentally meritorious and ranks the environmental impacts of those specific situations. LCA is somewhat fragmented and embryonic as we write, and we have chosen to describe aspects of several assessment systems that seem to us logical and useful. At present, and perhaps in perpetuity, there is no single best approach to life-cycle assessment, though one might hope that international study groups and standards organizations will begin to impose uniformity.

After the materials flows and environmental impacts of a product or a process have been assessed, the next step is to design a strategy for improvement. The techniques for doing so are generically called "design for environment" (DFE). DFE approaches can be designated *generic*, in that they refer to generally meritorious actions or considerations, and encourage their incorporation into product or process design activities in much the same vein as would be considerations of flow magnitudes or materials costs. Societal factors are involved in many of the decisions faced by industrial ecologists, and two chapters treat these issues.

The final part of the book deals with forward-looking topics. Many changes are on the horizon of the industry–environment relationship. Depending on how they are played out, they have the potential to modify many of the detailed recommendations of DFE, though not its overriding philosophy. Our hope is that the speculations we make will serve as food for thought and perhaps food for action.

SUGGESTED READING

Ausubel, J. H., and H. E. Sladovich, eds. *Technology and Environment*, Washington, DC: National Academy Press, 1989.

Clark, W. C., and R. E. Munn, eds. *Sustainable Development of the Biosphere*, Cambridge, UK: Cambridge University Press, 1986.

Hardin, G. The tragedy of the commons. *Science, 162* (1968): 1243–1248.

Henry, J. G., and G. W. Heinke. *Environmental Science and Engineering*. Englewood Cliffs, NJ: Prentice Hall, 1989.

Robey, B., S. O. Rutstein, and L. Morris. The fertility decline in developing countries. *Scientific American, 269*(6), (1993), 60–67.

Scientific American special issues: "Managing Planet Earth," Vol. 261, No. 3, September 1989, and "Energy for Planet Earth," Vol. 263, No. 3, September 1990.

World Commission on Environment and Development. *Our Common Future*. Oxford, UK: Oxford University Press, 1987.

EXERCISES

1.1 In 1983, the birth rate in Ireland was 19.0 per 1000 population per year, and the death rate, immigration rate, and emigration rate (same units) were 9.3, 2.7, and 11.5, respectively. Compute the overall rate of population change.

1.2 If the rate of population change for Ireland were to be stable from 1990 to 2005 at the rate computed in Ex. 1.1, compute the 2005 population. (The 1990 population was 3.72 million.)

1.3 Repeat Ex. 1.2 for the situation where the unrest in Northern Ireland is substantially modulated in 1998 and the emigration rate drops by 50%.

1.4 Using the master equation, the "Units of Measurement" section at the back of the book, and the following data, compute the 1990 GNP per capita and equivalent CO_2 emissions per equivalent U.S. dollar of GNP for each country shown in Table 1.4.

Table 1.4　1990 Master Equation Data for Five Countries

Country	Population[*]	GNP[†]	% of Global $CO_2^{\ddagger}$
Brazil	150	434,700	3.93
China	1134	419,500	9.12
India	853	262,400	4.18
Nigeria	109	27,500	1.14
United States	250	5,200,800	17.81

[*] Millions.

[†] Million equivalent U.S. dollars.

[‡] Percentage of total global greenhouse gas emissions, expressed in "equivalent CO_2 units". ECO_2 is computed by using gas heating effects and lifetimes to adjust emission fluxes for several of the infrared-absorbing gases. The global flux of ECO_2 is 13.15 Pg/yr.

Source: Data for this table were drawn primarily from *The 1993 Information Please Environmental Almanac*, compiled by World Resources Institute. Boston, Ma.: Houghton Mifflin, 1993.

1.5　Trends in population, GNP, and technology are estimated periodically by many institutions. Using the typical trend predictions in Table 1.5, compute the equivalent CO_2 anticipated for the years 2000 and 2025 for the five countries. Graph the answers, together with information from 1990 (previous exercise), on an ECO_2-vs.-year plot. Comment on the results.

Table 1.5　Master Equation Predicted Data for Five Countries

Country	Population[*] 2000	Population[*] 2025	GNP growth (%/yr) 1990–2000	GNP growth (%/yr) 2000–2025	Decrease in ECO_2/GNP(%/yr)
Brazil	175	240	3.6	2.8	0.5
China	1290	1600	5.5	4.0	1.0
India	990	1425	4.7	3.7	0.2
Nigeria	148	250	3.2	2.4	0.1
United States	270	307	2.4	1.7	0.7

[*] Millions.

Source: Data were drawn primarily from J. T. Houghton, B. A. Callander, and S. K. Varney, *Climate Change 1992*. Cambridge, UK: Cambridge University Press, 1992.

Technology and Industry: History and Recent Trends

2.1 THE BEGINNINGS OF INDUSTRIAL ACTIVITY

Industry is defined as the commercial production and sale of goods and services. By that definition, industry has existed since the first transactions, probably involving agricultural goods. By the second millenium B.C., copper and bronze working was supplementing agriculture in the industrial arena and artisans were crafting and selling jewelry and armaments. The ability to create pottery vessels was also an early achievement of civilization. By the time of Christ, a variety of metals capable of being worked at modest temperatures were being used to fashion small tools, weapons, and ornaments. None of these activities was performed on a scale large enough to have any significant global environmental impact, although there are indications that exploitation of local resources by agriculture contributed to the decline of a number of urban centers during this period.

In the Middle Ages, the growth in population led to the need for more agricultural land, and industrial impacts on the environment began with slash-and-burn agriculture and other exploitation of soils and forest resources in China, Greece, Crete, Roman North Africa, Mexico, and elsewhere. By about the beginning of the eleventh century, China was beginning to develop a technological capacity for the manufacture of such items as wrought iron, explosives, and textiles. Over the next few centuries, notable technological innovations occurred: the hinged ship rudder (ca. 1200), plate armor (ca. 1300), the casting of iron cannon (ca. 1550), and the production of sulfuric acid (ca. 1600). In all cases, however, the products were made by individuals or small groups in very labor-intensive ways. As a consequence, industrial activity was not widespread and was approached from the standpoint of the careful craftsperson, not the resource exploiter. Little prospect existed for global environmental despoliation, nor does archaeology or natural history reveal any.

2.2 THE INDUSTRIAL REVOLUTION

The Industrial Revolution began in about 1750 with the advent of several nearly simultaneous technological innovations. It was based on coal and iron. Coal provided the power for the newly invented steam engines and the energy to produce the iron. The abundant availability of iron permitted the manufacture of high-quality tools and machines, the building of improved bridges and ships, and, ultimately, the mechanization of agriculture and thus the transformation from agrarian life to urban life for a sizable portion of Earth's population. The Industrial Revolution was particularly successful in societies, first in Europe and then in North America, that were able to bring together within a small geographical region the necessary scientific and technical knowledge with substantial quantities of energy, work force, materials, and capital.

Perhaps the best measure of the success of the Industrial Revolution is the vast increase in labor productivity that resulted. Figure 2.1 shows a compilation of relative labor productivity for several countries from the time of the invention of the steam

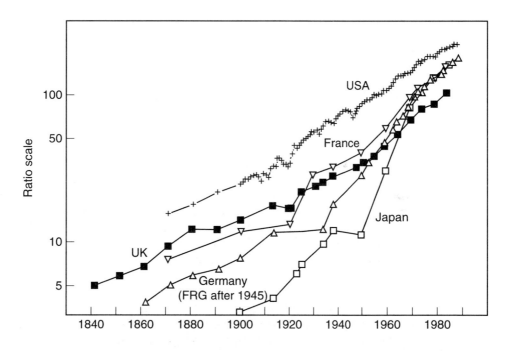

Figure 2.1   The growth in manufacturing labor productivity for five countries over the past century and a half. International comparisons of labor productivity are difficult, and it is perhaps most useful to look at the relative evolution of productivity for individual countries rather than to place reliance on comparitive numerical values. The ordinate ratio is in units of equivalent U.S. dollars per work hour. (Adapted with permission from A. Grübler, Preprint: Industrialization as a historical phenomenon, to be published in *Industrial Ecology and Global Change*, R. Socolow, C. Andrews, F. Berkhout, and V. Thomas, eds., Cambridge, UK: Cambridge University Press, 1994.)

engine to the present. The productivity increases are about a factor of 20 over this period.

   As industrialization developed, metals other than iron were needed and techniques were developed to recover them from their ores. Copper was particularly important because of its high electrical conductivity, and was produced in large quantity in the latter half of the nineteenth century. In the present century, the production of zinc, aluminum, and other metals from their ores has become common.

2.3 MODERN INDUSTRIAL OPERATIONS

Modern industrial technology has made major gains by implementing a succession of process changes in the major industrial sectors. In a productivity illustration that could be repeated for a number of industries with only minor variations (chiefly in the time scale), manufacturing processes for steel are shown in Fig. 2.2. The figure demonstrates that new processes tended to supplant the old on time scales of 30 or 40 years, each process being considerably more efficient than the one it replaced. Processes often have rather different characteristics, however. In the iron and steel industry, for example, the basic oxygen furnace (BOF) and electric arc furnace (EAF) are, to some extent,

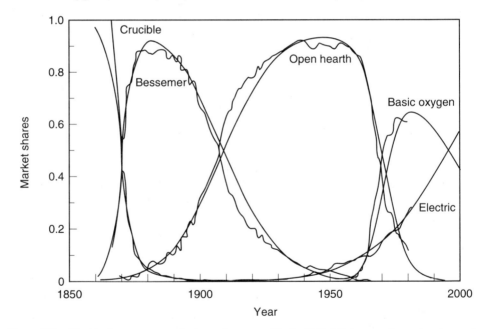

Figure 2.2   Changes in processes used to manufacture steel in the United States over the past century and a half. The jagged lines are historical data and the smooth curves are estimates from coupled equation models. (After N. Nakićenović, Dynamics of change and long waves, in *Life Cycles and Long Waves*, T. Vasco, R. Ayres, and L. Fontvielle, eds., pp. 147–192, Lecture Notes in Economics and Mathematical Systems. Berlin: Springer Verlag, 1990.)

complementary technologies. The BOF is used for making steel from pig iron, drawing energy from the reduction of carbon in the pig iron and thus requiring only about 22 MJ/metric ton of steel. However, the BOF can take only about 30% scrap in its charge and still work effectively. The EAF can use all scrap in its charge, but takes about 370 MJ/metric ton of steel in energy input. Overall, about 50% of all steel production is from recycled material, so energy-intensive recycling processes are required if this level of reuse is to be maintained or increased.

The advances of the Industrial Revolution did not come without price. Of the mass of materials and ore residues extracted from the ground for conversion into products, more than 90% was discarded without being used. Concomitantly, the first large-scale use of energy and the first large-scale emission of gases and particles into the atmosphere occurred. The relative efficiency of energy use is measured by a parameter called the *industrial energy intensity*, expressed in energy units per monetary unit of value added. Different countries show radically different patterns in energy intensity with time, as seen in Fig. 2.3. The general trend is toward decreasing energy intensity, with decreases of about a factor of 3 over the 30 years shown in the figure. In the cases of Brazil and Nigeria, however, energy intensity increased by about a factor of 2 from 1970 to 1985.

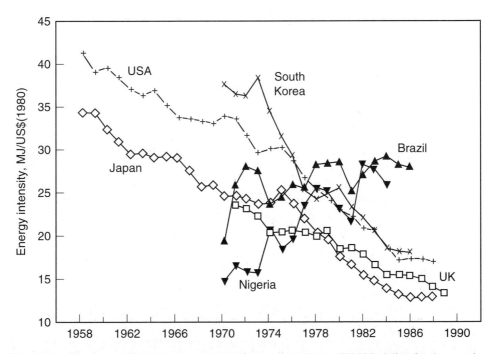

Figure 2.3   Changes in industrial energy intensity (expressed in MJ per 1980 U.S. dollar) for six countries over the past several decades. (Adapted with permission from A. Grübler, Preprint: Industrialization as a historical phenomenon, to be published in *Industrial Ecology and Global Change*, R. Socolow, C. Andrews, F. Berkhout, and V. Thomas, eds. Cambridge, UK: Cambridge University Press, 1994.)

The energy-intensity patterns of Fig. 2.3 are reflected in the *carbon-intensity* patterns for the same countries, shown in Fig. 2.4. Carbon intensity measures not energy used, but carbon atoms combusted to produce that energy; it thus plays out differently for countries only burning coal as opposed to countries with substantial hydropower. As it turns out, however, the same countries whose energy intensity decreased since the 1950s show substantial decreases also in carbon intensity, plotted against industrial value added. From the standpoint of minimization of waste generation, their industrial operations became two to three times as efficient over that period. In contrast, Brazil and Nigeria increased carbon use by about a factor of 2 without at the same time adding a significant amount of value to their industrial products. This figure is perhaps as good a simple summary as one can provide of how to practice industrial ecology and how not to practice it; it also reflects the profligate use of energy and resources that characterizes less developed, as opposed to more developed, economies.

Industrial emissions continue today at a high rate. It is interesting that the processes that began the Industrial Revolution more than two centuries ago—the generation of energy by the combustion of fossil fuels and the smelting of ores to recover metals—

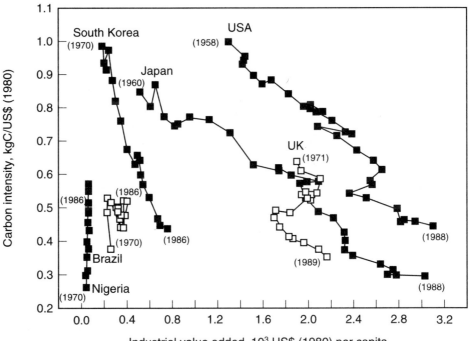

Figure 2.4   The evolution of industrial carbon intensity (expressed as kg C per 1980 U.S. dollar) as a function of the per capita level of industrialization. (Reproduced with permission from A. Grübler, Preprint: Industrialization as a historical phenomenon, to be published in *Industrial Ecology and Global Change*, R. Socolow, C. Andrews, F. Berkhout, and V. Thomas, eds., Cambridge, UK: Cambridge University Press, 1994.)

are still processes that have major impacts on the atmosphere of Earth, over long and short distance scales and over long and short time scales.

2.4   STATE OF DEVELOPMENT AND QUALITY OF LIFE

Some interesting and intricate relationships exist among levels of per capita income and indicators of environmental quality. A graphical representation of several of these is given in Fig. 2.5. The figure demonstrates a clear correlation between low income and the probability of being without safe water or adequate sanitation. Low-income people tend to live in countries whose industrial development is not far along; as a result, their generation of municipal wastes and energy-related pollutants such as carbon dioxide is low, but so is their ability to manage the residuals that are produced. Conversely, high-

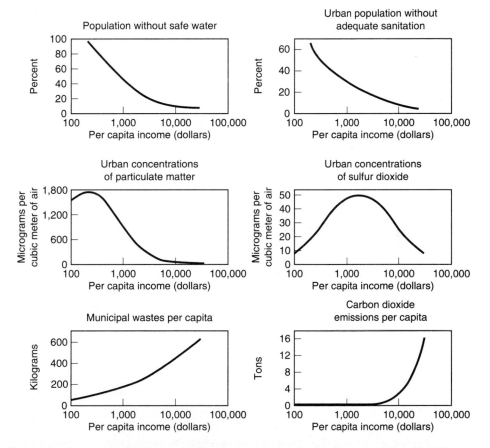

Figure 2.5   Environmental indicators at different country income levels. (Reproduced from *World Development Report 1992*. Copyright 1992 by The International Bank for Reconstruction and Development/The World Bank. Reprinted by permission of Oxford University Press, Inc.)

income populations have few problems with water supplies or sanitation, but produce large amounts of wastes and energy-related pollutants.

The measures of environmental quality that are related to industrial development rather than to personal activities tend to be least satisfactory not at low- or high-income levels but at intermediate ones. This apparent anomaly occurs because the rapid increase in average incomes that occurs in the early stages of industrial development is lagged by the implementation of sound environmental policies, usually because of an exclusive focus on economic growth. Hence, humans at each income level, low, moderate, and high, are responsible for, or subject to, unsatisfactory environmental problems under the current pattern of industrial development.

2.5 TRENDS IN TECHNOLOGY

An increasingly evident feature of industrial products is *dematerialization*, that is, using less material while providing the same or better service or function. Examples abound, as in the substitution of smaller, stronger, and lighter plastic composites for metal support beams, the design and manufacture of small home stereo systems that produce sound quality and volume equivalent to much larger, older systems, or the vast increase in computational power provided by today's ever smaller, lighter computers and workstations. As a consequence of such efforts, the use of engineering materials, as exemplified but not restricted to the cases of the metals shown in Fig. 2.6, has been even further below the 1970 projections of "experts". In countries where the use of smaller amounts of material coincided with efforts at energy conservation for residential and other uses, the pattern for energy consumption was similar to that for metal use for nearly two decades.

A second trend in technology that complements the use of lesser quantities of materials is the *substitution* of more environmentally suitable materials for less desirable but currently used alternatives. Technological innovation often permits enhanced performance to be realized as well. The situation for nonfossil fuel sources of energy is shown in Fig. 2.7. Here we see that rapid progress in photovoltaic and solar–thermal technologies suggests that those forms of energy generation will be economically competitive with fossil fuel and nuclear generation early in the twenty-first century, and that they will soon be economically superior to peak-load natural gas. Many assumptions are inherent in such predictions, of course, but they demonstrate the great progress that can be made in the substitution of more benign technologies for those that may have less desirable environmental impacts.

A third powerful trend in technology, remarked upon earlier in this chapter, is change in the pattern of energy intensity with time, called *decarbonization*. In the past three decades, high energy prices and enabling technology combined to reduce energy intensity by greater than half in the more developed countries. Most energy in the United States is supplied by the combustion of fossil fuels, so this decrease in energy intensity implies a decrease in pollution as well. Where government policies are used to

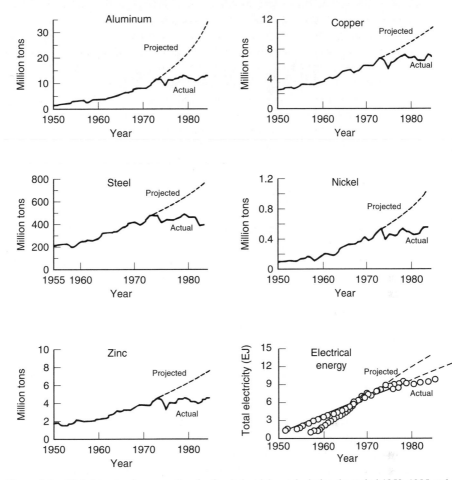

Figure 2.6   Global levels of consumption for five industrial metals during the period 1950–1985 and projections made in 1970 for 1970–1985 (J. E. Tilton, *Long-Run Growth in World Metal Demand: An Interim Report*, Mineral Economics and Policy Program. Golden, CO: Colorado School of Mines, 1987) and levels of U.S. electrical energy use for the same period (C. Starr, Implications of continuing electrification, in *Energy: Production, Consumption, and Consequences*, J. L. Helm, ed. pp. 52–71. Washington, DC: National Academy Press, 1990).

encourage energy conservation, as in Japan, energy intensity can be expected to continue this rapid decrease.

A final aspect of technology as an agent of change in industrial ecology is the computerization of information and technology. This process permits regular monitoring of technological operations, and also helps ensure that industrial processes that were originally designed on industrial-ecology principles can be maintained as designed, avoiding

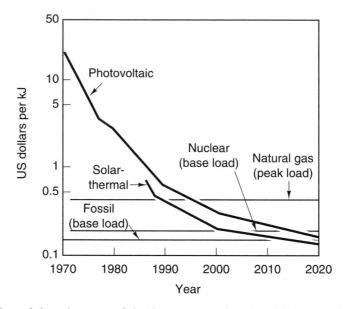

Figure 2.7   Costs of alternative means of electric power generation. Actual data are used for 1970–1990, estimates for 1990–2020. The photovoltaic estimate excludes storage costs and assumes high levels of solar insolation. The solar cell–thermal energy storage estimate includes storage costs and assumes high levels of solar insolation. The fossil fuel estimate is for natural gas and coal. Reference costs for three traditional sources of electric power are indicated. (Reproduced from *World Development Report 1992*. Copyright 1992 by The International Bank for Reconstruction and Development/The World Bank. Reprinted by permission of Oxford University Press, Inc.)

such problems as leaks, shutdown and startup problems, material-flow uncertainties, and the like.

These technology trends provide examples of changing the third term in the the master equation (Eq. 1.1, Chap. 1) for the better. A few cautions are advisable, however. One is that the pace of technological advance cannot be predicted, and is inevitably irregular. The second is that the diffusion of technology, once that technology is developed, is a strong function of government policy and societal preference. A recent policy initiative appropriate to this discussion is a section of the Convention on Climate Change signed in 1992 in Rio de Janiero at the United Nations Conference on Environment and Development:

> The developed country Parties ... shall take all practicable steps to promote, facilitate and finance, as appropriate, the transfer of, or access to, environmentally sound technologies and know-how to other Parties, particularly developing country Parties, to enable them to implement the provisions of the Convention. In this process, the developed country Parties shall support the development and enhancement of endogenous capacities and technologies of developing country Parties.

The uneven record of more developed country efforts to encourage industrialization in less developed countries by sporadic funding of technology transfer caution that a universal level of technology is more easily discussed than accomplished.

A third caution is the yet unmet challenge of developing policies that encourage the evolution of environmentally desirable technologies when it is impossible to predict a priori what those technologies will or should be.

2.6   TECHNOLOGY-ENVIRONMENT INTERACTIONS

2.6.1   Biomass Combustion

The combustion of "biomass" (fuelwood, brush, vegetation, grass, etc.) is a rich source of atmospheric emissions, and is a topic that is receiving much increased study. The species emitted comprise, besides carbon dioxide, a whole group of chemically reactive gases including carbon monoxide, oxides of nitrogen, methane, methyl chloride, various other hydrocarbons, and particulate matter. Different types of burning and different biomass materials produce variable amounts of other trace substances.

Biomass burning is sometimes a natural process, as in fires started by lightning, but is mostly anthropogenic. It is principally conducted in the tropics during the dry season to remove forest material and thus clear land for agriculture and grazing, to burn agricultural residues and dry savanna grasses, and, of course, for heating and cooking. Agricultural burning was still very common as late as the nineteenth century in Europe and North America, where it was probably the main source of air pollution at that time.

The current rate of biomass burning is rather uncertain, but is thought to be in the range of $2-5 \times 10^{15}$ g C/yr, mostly from tropical latitudes. Of far less quantitative importance are fires in temperate and boreal forests. (Boreal forests are those in far northern latitudes.) In the temperate regions, especially the United States, Canada, and Australia, extensive efforts have been made over the past century to control natural forest fires. For example, it is estimated that the area burned annually in the United States decreased by a factor of 10 between 1930 and 1980.

2.6.2   Crop Production

Crop production involves changes in land use and in soil chemistry. As such, it has the potential to affect the atmosphere. Perhaps the greatest potential effect has to do with carbon dioxide, which is released to the atmosphere when soil organic matter is oxidized by tilling.

Certain crops, especially rice, grow under anaerobic conditions and produce non-oxygenated gases as byproducts. Methane emission is of the greatest concern because its flux is relatively large. Not only has the total area in rice production increased over the past half century, but multiple cropping has increased the effective area under cultivation. As a consequence, rice production appears to have contributed significantly to the observed atmospheric increase in methane.

The extent to which our economic activity has managed to transform land from useful to degraded is seldom realized. The World Resources Institute estimates that nearly 2000 million hectares of land were so transformed between 1945 and 1990. The major causes include chemical degradation (nutrient loss, salinization, pollution) and physical loss (compaction, waterlogging, runoff).

Tilling of the soil may lead to substantial but not very well quantified emission of nitrous oxide, as implied from recent studies showing that 2.7% of the fixed nitrogen lost in drained, cultivated soils in south Florida appeared as N_2O. Fertilization of cultivated land is a common practice that results in the emission of N_2O. So far, field measurements suggest that this source of N_2O is not very large, but measurements have been made at temperate latitudes only; tropical soils may behave differently. This potential source for atmospheric N_2O is one that may be expected to grow with the globally increasing use of fertilizer containing nitrogen.

As the land used to grow crops has increased, and as agricultural practices have permitted land to be farmed more intensively, emissions to the atmosphere from these activities have increased apace. It is unfortunately the case that reliable data on world land use, from which areas devoted to crop production could be derived, are generally unavailable on a global scale before about 1950. It has been estimated, however, that during each of the periods 1860–1919 and 1920–1978, the land areas converted to regular cropping grew by about 400 million hectares, an area larger than the size of India. Whereas the land area *per person* devoted to the production of grain has decreased over the years as a consequence of population increases and improvements in agricultural efficiency, the use of fertilizer and energy per unit of production has increased sharply.

2.6.3   Domestic Animals

The animal kingdom is a source of several atmospheric trace gases. Methane is produced by fermentation in the guts of animals and by the digestion of wood by termites. Although difficult to quantify in detail, the fluxes from the undomesticated animal kingdom are small compared with other methane sources, whereas those from domestic ruminant animals are very significant, accounting for some 15% of the total methane source. In addition to methane, animals produce ammonia, hydrogen sulfide, and other products of excrement that affect air, water, and soil quality. The tendency for domestic animals to be gathered together in feedlots and pastures in the developed countries is thought to cause substantial local impacts, especially water pollution and the corrosion of metals, though these effects are very poorly quantified.

The number and mass of undomesticated animals on Earth have almost certainly decreased in recent times because of the steady expansion of human influence. The effects of this decrease on atmospheric emissions are, however, overwhelmed by the growth in domestic animal populations and the concomitant conversion of natural lands, especially forests, to pasture and agricultural lands under highly productive cultivation. Since 1950, the numbers of domestic animals has increased dramatically, with proportional increases in the emissions of methane and ammonia. (Cattle populations were estimated at 650 million worldwide in 1930 and 1350 million in 1975, for example.)

Because the consumption of higher levels of meat is a central characteristic of the diet of more highly industrialized countries, the rapid increase in the number of domestic animals may be expected to continue as long as the supply of grain does not limit their growth in numbers.

2.6.4   Coal Production and Use

The use of coal has traditionally been very directly related to industrialization and home heating, and has risen rapidly in the last century as nations increasingly joined the Industrial Revolution. (Figure 1.1, Chap. 1, shows the history of coal use for about the last 120 years. The scale used on the vertical axis is the amount of carbon dioxide emitted by the combustion process, a normalizing factor that makes it easy to compare the atmospheric effects of different energy sources.)

   The importance of coal combustion as a source of atmospheric trace gases will almost certainly grow with time, because the world's reserves of coal are very large in comparison with those of petroleum and natural gas. In addition, coal is generally the least expensive reliable source of energy. It is widely available, for example, in India and China, two of the world's most rapidly industrializing economies. However, coal's desirable characteristics are counterbalanced by the fact that it is the most detrimental energy source from the standpoint of atmospheric impacts. This unfortunate combination of properties means that coal combustion will be one of the most serious problems in planning for the long-term sustainable development of the biosphere. The result may be that future use of coal will be permitted only with a precleaning step to remove sulfur and other trace constituents, or by coal gasification. Should either of those be implemented, the cost advantage coal now possesses over other forms of fuel will diminish or disappear.

   Coal use results in large and diverse environmental effects.. At the earliest stage, trapped methane escapes during the mining and processing of coal, and mining activities and residues often result in the acidification of local surface waters. Later, during combustion, coal is a major source of CO_2, CO, hydrocarbons, NO and NO_2, SO_2, and soot. In addition, coal combustion produces hydrogen chloride, ammonia, and several trace metals, including mercury. In general, the combustion of coal produces less energy per unit of fuel weight than does that of petroleum or natural gas, and, as coal has a higher impurity content, its emissions per unit of produced energy are larger.

2.6.5   Petroleum Production and Use

Negligible amounts of petroleum were used as an energy source prior to the twentieth century, but petroleum use since then has grown rapidly (Fig. 1.1), especially from motor vehicles. The most dramatic feature of the figure is that in about 1975 petroleum combustion passed coal combustion as a source of carbon dioxide. This characteristic will probably continue for the next century, until world reserves of petroleum become sufficiently depleted that it ceases to be a major energy source.

The extraction of petroleum from its reservoirs in the planet's crust can result in large amounts of drilling waste, and the volatilization of gases, particularly methane. Petroleum's greatest environmental impact, however, occurs during its combustion. The primary gaseous emittant is carbon dioxide, but incomplete combustion produces carbon monoxide and a wide variety of hydrocarbon compounds as well. In addition to these carbon-based products, the high-temperature combustion of petroleum inevitably produces various oxides of nitrogen from the combustion of nitrogen impurities and from the oxygen and nitrogen in the air, which partially dissociate at the high combustion temperatures. NO and NO_2 are highly active catalysts in most atmospheric chemical chains, because they are involved in vital oxidizing reactions leading to ozone in the troposphere.

The sulfur content of petroleum varies widely and, as a result, so do the emissions of sulfur dioxide. In highly industrialized regions, these emissions can be a major cause of acidic precipitation and the subsequent degradation of ecosystems and materials. Sulfur dioxide and sulfates are also important species in aerosol and cloud chemistry.

Petroleum burned in inefficient combustors can be a prolific producer of carbon soot, which is involved in visibility degradation and in condensed-phase atmospheric chemistry. Carbon soot may also influence Earth's radiation budget by absorption of sunlight, a process especially noticeable during the early spring in the Arctic.

2.6.6   Natural Gas Production and Use

The third energy source whose emissions are quantified in Fig. 1.1, Chapter 1, is natural gas. This gas, which can also be recovered as a by-product of the production of petroleum, saw little use prior to the first quarter of the twentieth century, but its utilization for the production of energy has risen rapidly since that time. Natural gas tends to burn much more cleanly than coal or petroleum, with only carbon dioxide, and perhaps methane, being an emittant of great concern. Per unit of energy produced, the CO_2 emissions from natural gas are smaller than those from oil, which in turn are smaller than those from coal.

As with petroleum, Earth's supply of recoverable natural gas is limited. It may be largely used up by the middle of the next century, ceasing to be an important factor in atmospheric impact assessments.

2.6.7   Disposal of Residues

Solid and semisolid industrial residues have traditionally been discarded in landfills, the more modern of which are regularly covered with soil to minimize odor and the dispersion of debris. The leaching of various undesirable contaminants from landfills has the potential to degrade surface and subsurface water supplies, though modern landfills use liner and buffer techniques to guard against such problems. In many countries in the less developed world, however, waste disposal is effectively unregulated and significant adverse impacts can occur. This is especially unfortunate in the common situation where the same water reservoirs serve both as community water supplies and as waste depositories.

Methane gas (CH_4) is produced by the anaerobic decay of organic municipal and industrial residues in landfills. Worldwide emission estimates from this source have high uncertainty, but suggest methane releases to the atmosphere of order 0.5 g CH_4 per gram of biodegradable carbon disposed of on land. By counting all organic residue disposal worldwide, this source of methane is determined to be between 6 and 15% of the global CH_4 source. By far, most of this emission comes from the industrialized world. In the future, the contribution from less developed nations is expected to grow rapidly because of strongly increasing population and urbanization. Consequently, methane release from landfills may become one of the main sources of atmospheric methane in the next century, and it will be important to develop methods to use this methane as an energy source rather than to permit its escape into the atmosphere.

An alternative to burying residues is to incinerate them. Modern incinerators have low rates of emission of trace species, but carbon dioxide is emitted, of course, as may be small amounts of hydrogen chloride (HCl) (arising from the combustion of chlorine-containing plastics), heavy metals, dioxins, and other chemicals. The emission fluxes of CO_2, HCl, and heavy metals vary widely with the mix of incinerated materials and with the age and type of the incinerator facility.

2.6.8   Industrial Manufacturing Processes

The most diverse group of emissions into the environment comes from industrial activities, which have traditionally been a source of toxics to surface waters and groundwaters. Among the more notable constituents are heavy metals from plating operations, pesticides and herbicides from agricultural activities, organic solvents used for cleaning processes, and strong acids from chemical synthesis. Many of these constituents have long lifetimes in the environment, and the cessation of effluent emissions does not solve the problem in the short run, though it obviously contributes to longer-term solutions.

As regards atmospheric emissions from industrial activities, three classes of emittants can be singled out. The first is the family of chlorofluorocarbon compounds (CFCs). The CFCs have been the subject of recent international agreements (the Montreal Protocol and subsequent amendments) that are intended to prohibit their emissions by the end of the century. A second emittant of note is CO_2 from the manufacture of cement. A third industrial emittant of importance is atmospheric particles, which play major roles in reducing visibility, in some health-related impacts, and as a platform for various transformations of gaseous pollutants. Transition metals, generated by many of the materials processing industries, generally reside on emitted particles, though some are emitted as gases. For several of the metals, the fluxes are quite large, and soil and water quality impacts occur for such species as mercury and cadmium. Emission of the trace metals is widely variable depending on the industrial process, the raw materials used, and the degree of emission controls incorporated into the industrial process. Reduction in the emissions of larger particles can often be accomplished without great expense. In the case of the smaller particles, however, emissions control can be difficult and costly. This is unfortunate, as it is these small particles that cause most of the particle-related decreases in visibility.

Industrial gaseous emissions include many species, some rather poorly quantified. Hydrocarbons used as solvents, paints, and so on, are major emittants. A number of these compounds are highly volatile and are factors in the formation of photochemical smog in and downwind of urban areas.

2.7   THE EVOLVING DEVELOPMENT—ENVIRONMENT RELATIONSHIP

History offers a number of interesting and different examples of how society wrestles with the constraints of the master equation. At one extreme are the countries of the former Soviet Union, whose industrialization was accomplished during the last several decades at a very heavy environmental price. At the other are the current Organization for Economic Cooperation and Development (OECD) countries, who are managing to increase their industrial capacities and gross domestic products while substantially decreasing the emissions of pollutants (Fig. 2.8). If not completely adequate to the task, technology is in any case sufficient to begin the amelioration of problems. As was discussed before, technology and society must function in close partnership if such a change is to be achieved.

The ideal situation for the world in the next few decades will be if those countries and societies now undergoing rapid technological development avoid repeating the industry–environment mistakes of the Industrial Revolution. One can think of this approach as "clipping the peak" of Fig. 1.5 to achieve the development trajectories of the type shown in Fig. 2.9. In the cases of the more developed countries further along the development path, increased efforts can decrease their impact as well. To the degree that

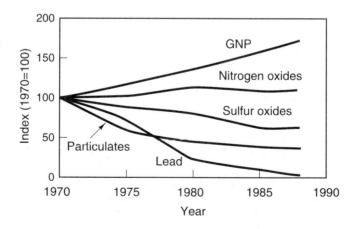

Figure 2.8   Recent relationships between gross national product and emissions of pollutants for the OECD countries. (Reproduced from *World Development Report 1992*. Copyright 1992 by The International Bank for Reconstruction and Development/The World Bank. Reprinted by permission of Oxford University Press, Inc.)

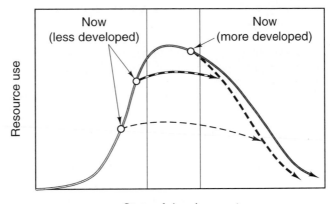

State of development

Figure 2.9   Desired redirections (dashed lines) of the typical historical development path, treated as the relationship between the state of technological development of society and its resulting environmental impact. "Now" refers to the approximate positions of countries at different levels of industrialization; compare with Fig. 1.5.

these potential development–environment trajectories can be achieved, benefits will accrue to both the more developed and less developed worlds.

For the longer term, one must consider whether it will be possible for technological improvements to occur and diffuse rapidly enough to override the demands placed on the planet by the increasing population. It will be a long time before it is clear which force will dominate, but the potential for difficulties is sufficiently great that it seems unlikely that truly sustainable development can occur without careful planning and prompt action.

SUGGESTED READING

Grübler, A. *The Rise and Fall of Infrastructures.* Heidelberg, Germany: Physica Verlag, 1990.

Faeth, P., ed. *Agricultural Policy and Sustainability: Case Studies from India, Chile, the Philippines, and the United States.* Washington, DC: World Resources Institute, 1993.

Herman, R., S. A. Ardekani, and J. H. Ausubel. Dematerialization, in *Technology and Environment*, J. H. Ausubel and H. E. Sladovich, eds., pp. 50–69. Washington, DC: National Academy Press, 1989.

"Managing Planet Earth", special issue of *Scientific American, 261* (3) 1989.

United Nations. *Convention on Biological Diversity*, Document 92-7807. Rio de Janiero, Brazil, 1992.

————. *Framework Convention on Climate Change*, Document A/AC.237/18. Rio de Janiero, Brazil, 1992.

World Bank. *World Development Report 1992.* Oxford, UK: Oxford University Press, 1992.

World Resources Institute (in collaboration with UN Environment Programme and UN Development Programme). *World Resources 1992–1993.* Oxford, UK: Oxford University Press, 1992.

EXERCISES

2.1 Energy consumption has often been expressed in quads, a quad being 10^{15} Btu. In 1989 the world's energy consumption was 385 quads. If 1 Btu = 1056 J, express the energy consumption in exajoules (EJ).

2.2 The total U.S. electrical energy use in 1985 was approximately 9.4 EJ. If the energy equivalents of a metric ton of coal and a 160 liter barrel of oil are 29.6 GJ and 610 GJ, respectively, find how many metric tons of coal would have been required to produce the electrical energy if coal combustion was the sole source. How many barrels of oil would have been required if oil combustion was the sole source?

2.3 Using the world population data from Fig. 1.4 and the global metals use data from Fig. 2.6, compute the per capita use of aluminum, copper, and iron in 1950, 1960, 1970, 1980, and 1985. (This will involve taking data from a graph, something practicing scientists and engineers do all the time.) The results show dematerialization, that is, the use of less material to achieve societal goals. Comment on the differences for the different metals.

2.4 Assume that China's development lags that of the OECD countries by 20 years, but otherwise conforms to the GNP and emissions curves of Fig. 2.8. The 1990 Chinese SO_2 emissions were 16.2 Tg and the GNP was U.S. $419,500 million. Compute the Chinese emissions of sulfur dioxide and the sulfur dioxide/GNP ratio for the years 2000, 2010, and 2020.

<table>
<tr>
<td>CHAPTER
3</td>
<td># A Survey of
Environmental Concerns</td>
</tr>
</table>

3.1 INTRODUCTION

The activities of industrial ecology are driven by the desire to avoid actual or perceived environmental impacts. Not all possible concerns can be addressed, of course, and some are more important than others. Thus, it is valuable for the industrial ecologist to have some understanding of the major environmental issues, lest action be directed toward goals of little significance. Accordingly, we present in this chapter brief summaries of what we and others consider to be the most important of the environmental concerns. We will note later in the book an important fact: that most sources of emission to the environment have multiple effects, and most effects have multiple causes. As a consequence, a single industrial product or process generally has the potential for impacts, even if often very small ones, on several environmental problems at once. Here, however, we will consider the effects themselves rather than to discuss the important but complicating subject of sources.

An important distinguishing characteristic among environmental problems is their temporal scale. If the problems, once begun, will endure a long time, they should command more attention than those capable of being reversed quickly once their cause is removed. By the same token, problems of large spatial scale are of more general concern than those whose effects are restricted to smaller spatial regions, although the latter may be quite important to those proximate to them. The characteristics of a long time scale and a large spatial scale tend to occur together, because long-lived contaminants are more likely to be dispersed over wide spatial areas during their lifetimes. Scale issues are compounded by sometimes lengthy lag times between forcing activities (e.g., increased emissions of CO_2) and the resulting change in natural systems (e.g., global climate change). We present our discussions ordered by spatial scale of impact; to a first approximation, this is also an order of importance.

One chapter in a book does not constitute a course in environmental science, of course, but will be a useful perspective later when we discuss decision making by product and process designers. Further information on environmental topics is available in the chapter references.

3.2 GLOBAL SCALE CONCERNS

3.2.1 Global Climate Change

Climate is most succinctly defined as the patterns of common meteorological conditions (temperature, precipitation, winds, etc.) over long time periods. (Thirty-year averages are often used as a modern measure of climate.) Climate has changed considerably over centuries or millenia throughout Earth's history, but concerns are now arising that climate changes may be proceeding extremely rapidly under the influence of humanity's activities. An example of the evidence leading to these concerns is the temperature record for the past century, as shown in Fig. 3.1. These data show that about 0.5°C of warming has occurred in 100 years, and that the decade of the 1980s was the warmest during that period. The pattern roughly parallels that of the use of fossil fuels and the resulting injection into the atmosphere of gases that can absorb radiation and produce warming.

Several research groups have used computer models to examine the effects of increases in the concentrations of radiatively active trace gases. Their calculations predict global temperature increases of 2–5°C by the year 2050. An average global temperature change of a few degrees does not seem very large until the change is put into perspective with past climate oscillations. For example, the Little Ice Age of

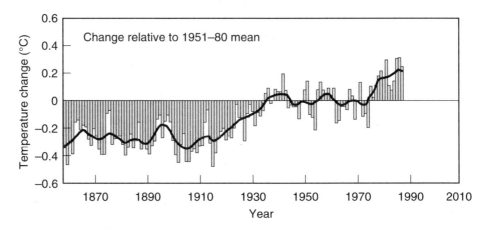

Figure 3.1   Annual deviation of the global mean (land and sea) temperature change over the past century relative to the average for 1951–1980. The curve shows the results of a smoothing filter applied to the annual values. (Policymakers' Summary, *Climate Change: The IPCC Scientific Assessment*, Cambridge, UK: Cambridge University Press, 1990.)

1400–1650 was about 0.5°C cooler than the present, and forced significant changes in agricultural practice and habitation. Should warming of one or two degrees occur, the planet would be at a temperature probably not seen for 120,000 years. Computer models predict that among the results of such warming would be a change in global mean sea level of about 20 cm by the year 2030 and about 45 cm by 2070, mostly due to thermal expansion of the oceans and the increased melting of mountain glaciers. Should such a change occur, it would cause major disruption to the large fraction of the world's population living in coastal regions, especially less developed nations in southeast Asia, and on islands.

There are vigorous scientific disputes about whether global climate change due to anthropogenic emissions—in particular, a general warming trend—has occurred, though most experts anticipate definite evidence for such occurrence within one or two decades. The warming will have as its cause a change in the atmosphere's radiation balance. This occurs because that portion of solar radiation that is not absorbed or scattered during its passage through the atmosphere reaches the ground and causes heating of soil, water, vegetation, and the adjacent air. Because any heated body radiates, emission of radiation from Earth's surface to the atmosphere follows. Earth is much cooler than the Sun, so most of its radiation occurs at long wavelengths, in the infrared portion of the spectrum. As with the incoming solar radiation, the outgoing radiation flux is reduced by absorbing atmospheric molecules. The energy of these infrared photons is, however, insufficient to cause chemical changes. Instead, the absorption merely increases the internal vibrational and rotational energy of each absorbing molecule. That excess energy is subsequently transferred to the atmosphere as kinetic energy (heat) by molecular collisions. Thus, any increase in the atmospheric concentrations of gases that absorb strongly in the 8–20 μm region where the principal radiation from Earth occurs will have a comparatively large warming effect. Among those species that do so are methane, nitrous oxide, and the chlorofluorocarbons CFC-11 ($CFCl_3$) and CFC-12 (CF_2Cl_2), together with the more abundant CO_2, H_2O, and O_3. Collectively, these absorbing molecules are called *greenhouse gases*; emission rates of these species must be addressed if action is to be taken to ensure the planet against future climate change.

3.2.2 Ozone Depletion

In 1985, researchers from the British Antarctic Survey showed that the total amount of ozone in the atmosphere over their observational site in Antarctica had shown a rapid decrease during the Austral spring (Fig. 3.2). Future trends in atmospheric ozone had already been a concern of scientists, who had predicted that the injection of chlorofluorocarbons into the atmospheric would gradually cause losses in stratospheric ozone over a period of decades as a result of a very efficient catalytic cycle for ozone destruction:

$$CFC + \text{high-altitude solar radiation} \rightarrow Cl + \text{other fragments} \qquad (3.1)$$

$$Cl + O_3 \rightarrow ClO + O_2 \qquad (3.2)$$

$$ClO + O \rightarrow Cl + O_2 \qquad (3.3)$$

$$\text{Net:} \quad O_3 + O \rightarrow 2\,O_2$$

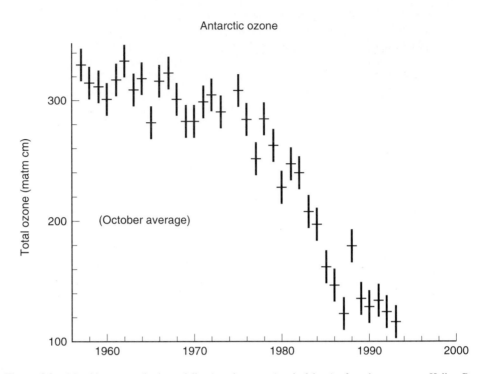

Figure 3.2   Monthly means (horizontal lines) and ranges (vertical bars) of total ozone over Halley Bay, Antarctica, for October of the years 1957 through 1993. A layer of gaseous ozone 1 mm thick at 1 atmosphere pressure and 0°C corresponds to 100 matm cm. (The data through 1984 are from J. C. Farman, B. G. Gardiner, and J. D. Shanklin, Large losses of total ozone in Antarctica reveal seasonal interaction, *Nature, 315* (1985): 207–210. Data from subsequent years are from NASA's total ozone monitor aboard the Nimbus 7 satellite.)

The chlorine species is efficiently regenerated in this sequence of reactions, so a single chlorine atom can destroy thousands of ozone molecules before finally being lost to a competing reaction. The effects of changes in the concentration of upper atmospheric ozone are expected to be directly felt by biological systems, because the penetration of ultraviolet solar radiation to Earth's surface is strongly limited by ozone absorption. The wavelength band 290–320 nm is the region to which biological organisms are most sensitive, and its intensity is greatly increased by ozone decreases. In addition, because the temperature structure of the atmosphere is strongly influenced by the amount of solar radiation absorbed by ozone, ozone losses may eventually result in changes in atmospheric circulation and hence in precipitation.

More recently, attention has also been given to several halogenated hydrocarbons containing bromine, widely used as fire extinguisher agents and fumigants. The reason for this concern is that bromine is an even more effective ozone depleter on a molecule-per-molecule basis than is chlorine, because bromine is lost to competing reactions much less rapidly.

3.2.3   Loss of Habitat and Reductions in Biodiversity

Although difficult to quantify with precision, there is general agreement among ecologists that the extinction of species of flora and fauna is occurring at a very rapid rate which, if continued, could constitute a global extinction event similar to those found periodically in the fossil record. The goal of maintenance of a high level of species diversity is not mere environmental radicalism. Rather, the entire life of the planet is dependent on the functioning of food webs from the simplest photosynthetic organism to the most complex. When a species becomes endangered, its gene pool is greatly reduced and its capacity to adapt to environmental changes becomes more limited. If a species is lost completely, its genetic material, produced by environmental adaptation over millions of years, is lost as well. At some point, breaks in food webs result in a chain reaction of problems for the global life-support system. A diversity of species provides other readily demonstrable values to humans, including new drugs and optimized biomass for energy processes. Many people also view loss of diversity as a moral issue.

The primary present cause of endangerment or extinction of species is loss or disturbance of natural habitats. These habitats have always varied as a function of climate and other external factors, but are now disappearing at a very rapid rate due to expanded urban areas, expanded crop production, and the like. Current data suggest that Earth's primary tropical forests could be essentially gone in four or five decades. These forests contain the richest diversity of species on the planet, so their loss would be a major blow. It is estimated that 0.2 to 0.3% of all species in the forests are lost per year at the present rate of deforestation; this constitutes the loss of about 4000 to 6000 species annually, perhaps 10,000 times the average natural extinction rate.

Loss of habitat has been called the "quiet crisis", because it occurs piecemeal, and because not only a suitable total amount of habitat needs to be preserved, but large contiguous tracts are needed for species and ecosystem survival. This characteristic has been especially noted on the African plains, where a number of large mammals, once abundant, are now threatened by habitat shrinkage and other pressures attributable to the acts of humans. Once a major change in land use has occurred, the habitat is irretrievably lost, because survival of native plant and animal life is no longer possible. Even where habitat is retained, disturbances caused by nearby human activities may have significant negative impacts. A complex version of this concern applies to migratory birds, which need suitable habitats in both their wintering and summering territories, and where the destruction of either could result in species loss.

A pressing need in the biodiversity/habitat research area is for extensive diversity surveys, so that the situation is accurately defined. It would also be advisable to modify present land use policies to encourage "cluster" development and preserve large tracts of land in a relatively undisturbed state. Industrial activities can often contribute to this end if they attempt to minimize alteration of undisturbed habitat and to encourage habitat diversity at industrial facilities by avoiding disruption of particularly sensitive biological regimes. Thus, new industrial facilities should preferentially utilize already altered land, such as urban or industrial areas. It is of interest that such actions might have positive social implications as well.

3.3 REGIONAL SCALE CONCERNS

3.3.1 Surface Water Chemistry Changes

Surface water chemistry is influenced by three factors: the natural environment containing the water body and its drainage basin, the effects of humanity's activities in directly altering that environment (such as applying fertilizer or discharging industrial residue streams), and the indirect effects of humanity's activities as communicated through the atmospheric transport and deposition of anthropogenic emittants. In any particular situation, all of the foregoing may contribute and any one may be dominant.

One of the most highly publicized environmental problems of the past two decades has been the acidification of lakes and rivers, generally (but not always) as a consequence of acidic precipitation. The degree to which bodies of water become acidified is strongly related to the chemical characteristics of the soil and rock in the immediate vicinity. If that material is capable of extensive buffering of incoming acid (as with rich organic humus in soil, for example), acidification is unlikely in the short term. However, freshwater in regions mostly in contact with underlying rock can be readily acidified. Most sites lie somewhere between the two extremes. A second common problem is eutrophication, a lack of accessible oxygen due to excess biological activity triggered by an oversupply of nitrogen and phosphorus, usually from heavy fertilization followed by runoff. Less common and frequently more localized insults include siltation as a result of impoper construction or forestry activity, pollution of water resources by toxic organics and heavy metals, and heavy salt loads as a result of overly intensive agricultural use (as in the Rio Grande in the American southwest).

Opposite trends in river water chemistry within a single country, as shown in Fig. 3.3, are evidence for differences in anthropogenic influence on small spatial scales. In the United States, for example, the Grand River has intensively cultivated lands in its drainage basin. These lands were receiving high and increasing applications of nitrogen fertilizer during the 1970s and 1980s. As a consequence, nitrate concentrations in the river increased by more than a factor of 2. Conversely, the fertilizer-related impact on the North Platte River was less intense, and improved municipal residue treatment is credited for lowering the riverine nitrate concentration during the same period. Elsewhere, certain activities such as mining often generate significant, albeit localized, negative impacts on surface water quality.

3.3.2 Soil Degradation

The extent to which arable land around the globe is being "used up", that is, rendered unusable for agricultural activity, is seldom recognized. Table 3.1 demonstrates that between 1945 and 1990, some 1970 million hectares of soil, or 17% of all vegetated land, have been degraded by human activity. In the past 45 years, an area about the size of China and India combined has been moderately to severely degraded, primarily as a result of agriculture, deforestation, and overgrazing.

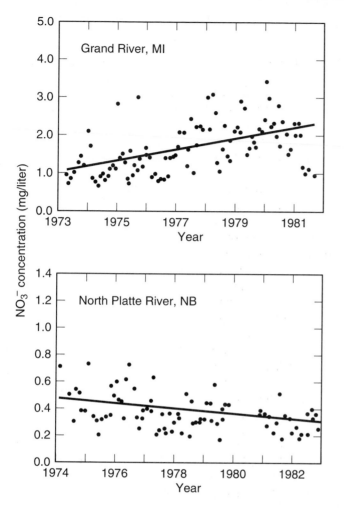

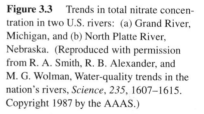

Figure 3.3   Trends in total nitrate concentration in two U.S. rivers: (a) Grand River, Michigan, and (b) North Platte River, Nebraska. (Reproduced with permission from R. A. Smith, R. B. Alexander, and M. G. Wolman, Water-quality trends in the nation's rivers, *Science*, *235*, 1607–1615. Copyright 1987 by the AAAS.)

A contributing factor is that unchecked soil erosion may result in a global food production potential loss of some 20%. Some 580 million people already live on marginal or fragile lands, driving even further degradation.

Although the causes of soil degradation and loss are complex, it is clearly unsustainable to continue using land in a linear Type I fashion, cycling it from productive to barren. Modified agricultural practices and new cultivars, and beyond that, economic and social conditions that support sustainable land use in the less developed countries until economic growth occurs, are clearly challenges for the industrial ecologist focusing on the agricultural and forestry sectors.

3.3.3   Precipitation Acidity

The single most important chemical species in clouds and precipitation is probably the hydrogen ion, whose concentrations can be indicated by specifying the solution acidity or pH value. The presence of atmospheric CO_2 assures that nearly all atmospheric water

Table 3.1 Global Anthropogenic Soil Degradation, 1945–1990

Degree	Definition	Area (M hectares)
Light	Mildly salinized, or eroded; widely spaced rills or hollows; > 70% of rangeland area still in native vegetation	750
Moderate	Moderate erosion or salinization; soil chemical and physical integrity compromised; rangelands have 30–70% native vegetation.	910
Severe	Frequent gullies and hollows; crops grow poorly or not at all; rangeland has < 30% native vegetation.	300
Extreme	No crop growth occurs and restoration is impossible. May be caused by, for example, water erosion (Central Italy), wind erosion (Somalia), or salinization (former Soviet Union near Iran).	9
Total		1969

Source: World Resources Institute, *World Resources 1992–1993*, pp. 111–126. Oxford, UK: Oxford University Press, 1992.

droplets will be at least somewhat acidic, and natural and anthropogenic nitrogen and (especially) sulfur species increase the acidity, that is, lower the pH value, to at least pH 5.0. Most rain near urban areas has pH levels nearer 4.0. Dew is generally less acidic, apparently because the acidity is neutralized by more basic constituents in particles deposited on the surfaces where the dew forms. Cloud and fog droplets are nearly always more acidic than rain, apparently because longer lifetimes and smaller drop sizes inhibit dilution of the acidic constituents. In some fogs, the pH of the droplets has been measured to be as low as 1.7, close to that of battery acid! It is no wonder that materials and organisms exposed to such fogs deteriorate rapidly.

The seriousness of the impacts of acid rain was a topic of intense investigation during the 1980s, one of the major topics of concern being the effect of precipitation on aquatic life in lakes. It is now clear that not all lakes are significantly affected, but only those where surrounding soils are shallow and most flow is through channels in bedrock or unconsolidated sediments highly resistant to weathering. Where that is the case, generally in high-latitude regions, loss of aquatic diversity can be significant. The sensitivity of a particular region can be expressed by the concept of *critical load*, that is, the areal flux of one or more pollutant species below which significant harmful environmental effects are not anticipated. Critical loads differ for different receptors, such as soils, lakes, or rivers, and for different geographical areas. In sensitive areas (those with little acid-neutralizing capacity), critical sulfur and nitrogen loads are of order 20 milliequivalents m^{-2} yr^{-1} for each. Where acid-neutralizing capacity is substantial, critical loads can be a factor of about 2 higher. When the critical loads are combined with computer models of atmospheric chemistry and diffusion, they specify the amount of acid gas emissions from upwind sources that can be permitted without the occurrence of significant environmental degradation.

Other impacts of precipitation acidity are less well-studied, but are also probably less important. Acid precipitation is definitely implicated in the degradation of susceptible materials, particularly metals and carbonate stones, the result being corrosion of building materials, statuary, and other components of the "engineered environment". However, acid rain has been determined to have generally negligible effects on forests and crops, in contrast to ozone and other constituents of smog that in high concentrations cause significant damage to vegetation of all kinds.

3.3.4   Visibility

Visibility may be defined as the greatest distance over which one can see and identify familiar objects with the unaided eye. The visibility involves two quite different factors: the degree to which light coming from the object is absorbed or scattered and the visual threshold of perception. The latter is, of course, unalterable, but the former can be heavily influenced by anthropogenic activities. In the atmosphere, solar radiation is absorbed and scattered by gas molecules, particles, and water droplets. The scattering by gases in the clean atmosphere provides an upper limit of about 300 km to the visual range. At moderate or high particle loadings, radiation scattering by particles is the primary limitation to visibility.

Standard visual range in the rural mountain/desert areas of the southwest United States currently averages about 130 to 190 km. In contrast, rural areas south of the Great Lakes and east of the Mississippi have median standard visual ranges of about 20–35 km. Because visibility is one of the measurements routinely made by airport weather observers, a long record of visibility from many different locations is available. Rudolf Husar and his colleagues at Washington University in St. Louis have used this information to create plots of average visibility for the eastern half of the United States for three 5-year periods spaced roughly a decade apart. The diagrams, reproduced in Fig. 3.4, show substantial visibility decreases in the Ohio River valley, the Tennessee–Kentucky region, and the Virginia–Carolina area, especially in summer. In the spring and fall, a noticeable decrease occurred across the entire eastern United States.

The two principal causes of decreased visibility are the emission to the atmosphere of small particles (i.e., those with sizes of the same order as that of visible light, about 0.2–0.7 μ) and the emission of reactive gases that are subsequently converted to small particles. The particles themselves are usually a consequence of combustion processes, such as the burning of petroleum by diesel engines or the combustion of coal. Convertible gases tend to be those that are water-soluble, with sulfur dioxide the most important and the oxides of nitrogen and nonmethane hydrocarbons also worth mention. Comparing the natural visibility levels with current visibility levels indicates that human-made contributions account for about one-third of the average visibility limitation in the rural West and over 80% of the average visibility limitation in the rural East. (Atmospheric water vapor and naturally generated particles provide the remaining visibility loss.)

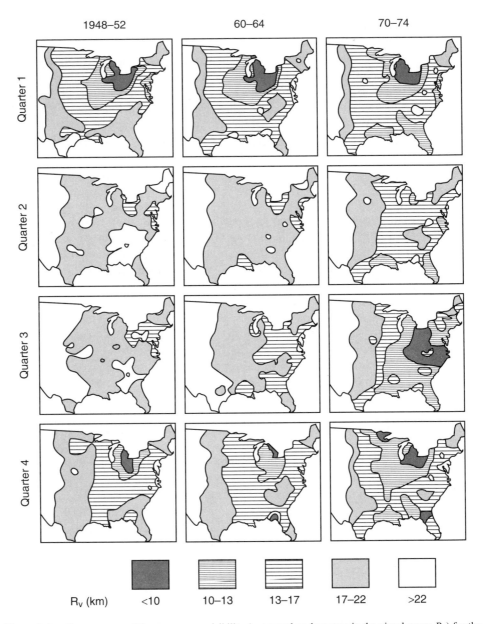

Figure 3.4 Contour maps of 5-year average visibility (expressed as decreases in the visual range R_v) for the eastern United States, centered on the years 1950, 1962, and 1972. The maps are given for the four seasons, from winter (quarter 1) to fall (quarter 4). (Adapted with permission from R. B. Husar, J. M. Holloway, D. E. Patterson, and W. E. Wilson, Spatial and temporal patterns of eastern United States haziness: A summary, *Atmospheric Environment, 15*, 1919–1928. Copyright 1981 by Pergamon Press, plc.)

3.3.5   Herbicides and Pesticides

Herbicides and pesticides are designed expressly to be biologically damaging, and as such are of significant concern from an environmental standpoint. The level of concern is a function of such factors as the toxicity of the product, its longevity, and its method and intensity of application. Improvements in each of these areas have been accomplished in recent years, and continued efforts are advisable.

Although the pesticide and herbicide industries are heavily regulated, the necessity to produce food for growing populations creates great pressure for the continued control of animal and plant pests, a task complicated by the rapid adaptation of pest populations to common control agents. Increasing evidence of widespread groundwater contamination by, and resulting regulatory controls on, agricultural chemicals is encouraging the development of less toxic means for achieving this control. Insect behavior, for example, is quite responsive to chemical signals. The use of nontoxic attractants, disruptants, and the like, perhaps in combination with revised practices such as "no-till" farming and minimal application of agricultural chemicals, has the potential to ameliorate many problems with minimal environmental effects.

3.4   LOCAL SCALE CONCERNS

3.4.1   Photochemical Smog

In the mid-1940s, it became obvious that production of ozone was taking place in the troposphere as well as in the stratosphere. After heavy injury to vegetable crops had occurred repeatedly in the Los Angeles area, it was shown that the plant damage could be produced by ozone, which was known to be a prominant constituent of "photochemical smog". The overall smog-reaction mechanism was eventually identified by Arie Haagen-Smit of the California Institute of Technology as

$$NMHC + NO + h\nu \rightarrow NO_2 + \text{other products} \tag{3.4}$$

where NMHC denotes various reactive nonmethane hydrocarbons (ethylene, butane, etc.), and $h\nu$ indicates a quantum of solar radiation of wavelength less than about 410 nm. Ozone formation by this mechanism is possible because solar radiation dissociates the NO_2 formed in Eq. (3.4):

$$NO_2 + h\nu \ (\lambda \leq 410 \text{ nm}) \rightarrow NO + O \tag{3.5}$$

and the recombination of O with molecular oxygen then produces ozone:

$$O + O_2 \rightarrow O_3 \tag{3.6}$$

The rate of formation of ozone and other toxic products of smog chemistry can be controlled by controlling the atmospheric concentrations of NO, NO_2, and NMHC. Because of the cyclic nature of the chemistry, NMHC control may be more effective for some regions, NO_x (NO + NO_2) control for others.

3.4.2 Groundwater Pollution

High quality groundwater is essential to the health and welfare of a large fraction of Earth's population. Although groundwater *supplies* continue to be of concern, especially in overbuilt areas or in semidesert regions where they are frequently mined for agricultural use, groundwater *quality* is increasingly a focus of inquiry. Among the many sources of contaminants to groundwater are sewage disposal (including septic tanks and cesspools), agricultural activities, solid-residue disposal in landfills, disposal of liquid residue in deep wells, petroleum leakage, and washoff of road salt, pesticides, and other dispersed chemicals. As materials from these sources permeate the soil, they are filtered, buffered, and diluted, thereby losing much of their potency. This permeation, which often occurs over time scales of weeks to months, is quite effective at cleansing the percolating liquid as long as the contaminant levels are not excessive and as long as the trace contaminants are susceptible to scavenging by soil. The case of synthetic organic chemicals only sparingly soluble in water may be less benign, however. For these species, effective bacterial or physical mechanisms of neutralization have not been developed within the soil system, and they will be released unchanged to the groundwater if their lifetimes exceed the diffusion time.

Notwithstanding occasional serious contamination situations, most groundwater is so effectively filtered by soil that it meets or exceeds water quality guidelines. Exposure of humans to groundwater of unsatisfactory chemical content is uncommon. It is important, however, to attempt to maintain the natural systems in undiminished efficiency. Such a goal implies avoiding the overuse of groundwater supplies, especially the "mining" of irreplaceable groundwater resources. Thus, water that has had much less cleansing and processing by the soil will not have to be utilized and, in coastal areas, saline water will not intrude significantly into groundwater reservoirs as they are drained of their freshwater.

3.4.3 Radionuclides

Human beings are exposed to radioactivity and ionizing radiation from many natural sources. The most important common inputs are from cosmic and Earth background radiation. Next are routine medical and diagnostic exposures, and, for those who fly frequently, from enhanced cosmic radiation while aboard aircraft. All of these sources are well-studied and none is thought to constitute a significant health problem. Even less impact to the average population is estimated from uses of radioactive materials in commercial products such as smoke detectors.

A radiation source that is unimportant for most people but substantial for a few is indoor radon. Radon is emitted from soil or building materials containing high concentrations of radium and other radon precursors. Radon-emitting soils are seldom a problem if their presence is recognized, because provision for frequent air changes in basement areas is generally sufficient to keep concentrations well below the levels considered dangerous to health. However, the occurrence of radon-emitting soils varies widely with geography and the emission rate can be very different over distances as small as a few hundred yards. Testing of individual buildings for radon is thus advisable in regions where the soil characteristics suggest the possibility of a significant radon-source flux.

3.4.4   Toxics in Sludge

The treatment of industrial or municipal wastewater produces a moist solid mass called *sludge*. Some means of disposal is then required for the sludge, which may contain a variety of undesirable anthropogenic pollutants. Much of this material is removed or depleted during processing, however. Different disposal options have been utilized over the years, including ocean dumping (now illegal in the United States), landfilling, incineration, and land application as a fertilizer. In fact, sewage sludge is rich in organic matter and nutrients, and "clean" sludge can be an excellent fertilizer.

Sludge is, by legal definition at least, potentially hazardous, so its disposal is usually under fairly rigorous supervision, especially in the more developed countries. As such, the potential for exposure (especially of humans) is not very high. The same is not necessarily true of the ecosystems near the disposal site, however, and extensive monitoring and site maintenance may be advisable.

3.4.5   Oil Spills

Oil spills are major news event because of their photogenic properties and the image of environmental disdain that they project. They are usually the result of damage to the hull of a petroleum tanker ship or truck, but can result from accidents during petroleum transfer activities. The largest portion of spilled petroleum is crude oil, which is less toxic than refined petroleum products like gasoline.

Notwithstanding their image, oil spills are regarded by experts as among the more moderate risks to the environment. This judgment is based on the fact that spilled oil degrades, loses toxicity, and generally does not persist in a biologically active state in the environment. In addition, improved spill-response techniques and more rugged tanker construction continue to reduce oil spill losses substantially. Moreover, there is evidence that the volume of oil released from natural seeps is in many cases as extensive as from anthropogenic oil spills.

3.4.6   Toxics in Sediments

Bays, harbors, and other receivers of flowing water build up layers of sediment as small particles and other debris flow into them. The sediment invariably contains a variety of sparingly soluble or insoluble contaminants that have entered the water stream from industrial sources and surface runoff (especially from agricultural sources, atmospheric deposition, and the like). Once deposited, materials in the sediment are not necessarily sequestered, because bioturbation (sediment disturbance by aquatic life), dredging, and ship traffic tend to remobilize long-lived sedimentary constituents.

The materials of special concern in sediments tend to be those that are both toxic and of anthropogenic origin, such as pesticides and herbicides (especially those used before biodegradability became recognized as an important attribute), aromatic organic molecules, and heavy-metal compounds. The presence of the latter in sediments is indicated by, for example, a core taken from the mouth of the Mississippi River. An analysis was performed separately for "pollutant lead" and total lead, the former being based on the difference between the measured level and that attributable to natural-sediment lead concentrations. The data show (Fig. 3.5) that pollutant lead contributions were at a maximum in about 1970 (just before the Clean Water Act began to be implemented and more or less coincident with the maximum emission of lead compounds from gasoline). This figure shows how dramatically emissions controls can be reflected in evolving environmental reservoirs.

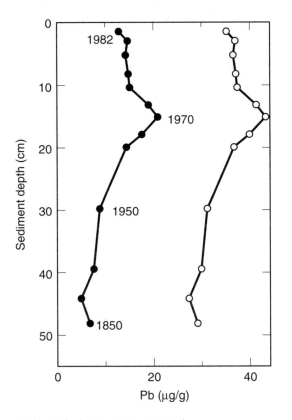

Figure 3.5   Lead concentration as a function of depth in sediments near the river mouth. The closed circles refer to "pollutant lead" (see text) and the open circles to total lead. The top of this sediment record was formed during 1982 and the bottom in about 1850. Two intermediate calibrated dates are indicated: 1950, when lead emissions from gasoline were rather low following the World War II, and 1970, when lead use in industry and gasoline was high and when the lead peak in the sediments occurred. (Adapted with permission from J. H. Trefry, S. Metz, R. P. Trocine, and T. A. Nelson, A decline in lead transport by the Mississippi River, *Science, 230,* 439–441. Copyright 1985 by AAAS.)

3.4.7   Hazardous Waste Sites

The disposal of hazardous waste is one of the most public and contentious of environmental issues. Hazardous waste sites are the locations where toxic materials, including benzene, polychlorinated biphenyls (PCBs), lead, arsenic, cadmium, and pesticide residues, are confined. If the lifetimes of the chemicals are longer than the time needed for their escape to the ambient air or to adjacent groundwater, the emissions have the potential to cause significant harm, especially to proximate ecological systems.

Both inactive and active sites are of potential concern. Sites in active use can be much more easily monitored. Inactive sites are characterized by capping and a presumably lower potential for emissions, at least on a large spatial scale, but their inaccessibility for inspection creates the possibility of poorly controlled emissions going unnoticed for extended periods of time. Older inactive landfills often had either no liners or ineffective ones, so that flowing groundwater may be able to readily mobilize the landfilled chemicals.

The U.S. Environmental Protection Agency (EPA) lists more than 1000 hazardous waste sites in the United States. Worldwide there are many, many more. The effectiveness of hazardous waste-site management varies significantly among countries: It has been estimated that only a few percent of U.S. sites are likely to have significant detrimental effects on their surroundings, for example, but many less developed countries have virtually no site-control management systems in operation. The long lifetimes of many of the landfilled chemicals and the spotty nature of landfilling operations suggest that monitoring of both active and inactive waste-disposal sites is advisable and that vigorous efforts should be made to reduce the volume and toxicity of materials entering active hazardous waste sites.

3.5 DISCUSSION

This brief summary of environmental problems related to industrialization and society readily demonstrates that the problems are diverse and complex. The spatial scales involved range from quite local to global, the time scales from less than a day to well over a century. Furthermore, many of the approaches taken historically to deal with these concerns have followed a pattern of transferring problems rather than eliminating them. For example, sulfur is not desired as a gaseous emittant from fossil fuel combustion, so one solution was to scrub the sulfur from the exhaust gases and create a sulfur-rich liquid-residue stream. This residue stream was then processed into sludge and the sludge landfilled. Thus, a problem was moved from air to water to soil, but never really solved. Another aspect of environmental concerns that was stated earlier is that many problematic species have multiple sources and many environmental concerns have multiple causes. These characteristics have been among those that have confounded historical approaches to perceived environmental problems, in which single-issue approaches have created or exacerbated some problems as a consequence of solving others.

These historical perspectives are important to the industrial ecologist because they emphasize the difficulty of treating environmental problems once created. The rational approach to the industry–environment interaction is not to continue single-issue amelioration, or problem transfer among environmental regimes, but to minimize or eliminate altogether such problems from the beginning. Perhaps the simplest statement of the goal of industrial ecology is that it is designed to eliminate the causes of environmental problems rather than to work to solve them after they become apparent.

SUGGESTED READING

Graedel, T. E., and P. J. Crutzen. *Atmospheric Change: An Earth System Perspective.* New York: W. H. Freeman, 1993.

Kämäri, J., M. Forsins, and M. Posch. Critical loads of sulfur and nitrogen for lakes. II. Regional extent and variability in Finland. *Water, Air, and Soil Pollution, 66* (1993): 77–96.

Masters, G. M. *Introduction to Environmental Engineering and Science.* Englewood Cliffs, NJ: Prentice Hall, 1991.

(A number of useful citations are given in Exercises 3.5–3.7.)

EXERCISES

3.1   Computer models predict that an average planetary warming of 2°C would raise the global mean sea level by 45 cm. What would be the consequences of such a change for Bangladesh, Marshall Islands, Argentina, and Japan?

3.2   The total land area of Earth is 5.10×10^8 km^2. If tropical rain forest area in 1990 was 7.7×10^6 km^2, what fraction of the surface was tropical rain forest? What will be the fraction in 2020 if the area is decreased by 12% per decade, as predicted in some studies?

3.3   Figure 3.3 shows that the 1975 nitrate level in the Grand River was 1.5 mg/l. If the flow of the river is 50 m^3/s at a monitoring site, what is the nitrate flow past the same site?

3.4   The average concentrations of sulfate and nitrate ions in precipitation falling on Hubbard Brook, New Hampshire in 1970 were 2.4 and 1.4 mg l^{-1}, respectively. If the average rainfall is 135 centimeters per year, compute the annual input of acid (equal to the sum of sulfate and nitrate ions) to the area in units of meq m^{-2}/yr for 1970. If the critical load for the area is 30 meq m^{-2}/yr, is there reason for concern?

3.5   Prepare a five-page report on one aspect of local-scale environmental concerns. The report should have the following outline: explanation of the concern, history of the problem (as supported by data), seriousness of the situation at present, projections for the future, and potential for dealing with the problem. The report may be done for your own local geographical area or for another of your choice. The library and your instructor will be good sources for reference material, in addition to the following. **Smog**: Committee on Tropospheric Ozone Formation and Measurement, *Rethinking the Ozone Problem in Urban and Regional Air Pollution*, Washington, DC: National Academy Press, 1991; J. H. Seinfeld, Urban air pollution: State of the science, *Science, 243* (1989): 745–752. **Radionuclides**: C. R. Cothern and J. E. Smith, Jr., eds., *Environmental Radon*, New York: Plenum Press, 1987. **Oil Spills**: National Research Council, *Oil in the Sea: Inputs, Fates, and Effects*, Washington, DC: National Academy Press, 1985; M. Holloway, Soiled shores, *Scientific American, 265* (4) (1991): 103–116. **Hazardous waste sites**: J. Bernstein, Report from Aspen, *The New Yorker* (November 25, 1991): pp. 121–136.

3.6   Prepare a five-page report on one aspect of regional scale environmental concerns, using the outline format of Ex. 3.5. The report may be done for your own regional geographical area or for another of your choice. The library and your instructor will be good sources for reference material, in addition to the following. **Surface-water chemistry**: R. A. Smith, R. B. Alexander, and M. G. Wolman, Water-quality trends in the nation's rivers, *Science, 235* (1987):

1607–1615. **Groundwater**: R. A. Freeze and J. A. Cherry, *Groundwater*, Englewood Cliffs, NJ: Prentice Hall, 1979. **Precipitation acidity**: O. P. Bricker and K. C. Rice, Acid rain, in *Annual Review of Earth and Planetary Sciences, 21* (1993): 151–174; P. M. Irving, ed., *Acidic Deposition: State of Science and Technology*, Washington, DC: U.S. Government Printing Office, 1991; S. E. Schwartz, Acid deposition: Unraveling a regional phenomenon, *Science, 243* (1989): 753–763. **Visibility**: Committee on Haze in National Parks and Wilderness Areas, *Protecting Visibility in National Parks and Wilderness Areas*, Washington, DC: National Academy Press, 1993; R. B. Husar, J. M. Holloway, D. E. Patterson, and W. E. Wilson, Spatial and temporal pattern of eastern U.S. haziness: A summary, *Atmospheric Environment, 15* (1981): 1919–1928; P. M. Irving, ed., *Acidic Deposition: State of Science and Technology*, Washington, DC: U.S. Government Printing Office, 1991.

3.7   Prepare a five-page report on one aspect of global-scale environmental concerns, using the outline format of Ex. 3.5. The library and your instructor will be good sources for reference material, in addition to the following. **Global warming**: J. T. Houghton, G. J. Jenkins, and J. J. Ephraums, eds., *Climate Change: The IPCC Scientific Assessment*, Cambridge, UK: Cambridge University Press, 1990; P. D. Jones and T. M. L. Wigley, Global warming trends, *Scientific American, 263* (2) (1990): 84–91; R. M. White, The great climate debate, *Scientific American, 263* (1) (1990): 36–43. **Ozone depletion**: O. B. Toon and R. P. Turco, Polar stratospheric clouds and ozone depletion, *Scientific American, 264* (6) (1991): 68–74; S. Solomon, Antarctic ozone: Progress toward a quantitative understanding, *Nature, 347* (1990): 347. **Soil Degradation**: M. Huston, Biological diversity, soils, and economics, *Science, 262* (1993): 1676–1680. **Loss of habitat**: R. Repetto, Deforestation in the tropics, *Scientific American, 262* (4) (1990): 36–42; D. Skole and C. Tucker, Tropical deforestation and habitat fragmentation in the Amazon: Satellite data from 1978 to 1988, *Science, 260* (1993): 1905–1910. **Biodiversity**: J. Terborgh, *Diversity and the Rain Forest*, New York: Scientific American Library, 1992; R. H. May, How many species inhabit the Earth? *Scientific American, 267* (4) (1992): 42–48; C. C. Mann and M. L. Plummer, The high cost of biodiversity, *Science, 260* (1993): 1868–1871.

CHAPTER

4

An Overview of
Risk Assessment

4.1 UNCERTAINTY AND RISK

We live in a problematic world, where many things are not definitively known to us and may not ever be. Understanding or predicting the consequences of a specific action or circumstance is especially difficult in the case of environmental issues, which exhibit several forms of uncertainty. In many cases, such as the toxicology of a new organic compound, there are few data but no practical reason why such data cannot be generated. Many natural systems such as the atmosphere or oceans, however, appear to be inherently chaotic, and their evolution through time is uncertain. Moreover, there is a broad middle ground where as a practical matter it will be virtually impossible to develop sufficient data to dispel significant uncertainty even though there may not be any theoretical reason why this cannot be done.

The inevitability of uncertainty in dealing with environmental issues is compounded for the industrial ecologist who must integrate a changing technology with two sets of complex systems, the environmental and the economic. In most cases, the uncertainty is defined and managed through evaluating and quantifying risk.

Risk can be defined as the probability of suffering harm from a hazard. Risk is not foreign to us: we face, and intuitively evaluate, a myriad of risks every day. Most of them are not even considered consciously ("Should I drive to work today on a crowded superhighway?"). Others we may consider but rationalize ("I'll smoke for another month, then give it up."). The risks may be quite substantial, as can be seen from Table 4.1, which presents annual and lifetime mortality rates associated with common activities based on data from The Netherlands. For comparison purposes, standards for environmental cleanups in the United States frequently are based on a 10^{-6} lifetime risk, or

Table 4.1   Annual Mortality Rate Associated with Certain Occurrences and Activities in the Netherlands

Activity/Occurrence	Annual Mortality Rate	Lifetime Mortality Rate
Drowning as a result of a dike collapse	1×10^{-7} (1 in 10 million)	1 in 133,000
Bee sting	2×10^{-7} (1 in 5.5 million)	1 in 73,000
Being struck by lightning	5×10^{-7} (1 in 2 million)	1 in 27,000
Flying	1×10^{-6} (1 in 814,000)	1 in 11,000
Walking	2×10^{-5} (1 in 54,000)	1 in 720
Cycling	4×10^{-5} (1 in 26,000)	1 in 350
Driving a car	2×10^{-4} (1 in 5,700)	1 in 76
Riding a moped	2×10^{-4} (1 in 5,000)	1 in 67
Riding a motorcycle	1×10^{-3} (1 in 1,000)	1 in 13
Smoking cigarettes (one pack per day)	5×10^{-3} (1 in 200)	1 in 3

Source: Ministry of Housing, Physical Planning, and Environment, *National Environmental Policy Plan: Premises for Risk Management*, p. 7, The Hague, The Netherlands, 1991.

one additional fatality in a million over a lifetime. Thus, walking, cycling, or driving a car are all far more risky (higher probability of resulting in mortality) than the standards the United States chooses to impose on cleanups of contaminated sites, a significant efficiency consideration when it is recognized that cleanup standards are the primary determinant of cost for virtually any cleanup.

Risk as defined earlier seems deceptively objective; risks are, in fact, intensely subjective. People worry about some risks all out of proportion to the actual potential for harm, and ignore others much more dangerous. A study devised by psychologist P. Slovick of Decision Research, Inc., demonstrated this tendency. Slovick composed a list of 30 activities and technologies, and had them ranked for perceived risk by a group of experts and by a group of informed lay people. Whereas some risks were ranked similarly by the two groups (motor vehicles and food coloring, for example), others were ranked dramatically differently: nuclear power was ranked first by lay people and twentieth by the experts, and X-rays were ranked twenty-second by lay people and seventh by the experts.

The reasons for the discrepancy between expert and public risk perception appear to be that the public perceptions are biased by a number of factors, including the following:

- The extent to which the risk is seen to be controllable, either by the individual or by society.
- Whether the risk is "dreaded" (such as cancer).
- The extent to which the risk is involuntarily assumed.

- The extent to which the risk is observable, and known to those exposed to it.
- The immediacy of the effect.
- The identity of the victims (are they sympathetic, such as children, and identifiable as individuals, as opposed to statistical groupings).
- The extent of media attention.

As Fig. 4.1 demonstrates, risks that are "dreaded", uncontrollable, and not observable by those exposed to them are most likely to generate calls for government intervention. Many environmental problems possess these attributes.

An additional consideration is that just as people discount money—"a dollar today is worth more than a dollar tomorrow"—they also apparently discount lives. Surveys have shown that many people value one life saved today as equivalent to three saved 10 years from now, or 11 saved 50 years from now, or 44 saved 100 years from now. This perspective has clear and negative implications for the support that industrial ecology activities designed to reduce future risks may receive from the public today.

The differences in risk perception extend, of course, into the environmental arena. Table 4.2 shows the results of an exercise similar to the relative risk survey in which U.S. EPA managers and a group of citizens were independently asked to prioritize envi-

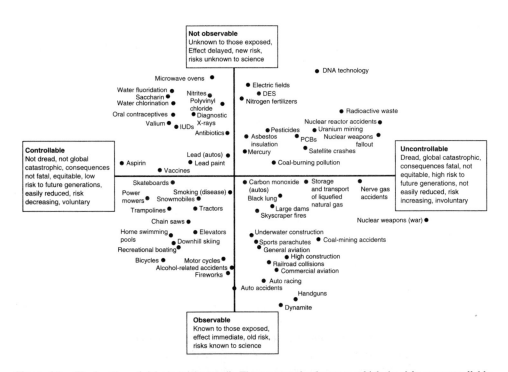

Figure 4.1   The location of risks in "risk space". The axes are the degree to which the risks are controllable and observable. (Reproduced with permission from M. G. Morgan, Risk analysis and management, *Scientific American, 269* (1): 32–41. Copyright 1993 by Scientific American, Inc.)

Table 4.2   Comparitive Rankings of Environmental Problems

Public Perception of Risk (Priority Ranking)	Corresponding U.S. EPA Ranking (*Unfinished Business*, 1987)
1. Chemical waste disposal	16. Hazardous waste sites—active 17. Hazardous waste sites—inactive
2. Water pollution	9. Direct point source discharges 10. Indirect point source discharges 11. Nonpoint source discharges
3. Chemical plant accidents	21. Accidental releases—toxics
4. Air pollution	1. Criteria air pollutants 2. Hazardous air pollutants
5. Oil tanker spillage	22. Accidental releases—oil
6. Exposure on the job	31. Worker exposure
7. Pesticide residues on food	25. Pesticide residues on food
8. Pesticides in farming	26. Application of pesticides 27. Other pesticide risks
9. Drinking water	15. Drinking water
10. Indoor air pollution	5. Indoor air pollution 30. Consumer product exposure
11. Biotechnology	29. Biotechnology
12. Strip mining	20. Mining waste
13. Nonnuclear radiation	6. Radiation other than radon
14. The greenhouse effect	8. Global warming

The following U.S. EPA problem areas were not ranked in the public poll:
3. Other air pollutants; 4. indoor radon; 7. stratospheric ozone depletion; 12. contaminated sludge;
13. estuaries, surface waters, and oceans; 14. wetlands; 18. nonhazardous waste sites—municipal;
19. nonhazardous waste sites—industrial; 23. releases from storage tanks;
24. other groundwater contamination; and 28. new toxic chemicals.

ronmental problems. It is obvious that many topics seen as very important by one group were much less important to the other.

It has been argued, usually by scientists and technologists, that the public is unrealistically risk-adverse and "risk-illiterate", and cannot evaluate risk rationally. Though this position can be supported by evidence, it is in many ways immaterial: in societies such as the United States or European democracies, for example, public perception controls public policies. The problems of unrealistic risk aversion and the inability to objectively evaluate risk become especially difficult when two potential risks must be balanced against one another: In some cases, there may be simply a refusal to recognize that any trade-offs exist. The "Delaney Clause", passed by the U.S. Congress in 1954 in response to the public phobia about cancer, is a case in point. The legislation essentially bans the slightest use of any pesticide suspected of causing cancer even at high dose levels regardless of how much safer its use may make the food supply.

4.2   TYPES OF ENVIRONMENTAL RISK

Environmental risk is usually defined in terms of damage to biological systems, and, generally, to humans. The most dreaded risks are those related to toxins (poisons), which act directly to degrade the organism that is exposed. Acute toxins have effects that are manifested rapidly (e.g., minutes, hours, or even a few days) after exposure to a hazard. Chronic toxins have effects that become manifest only after a significant amount of time—weeks, months, or even years. Many carcinogens (substances causing cancer) are chronic toxins, and low-level exposure to many heavy metals, such as lead, produces chronic, rather than acute, effects. Another group of substances of substantial concern are the mutagens, substances that can cause inheritable changes in DNA.

A second type of environmental risk, broadly defined and certainly generally accepted, has to do with impacts that degrade widely appreciated aesthetic properties. Included in this group are loss of visibility due to air pollution, petroleum globules in surface waters and on beaches as a result of oil spills, and the corrosion of statuary, buildings, and other valued materials as a consequence of acid rain.

The third type of environmental risk involves damage to important systems of the planet. These systems might be biological, as with species diversity or food supplies, aquatic, as with the circulation systems of the oceans, or atmospheric, as with global warming. As discussed in the previous chapter, large uncertainties are connected with evaluating these risks, but potentially very important consequences could result if any of the planetary systems are significantly changed.

The three types of risk, and their relationship to the impacts discussed, are listed for convenience in Table 4.3.

The recognition of different types of risk brings to the fore the issue of risk comparability, because risks as disparate as human carcinogenicity, loss of biodiversity, and loss of a unique ecosystem are not readily reduced to any common quantitative denominator. For this reason, the Dutch National Environmental Policy Plan explicitly declares that three categories of impacts—effects on biological systems, including humans; effects expressed in monetary terms; and effects on quality of life—cannot be considered as interchangeable. This declaration begs the issue, however; unless the impacts are somehow ranked on the same scale, some will usually be regarded as more worthy of attention than others.

Table 4.3   The Division of Major Areas of Environmental Concern into
Types of Risk

Damage to Biological Systems	Aesthetic Degradation	Damage to Planetary Systems
Acid rain (aquatic effects)	Acid rain (Corrosion of materials)	Biodiversity loss
Air toxics (including smog)	Oil spills (visual effects)	Changes in ocean circulation
Groundwater degradation	Visibility loss	Global warming
Hazardous waste sites	Loss of opportunity for wilderness	Ozone depletion
Herbicides, pesticides	experiences	Loss of arable land
Oil spills (wildlife effects)		
Surface-water degradation		
Radionuclides		
Toxics in sediments		
Toxics in sludge		
Loss of habitat		

4.3   RISK ASSESSMENT

Risk assessment may be defined as a method for evaluating risks so that they may be
better reduced, prioritized, avoided completely, or rationally managed. The assessment
may range from the informal and qualitative (in our personal lives) to the relatively rigorous and quantitative (in environmental regulation). In all cases, however, risk assessments deal with uncertainty and rely on assumptions and methods that may or may not
be reasonable under the circumstances. If well done, they are best estimates; if poorly
done, they are useless or, worse, lead to inappropriate or harmful choices.

1. In the *hazard-identification* stage, the first of five stages in a risk assessment, one
    determines whether a particular situation or material indeed poses a nontrivial risk.
    In environmental situations, this determination may not be straightforward,
    because it is often based on extrapolation from animal bioassays or more limited
    tissue- or cell-culture tests, epidemiological studies, or application of generic
    hazard-assessment procedures to specific conditions. Some of the uncertainties
    that may underlie a toxicological experimental design are presented in Table 4.4.
    In principle, these factors are multiplicative, so that the uncertainty in the risk-
    assessment result can be many orders of magnitude.

2. The second stage of risk assessment, taken if a risk is determined to be present,
    is to *evaluate the dose*. This step may also require defining microenvironments.
    For example, assessing the risk of formaldehyde released by certain building
    materials incorporated into mobile homes would focus on the indoor air microen-
    vironment. It is generally the case that the concentration of a causitive substance
    is determined by measurement in the environment of interest, for example, nitro-

Table 4.4   Contributions to Uncertainty from Toxicology Studies

Source	Estimated contribution
Extrapolation of animal toxicology data from high doses to low doses	Factor of 10^5–10^6
Basing risk estimate on a curable cancer, such as skin cancer	Factor of 10^4
Relating animal data to human impact	Factor of 10^3–10^4
Effects of doses of more than one chemical at once	Factor of 10^1–10^2
Basing risk estimate of response in animal tissue with no corresponding human tissue	Factor of 1–10
Statistical noise—extrapolating from only a few animal impacts in a large test animal population	Factor of 2

Source: Abstracted from C. R. Cothern, Uncertainties in quantitative risk assessment—Two examples: Trichloroethylene and radon in drinking water, in C. R. Cothern, M. A. Mehlman, and W. L. Marcus, eds., *Risk Assessment and Risk Management of Industrial and Environmental Chemicals*, pp. 159–180, Princeton: Princeton Scientific Publishing Company, 1988.

gen oxides in smoggy air or a pesticide in groundwater. In the case of a biological system risk assessment, the dose is the portion of the concentration that is bioavailable (some heavy metals in water may be sorbed to particles and not bioavailable, for example).

3. The third step in the risk assessment is to *determine the probability of an undesirable impact as a result of the delivered dose.* Such assessments may consider the whole organism or they may focus on specific organs or internal biological systems. (Figure 4.2 illustrates the routes of exposure that can be considered in such an analysis). There may also be an effort to focus on sensitive subgroups within the population; children or the elderly, for example, frequently exhibit increased sensitivity to some risk agents.

Table 4.4 indicates that a major uncertainty in risk analysis is the methodology used to extrapolate the results of bioassays at very high concentrations of the toxicants to very low concentrations: This choice in itself can determine the outcome of the risk assessment. Figure 4.3, for example, compares four different methods of extrapolating animal data for trichloroethylene ingestion exposure down to very low concentrations. Whether the low concentrations pose any risk at all turns out to be dependent on the choice of extrapolation method. There are numerous mathematical approaches to extrapolation, reflecting the underlying uncertainty about carcinogenesis in biological systems. The *log probit model*, for example, assumes a Gaussian distribution of the logarithms of the test doses. With this model, the probability P of an effect from dose d is given by

$$P(d) = \Gamma(\alpha + \beta \log_{10} d \,) \tag{4.1}$$

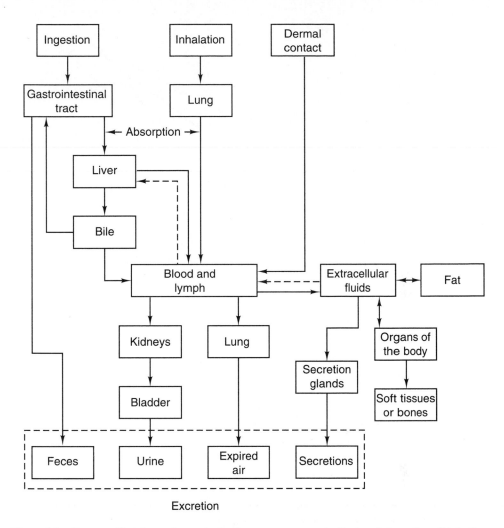

Figure 4.2   Dose manifestation pathways in dose/response assessments: Routes of absorption, distribution, and excretion of toxic chemicals in a mammalian system. (Office of Policy Analysis, *Principles of Risk Assessment: A Technical Review.* Washington, DC: U.S. Environmental Protection Agency, 1987.)

where α and β are constants, and Γ is the standard cumulative Gaussian distribution. Obviously, as dose approaches zero, the number of affected individuals does so as well.

The *multistage model*, on the other hand, assumes that even at very low concentrations, dose and effect are linearly related: in essence, that there are no safe exposures. The generalized form of this model is

$$P(d) = 1 - \exp\left[-(\gamma_0 + \gamma_1 d + \gamma_2 d^2 + \cdots + \gamma_r d^r)\right] \tag{4.2}$$

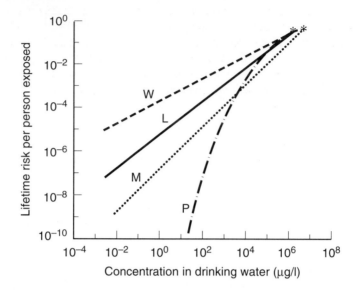

Figure 4.3   Bioassay data for ingestion of trichloroethylene in water (the asterisks in the upper right corner) and extrapolations by the logit (*L*), multistage (*M*), log probit (*P*), and Weibull (*W*) models. (Adapted with permission from C. R. Cothern, Uncertainties in quantitative risk assessment—Two examples: Trichloroethylene and radon in drinking water, in C. R. Cothern, M. A. Mehlman, and W. L. Marcus, eds., *Risk Assessment and Risk Management of Industrial and Environmental Chemicals*, pp. 159–180. Copyright 1988 by Princeton Scientific Publishing Company.)

which, for small values of $P(d)$ and assuming that a chemical of concern affects only one stage of a multistage biological system, can be approximated by a *linear model* relating dose and probability of effect:

$$P(d) = \gamma_0 + \gamma_1 d \tag{4.3}$$

The γ's are constants and, in the expanded model, r is the number of stages in a biological system affected by the chemical.

In Fig. 4.3, the linear and log probit models capture nicely one of the fundamental arguments in many risk assessments of carcinogens: Is there a threshold below which carcinogenesis does not occur? If there is, a variation of the log probit model is appropriate; if not, a linear model is more reflective of actual risk.

A major complicating factor in risk analysis from an ecological standpoint is that it has generally focused on human carcinogenesis as the endpoint, thus drawing regulatory and public attention away from other human health risks as well as virtually all risks to other species and to ecosystems in general. An additional confounding effect arises because risk analysis has been applied almost entirely to the characterization of industrial substances while ignoring the ubiquitous toxins of the natural world. Some have argued that this approach has seriously skewed public perception of the true risks posed by synthetic materials. The argument is that if risks are only assessed for industrial (i.e., synthetic) materials, it implies to the public that natural materials are safer, even if this is not the case. A useful perspective is provided by the relatively large amounts of naturally occurring pesticides that are produced by plants in self-defense mechanisms against insects and fungi and are then ingested in our daily intake of food. As Table 4.5 shows, the average doses of these chemicals are far larger than those of synthetic pesticides and other environmental pollutants.

4. The fourth stage in risk assessment involves *determining the exposure*, that is, what is the population that is exposed to the dose? Once that population is established, the total risk impact is then computed from

$$I = NP(d_i) \tag{4.4}$$

where N is the number of exposed people.

The four-step approach is well-established as a tool for human risk assessment, though the choice of an appropriate dose/response relationship is a continuing point of contention. Comprehensive risk assessment intends to apply similar methods to the evaluation of aesthetic and planetary risk. Few efforts have been made to put these three types of risk on a comparative basis, but it is instructive to consider how this might be done.

In the case of aesthetic risk, dose/response relationships have been established for such impacts as sulfur gas emissions/visibility degradation and acid rain pH/carbonate stone loss. One can estimate the number of people for whom visibility degradation represents loss of quality of life, and thus Eq. (4.4) could be

Table 4.5   Average Exposures to Natural and Synthetic Pesticides

HERP (%)*	Average Daily Human Exposure	Human Dose Produced
0.1	Coffee (3 cups)	Caffeic acid (24 mg)
0.03	Spices	Safrole (1.2 mg)
0.03	Orange juice (0.8 glass)	d-Limonene, 4.3 mg
0.002	DDT	DDT (13.8 μg)
0.0009	Brown mustard (68 mg)	Allyl isothiocyanate (63 μg)
0.0008	DDE	DDE (6.9 μg)
0.0006	Celery (0.4 stalk)	8-Methoxypsoralen (13.2 μg)
0.0002	Toxaphene	Toxaphene (600 ng)
0.00009	Mushroom	p-Hydrazinobenzoate (28 μg)
0.000001	Lindane	Lindane (32 ng)

*HERP is the human exposure/rodent potency index. It represents the percentage of the rodent potency dose received by a human being during a given lifetime exposure.

Source: Values are abstracted from Table 3 of L. S. Gold, T. H. Stone, B. R. Stern, N. B. Manley, and B. N. Ames, Rodent carcinogens: Setting priorities, *Science, 258* (1992): 261–265.

invoked as before. Few people, however, would rank visibility degradation at the same level of seriousness as carcinogenesis, so it is probably appropriate to add a weighting factor ω to the equation:

$$I = \omega NP(d_i) \tag{4.5}$$

where ω would be expected to be small (< 1) but must be determined by societal consensus.

In the case of global impacts, the number of people is not in question—it is the population of the planet—but the dose/response relationships and weighting factors are not at all established. It is clear that the weighting factors should be high for serious irremedial impacts, since the whole sustainability of the planet is at risk. Further, since the risk of global impacts often extends for several generations, integration over time is necessary. Equation (4.5) then becomes in principle

$$I = \omega \int_{t_o}^{t_L} N(t)\, P(d_{i,t})\, dt \tag{4.6}$$

where both the dose and the affected population are time dependent and the integration is performed from the present time (t_o) through the lifetime (t_L) of the substance in question.

Establishing defendable dose-response functions for planetary systems impacts is complicated in part because the time and distance scales remain well outside the norm of standard human experience. A perspective on this issue is provided by attempting to construct for global warming a dose manifestation diagram similar to that for mammalian systems. As seen in Fig. 4.4, the increase in green-

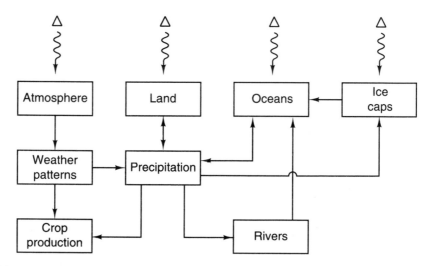

Figure 4.4   Dose manifestation pathways in dose/response assessments: Routes of heat absorption and distribution in the Earth system.

house gases threatens to add heat to four components of the system: the atmosphere, the land, the oceans, and the ice caps. Added heat to land and ocean can be expected to increase evaporation and thus precipitation. Adding heat to the atmosphere may change weather patterns, and thus alter precipitation location as well as amount. Among the likely results are modifications to the ice caps, which are major reservoirs for surface water. Other effects of changed precipitation are likely on river flows and crop production. Uncovering the relationships between the dose of heat and the response of various parts of the system will be enormously difficult. In principal, however, risk assessment is approached in the same way as a mammalian cancer concern, and should be so regarded.

5. The final stage in any risk assessment is to *characterize the risk*; that is, the results are summed up in a statement that attempts to quantitatively define the risk posed by the risk agent under the assumptions of the assessment. Ideally, the risk characterization should include relevant statistical information concerning the uncertainty of the estimate, such as variance and confidence level.

4.4   RISK MANAGEMENT

It is not surprising that the evaluation of risk through formal risk-assessment methodologies is frequently highly contentious. Moreover, the levels of uncertainty and methodological and data weaknesses are such that no risk assessment can be made without numerous assumptions, virtually all of which are also highly contentious. So why do it? The reason is that the assessment is needed in order to set the basis for *risk management*, the steps that are taken to respond to the perceived risk.

Can we expect society to manage in a reasonable way the results from risk assessments? It is far from obvious that thoughtful responses to risk will be a frequent occurrence. Clearly, the human desire is for risk-free situations, notwithstanding the obvious fact that many of life's activities involve a measure of risk. In fact, the very word risk carries with it the meaning that the prospect of an unwanted happening is problematical rather than assured. Risks are, of course, to be avoided if such avoidance does not involve actions that result in even less-desired consequences. Thus, one generally chooses to cross a busy street at a corner with a traffic light, even if a short walk to that corner is required. To put the discussion in the context of industrial ecology, if one has need for a product, and the responsible manufacture of that product imposes some degree of impact on resources and the environment, prioritization and choices among alternatives must enter the picture. A crucially important concept for making logical decisions about industry and about life is thus that risk cannot be avoided; it must be accepted and prioritized.

SUGGESTED READING

Cohrssen, J. J., and V. T. Covello. *Risk Analysis: A Guide to Principles and Methods for Analyzing Health and Environmental Risks*, U.S. Council on Environmental Quality. Washington, DC: National Technical Information Service, 1989.

Cothern, C. R., M. A. Mehlman, and W. L. Marcus, eds. *Risk Assessment and Risk Management of Industrial and Environmental Chemicals.* Princeton: Princeton Scientific Publishing Company, 1988.

Cropper, N. L., and P. R. Portney. Discounting human lives, *Resources, 108* (Summer 1992): 1–4.

Freudenburg, W. R. Perceived risk, real risk: Social science and the art of probabilistic risk assessment. *Science, 242* (1988): 44–49.

Gold, L. S., T. H. Slone, B. R. Stern, N. B. Manley, and B. N. Ames. Rodent carcinogens: Setting priorities. *Science, 258* (1992): 261–265.

Science special issue on risk assessment, *236,* April 17, 1987.

Zeckhauser, R. J., and W. K. Viscusi. Risk within reason. *Science, 248* (1990): 559–564.

EXERCISES

4.1 The U.S. standard for maximum levels of cadmium in drinking water to ensure against chronic risk is 1.1 mg/1. If cadmium follows the same dose-response patterns as invoked in Fig. 4.3 for trichloroethylene, but is three times less toxic, what extrapolation method was used to set the standard? (Recall that U.S. standards are set to have risk no higher than one lifetime response per million persons exposed.) What would the standard have been had other extrapolation methods been used?

4.2 For each of the following scenarios: (a) Define the probable risk qualitatively. (b) Predict the public reaction. (c) Determine whether the risk, if any, warrants a change in existing practices or technologies.

Scenario 1. A truck carrying low-level radioactive material overturns on an exit ramp near a school. A poorly packed canister splits open, but it is not apparent that any material actually escapes.

Scenario 2. On a late-night television talk show, a guest with an inoperable brain tumor claims that it was caused by a newly introduced portable telephone/television, and announces his intention to sue for substantial damages.

Scenario 3. The U.S. Food and Drug Administration announces that 2 of 100 shipments of kumquats from Erebus have tested positive for residues of a pesticide that has caused elevated occurrences of benign tumors in rats, but not in mice or hamsters, in standard toxicological assays.

Scenario 4. A startup firm, Superconductors 'R' Us, announces that it has developed a malleable, formable material that remains superconducting at room temperatures and that contains the highly toxic element thallium. The corporation announces plans to license the material widely across industries for use in a broad variety of energy-saving devices. Initial U.S. Department of Energy and U.S. EPA estimates indicate that energy-use reductions of up to 50% may be possible in highly electrified countries.

PART II: RELEVANT EXTERNAL FACTORS

CHAPTER
5

Relationships of Society to Industry and Development

5.1 WANTS AND NEEDS: THE DRIVING FACTOR

Industrial systems operate within societies and their economic structures, rather than distinct from them. This relationship produces benefits, such as the creation and expansion of markets; it also produces liabilities, such as environmental impacts. As a consequence, industrial systems are constrained by governmental policies and regulations, and, more broadly, by social morays and economic and technological conditions. To the extent that the benefits of industry–society interactions are recognized and encouraged, they can serve to further industrial operation and industrial ecology.

All industrial activity is a response to society's needs and wants. The terms *needs* and *wants* have a variety of meanings, but from the standpoint of the industrial system, both generate demand for products. Sample needs and wants of different societal constituencies at different spatial scales are shown schematically in Table 5.1, transportation being used as an illustration. At the local level, the desire of government for development leads to the construction of rail lines and highways, thus allowing producers ready access to markets and to labor supplies. Individual consumers use these facilities not only for commuting to work, but also for such private tasks as shopping. In many cases, it is immaterial to the customer exactly how the transportation is provided. If public transportation is sufficiently convenient and inexpensive, it will be used. Otherwise, customers will prefer to operate private automobiles even though ownership is often expensive and adds complexity to daily living.

As spatial scales become greater, planning assumes longer time horizons, the movement of goods and services is a central focus, and individual transportation becomes less and less appropriate. Transportation as a component of security and competitiveness assumes interest at the national scale. At the international scale, factors

Table 5.1 The Needs and Wants of Principal Constituencies for Transportation, Evaluated at Different Spatial Scales

Constituency	Spatial Scale		
	Local	National	International
Government	Regional development	National security	Trade competitiveness
Primary producers	Dedicated systems	Dedicated systems	Market diversity Stability of demand
Secondary producers	Labor supply	Product distribution Market access	Exports Market presence
Consumers	Commuting Shopping	Recreation Business	Vacation Business

Source: Adapted from a matrix devised by T. E. Graedel and S. Rayner.

such as the opening of markets and the provision for shipment of large quantities of manufactured goods become elements of tranportation planning. The message of this table is that there are many facets of the industry–society relationship, and many levels of motivation and constraint. The sum of these forces produces society's needs and wants.

Once needs and wants are identified, how does industry act within society to respond to the resulting demands? A concept for such interactions is shown in Fig. 5.1.

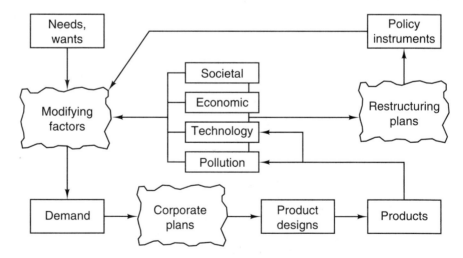

Figure 5.1 Interactions between industrial activities and societal systems. The wavy-line boxes identify planning stages necessary to connect forcing functions and responses. (Adapted from a matrix devised by C. J. Andrews, T. E. Graedel, and S. Rayner.)

The flows of information in the figure begin with the needs and wants in the upper left of the diagram. These motive forces are modified by various societal factors, economic constraints, concerns regarding hazards and environmental impacts, and the state of technology. The result is a demand for specific goods and services. Industrial corporations, responding in their own ways to available information, design, evaluate, and produce these goods and services.

5.2 SOCIETY AND SUSTAINABLE DEVELOPMENT

"Sustainable development" has been defined as "Development that meets the needs of the present without compromising the ability of future generations to meet their own needs" (World Committee on Environment and Development, 1987). Although it is generally recognized that societies are moving toward a greater appreciation of sustainable development, that movement is not instantaneous. At a 1992 meeting of the Economic Commission for Europe, the technological transformation was described as a five-stage process:

Stage 1. Ignorance: The environmental problems are unknown.

Stage 2. Lack of interest: The environmental problems are known, but people do not care about them.

Stage 3. Reliance on technology: People hope that new technology will solve all environmental problems.

Stage 4. Toward sustainability: Conversion of society in the direction of more environmentally adapted developments has begun.

Stage 5. Absolute sustainability: The ecological cycle has been brought full circle.

No society or community has reached the final stage and perhaps none will ever do so. Nonetheless, many groups are now progressing through these stages. Especially as those efforts are encouraged by governmental actions, their progress will continue.

The societal modifying factors in Fig. 5.1 evolve gradually as society moves through the stages of sustainable development. These factors are far from uniform when one society is compared with another. The most obvious example of nonuniformity is the difference in typical environmental approaches adopted by less developed and more developed countries. The latter increasingly see environmental degradation as an important limitation on industrial activity, at least where industrial activity takes the form of unconstrained industrial development. In contrast, the priority in most less developed countries is toward development that will transform their economic base from an emphasis on agriculture and natural resources to an emphasis on industrial production, the goal being overall improvement in quality of life. With this latter focus, environmental concerns generally receive less attention.

Although societal differences make global implementation of industrial ecology more complex and challenging than might be the case in a more uniform world, Ashok

Gadgil of Lawrence Berkeley Laboratory in California points out that under some circumstances the interests of less developed and more developed countries can have substantial overlap. His chief example is one of promotion of energy efficiency. Whereas energy efficiency may not be high on the priority list for officials of less developed countries, an ample supply of energy is near the top of that list. However, energy-generation facilities are very costly, and raising capital is difficult. It turns out that a factory for the production of compact fluorescent lamps is more than a hundred times cheaper than building power plants, if compared on the basis of units of energy conserved or produced. Many similar examples may emerge as materials conservation and reuse continue to be studied. Thus, environmentally sound strategies can often be sound development strategies as well.

As demonstrated in previous chapters, the human population since the beginning of the Industrial Revolution has been growing at a roughly exponential rate. Prior to that time, environmental constraints on growth were local, not regional or global, so intra- and interspecies competition involving humans did not challenge the perception of the desirability of growth. As a result, it is not surprising that most, if not all, human institutions and intellectual disciplines implicitly assume that continuing rapid growth is both feasible and desirable. They embody the assumptions and attitudes appropriate for an exponential growth phase, not the carrying-capacity phase. Thus, most economic approaches favor an aversion to implied limits to growth, especially reduced growth in capital stock. Such an economic philosophy is reflected in the way in which the capital market operates on the basis of an implied cost of capital—a discount rate—that effectively makes resources "worth more" if they are used today, as opposed to being saved for future generations.

In law, the obvious example of the "now is better" approach is the failure of the legal system to recognize any rights of parties not present in a dispute, thus disenfranchising future generations. Issues of family planning and population control are clouded by such legal and cultural factors as women's rights, religious beliefs, and government policy. In addition, progress is slowed by lack of a formal international legal structure for environmental issues, notwithstanding the promising example of the Montreal Protocol for Stratospheric Ozone Protection.

The fundamental point is that, over time, implementation of industrial ecology and migration toward sustainable development will involve significant and difficult cultural, religious, political and social change. Indeed, it is likely that these societal systems, not the technological and economic systems upon which so many people focus, will be the most difficult to integrate into sustainable development efforts.

5.3 IMPLICATIONS FOR THE CORPORATION

The private firm is a critical element of the progress toward sustainability, and will be inevitably impacted significantly as industrial ecology principles begin to be implemented. Among the impacts that bear emphasizing are the following:

1. The private corporation, as the expert in its technology, needs to become a partner in the development of new regulatory structures. This will require corporations to develop positions that are acceptable to all stakeholders in the society—including regulators and environmentalists—and eschew the adversarial, negative approaches that have too frequently characterized past behavior.

2. New organizations and information flows will have to be developed as corporations internalize environmental issues. In particular, the legal and government affairs organizations within corporations will have to become sophisticated in developing and presenting positions that integrate environment and technology in a responsible manner.

3. Corporations need to participate in efforts to design and implement full-cost accounting, so that environmental costs are automatically incorporated into economic decisions.

4. Corporations need to view society as a whole, and the community within which they operate, as full partners in their activities. They need, in fact, to view communities (in the broadest sense of the word) as customers of their services rather than just purchasers of their products. Environmental issues may, over time, shift the fundamental raison d'être of private corporations away from a sole focus on shareholders and profits toward a broader social role.

5.4 OPTIONS FOR TECHNOLOGY–SOCIETY RELATIONSHIPS

The preceding factors do not negate the critical role of technology in implementing industrial ecology. Table 5.2 presents four approaches sometimes taken in approaching the interactions among industry, environment, and society. These options should be considered in light of Fig. 1.3, Chap. 1, which illustrates graphically the different historical relationships between the supportable human population and the state of technology.

The first option in Table 5.2 is *radical ecology*. It represents essentially a program of preindustrial, low (even anti-) technology, pastoralism. It rejects the use of modern agriculture, electronics, medicine, transportation, and other benefits of technology. Although this option clearly responds to the environmental costs associated with the Industrial Revolution to date, there are significant costs to its implementation, including a greatly enhanced susceptibility to famine, disease, and "acts of God". Adoption of this option seems likely to result in a world that could not support the current population, much less that of the future. Were technology to be suddenly removed from society, the result would almost certainly be a swift and chaotic collapse of the human population. The moral and ethical implications of such a population reduction are obviously considerable.

The fourth option, *continuation of current trends*, is also a fundamentally flawed, high-cost choice. The most obvious problem with this path, which is essentially a continuation of exponential growth until terminated by extreme environmental pressure, is that it can be followed only in the short term and only by imposing substantial costs on future generations and global environmental systems generally. Continually escalating

Table 5.2 Options for Technology–Society Interactions

Approach	Effect on Technology	Implications
Radical ecology	Return to low technology	Unmanaged population crash; economic, technological, and cultural disruption
Deep ecology	Appropriate technology, "low-tech" where possible	Lower population, substantial adjustments to economic, technological, and cultural status quo
Industrial ecology	Reliance on technological evolution within environmental constraints; no bias for "low-tech" unless environmentally preferable	Moderately higher population, substantial adjustments to economic, technological, and cultural status quo
Continuation of status quo	Ad hoc adoption of specific mandates (e.g., CFC ban); little effect on overall trends	Unmanaged population crash; economic, technological, and cultural disruption

Source: Adapted from B. R. Allenby, Industrial ecology: The materials scientist in an environmentally-constrained world. *MRS Bulletin, 17* (3) (1992): 46–51.

material flows and rapid growth in capital stock, energy, and resource consumption simply cannot be maintained. Ad hoc responses to individual environmental symptoms rather than a systems-based, comprehensive approach to environmental perturbations have been shown to be inadequate to achieve a sustainable global system. The most likely outcome of continuation of the status quo is, ironically, similar to that resulting from the radical ecology: a dramatic and uncontrolled reduction in human population, with a significant risk of political, economic and social disruption.

The final two options, *deep ecology* and *industrial ecology*, share a recognition that environmental considerations and constraints must be internalized into human cultures and economic activity at all levels. They differ, however, in their view of the role of technology in the transition to a sustainable world.

Deep ecology tends to view technology with suspicion, in part because of the recognized impact of technology on environmental perturbations. Accordingly, technology is given little role in the evolution of a sustainable world; rather, the proponents advocate a return to low-technology options, such as bicycles for transportation. The result would be a lower level of support for populations and a reduction in Earth's carrying capacity, but such a reduction might be capable of being managed well enough to produce a gradual and controlled transition. In the deep ecology view, technology is something to be controlled, not exploited; it remains part of the problem rather than being viewed as a necessary component of the solution. Nonetheless, given the vested interest many institutions, states, and consumers have in an industrialized economy, this option may well be politically unattainable without great difficulty.

Industrial ecology, on the other hand, recognizes the need for continued technological evolution and sees the development of environmentally appropriate technologies as a critical component of the transition to a sustainable world. Given that the global technology state and the level of human population are inextricably linked, if the goal is to maintain current population levels (or even allow for population growth), evolving an appropriate technology state is a crucial requirement. This new technology must be innovative at its core, and include the development of efficient products, processes, and services, improved energy efficiency, use of advanced materials, miniaturization, and advanced information management. It is important to recognize, however, that industrial ecology is not merely naive technological optimism. The data on human perturbation of complex fundamental natural systems are sparse and uncertain, and do not support a facile certainty that the human species can migrate itself to a relatively stable, desirable carrying capacity without significant economic, social, and cultural dislocations, or even precipitous population fluctuations. Very much in the spirit of William James's pragmatism, however, industrial ecology adopts as an operative assumption the possibility of a reasonably smooth transition to a stable carrying capacity: to assume otherwise, as some do, too frequently leads to extremism, despair, or inaction. Rather than unbridled optimism, there is an element of realism (if not fatalism) in this assumption: Progress from the present point in human history must occur within the degrees of freedom that actually exist, not those that wishful thinking would create. A world population approaching six billion, all desiring a better material life-style, cannot be ignored.

The four societal options are now seen to identify themselves with the possible outcomes shown on the population curves of Fig. 1.3. In Fig. 5.2, we add societal option labels to those curves. It is clear that as a society, it is much more attractive to consider the paths of industrial ecology or deep ecology rather than the more extreme alternatives. In addition, because it is difficult to conceive of a global decision to reduce population in an orderly fashion, deep ecology may not be a realizable scenario. Thus, there is a fundamental systems truth to emphasis on industrial ecology. Although it is generally recognized in the biological context, it is a characteristic of all complex systems that they can evolve in response to external changes in the environment. Indeed, in many cases, this ability to evolve is critical to continued survival of the system itself. There is no question whether human society will respond to the significant forcing of environmental constraints. Rather, the question is whether that society will choose to do so in an orderly, rational manner. Our thesis is that we should indeed do so, and that industrial ecology is the best guide to the form of that response.

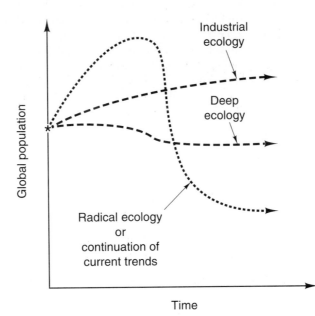

Figure 5.2   The alternative human population scenarios suggested by different options for technology-society interactions. The asterisk marks the present global population of approximately 5.3 billion.

SUGGESTED READING

Allenby, B. R. Evolution of the private firm in an environmentally constrained world. Paper presented at the Workshop on Engineering Within Ecological Constraints, National Academy of Engineering, Washington, DC, April 17–19, 1994.

Environmental Advisory Council. *Eco-Cycles: The Basis of Sustainable Urban Development*, Report Jo 1968. Stockholm: A. Allmänna Förlaget, 1992.

Gadgil, A. Integrating environmentally friendly technologies in the developing process, in *Industrial Ecology and Global Change*, R. Socolow, C. Andrews, F. Berkhout, and V. Thomas, eds. Cambridge, UK: Cambridge University Press, 1994.

Stern, P. C. A second environmental science: Human-environment interactions. *Science, 260* (1993): 1897–1899.

Turner, B. L., W. C. Clark, R. W. Kates, J. F. Richards, J. T. Matthews, and W. B. Meyer, eds. *The Earth as Transformed by Human Action*. Cambridge, UK: Cambridge University Press, 1990.

EXERCISES

5.1   Technological transformation is described in this chapter as a five-stage process. At which of these stages do you put the world as a whole? At which do you put your country? Predict the year each of the higher stages will be achieved, and defend your reasoning.

5.2 Redo Table 5.1 for communications rather than for transportation. If you are an executive in a corporation that manufactures communications equipment, what business opportunities are suggested by your table?

5.3 Increasing urbanization of the world's peoples is everywhere evident. For example, the percentage of the global population living in cities with greater than one million inhabitants increased from about 29% in 1950 to about 43% in 1985. Discuss what this trend implies for flows of products and residues: food, water, sewage, and so on. Overall, is urbanization environmentally advantageous or disadvantageous?

CHAPTER

6

Governments, Laws, and Economics

6.1 GOVERNMENTAL STRUCTURES AND ACTIONS

Industrial ecology includes among its many facets an understanding of the potential role of government as a factor in implementation. This factor requires the practitioner to understand the political and legal environment within which she or he operates, as well as the potential for influencing the development of rational public policies. Often, these activities are interrelated with economic issues.

Governments at all levels purchase products for their own use, so are direct customers, but they may also be thought of as "indirect customers". In this role, they express not the direct demand of consumers, which is the province of the marketplace, but the indirect demand of consumers as formulated in legislation, regulation, and less formal policies and practices. More than is commonly realized, governments as well as customers shape markets, and they have it within their power to create new markets. For example, the U.S. Clean Air Act Amendments of 1991 included a requirement that employers in geographical areas with poor air quality decrease the amount of commuting performed by their employees by at least 10%. This requirement is tantamount to generating a new market for telecommuting services, related hardware and software, new van and bus services, and the like. However, few companies who could have taken advantage of this new market have done so, having failed to recognize governmental environmental actions as critical market dimensions in business planning.

Governments establish technology policies actively—Japan and Germany are examples of this category—and passively, as was generally the case of the United States during the Reagan and Bush administrations. Policies can encourage environmentally preferable technologies and industrial ecology principles: Product takeback legislation, if properly implemented, is one example. Alternatively, policies can impede such progress: Laws that restrict the ability of government entities to purchase products containing refurbished subassemblies or components or that grant subsidies or tax breaks to extractive industries are examples of the latter. In general, government technology policies that assist the diffusion of new technologies will also indirectly increase the use of environmentally preferable technologies, because newer technologies tend to be more efficient in the use of resources and energy.

There are a number of dimensions that can significantly affect the ability of nation-states to respond to environmental challenges. Among these are the following:

1. *Form of government.* In general, democracies such as those in western Europe and the United States will be more responsive than more totalitarian governments such as those that used to exist in eastern Europe.

2. *Wealth.* Wealthier countries have more resources with which to respond to environmental challenges than do poorer countries, and the former may be able to place more relative value on environmental benefits.

3. *Size.* Even very progressive small countries such as Denmark or The Netherlands cannot overlook the fact that much of their industrial production is exported, and thus subject to standards and requirements beyond their direct reach.

4. *Focus.* Countries emphasize different aspects of environmental protection. For example, the United States is a leader in remediation, but lags behind Japan in energy efficiency and behind Germany and The Netherlands in developing consumer product takeback approaches.

5. *Culture.* There is a distinct contrast between Japan, for example, with its parsimonious approach to resources and energy born of its island status, and the former Soviet Union, for example, which possessed a greater natural resource base, a focus on industrialization at any cost, and an accompanying cavalier attitude toward conservation.

International trade forces the political system to think not just of the home nation-state but of all nation-states as stakeholders in environmental issues, and to recognize that approaches among even seemingly similar countries may be quite different for historic and cultural reasons. Nation-states that are environmental leaders will frequently be models for action in other countries. Countries playing a leadership role include Canada, Denmark, Germany, Japan, Sweden, the United States and, perhaps the most sophisticated, The Netherlands.

An example of the activist governmental policies that will become increasingly common is that of the city of Göteborg, Sweden, which in 1990 decided to explicitly consider environmental factors in its procurement practices. The municipality determined that the environmental impact of a product must be taken into account over its entire life cycle, and that preferred products would be those with the following characteristics:

• During a product's manufacture and lifetime, it shall give rise to as little environmental damage as possible.

- Products must not be harmful to the consumer/user (from an occupational health and safety viewpoint).
- When it becomes waste, a product shall be biodegradable or reusable.
- Product manufacture and use are to be based on economical use of raw materials and energy resources.

In order to have sufficient information on which to base its decisions, Göteborg requires that all prospective suppliers provide their products with an environmental label, stating whether hazardous chemicals are used, giving details of packaging, recycling and reuse information, and so forth.

THE NETHERLANDS APPROACH TO ENVIRONMENTAL POLICY

People tend to think of Germany, the United States, and perhaps Sweden as governments most advanced in environmental regulation. The most sophisticated, comprehensive approach to the integration of environment and technology, however, has without question been that of The Netherlands, as set forth in the National Environmental Policy Plan (1989) and the National Environmental Policy Plan Plus (1990). A third plan document is scheduled for release in the near future.[1] The plans, and the implementing Government Memoranda, are all explicitly based on the goal of attaining sustainable development in The Netherlands within one generation, with sustainable development being defined as in the Brundtland Report: "Development that meets the needs of the present without compromising the ability of future generations to meet their own needs." The approach is comprehensive: Although the lead is taken by the Ministry of Housing, Physical Planning and Environment, the need to include all other sectors, especially transportation and housing, is explicitly recognized, by including the relevant Ministries and Councils in the planning process (the plans, for example, were submitted to the Second Chamber of the States General of the Netherlands by not just the Ministry of Housing, Physical Planning and the Environment, but on behalf of the Minister of Economic Affairs, the Minister of Agriculture and Fisheries, and the Minister of Transport and Public Works). Target activities for the plans include agriculture, traffic and transport, industry and refineries, gas and electricity supply, building trade, consumers and retail trade, environmental trade, research and education, and "societal organizations" (environmental groups, unions, etc.).

At all stages of analysis, the economic impacts of proposed changes are explictly considered. Moreover, the emphasis is clearly on collaboration with industry and other stakeholders in developing and implementing specific proposals, the goal being scientifically and technically correct decisions. This collaborative approach is

---

[1] The plans are available from the Ministry of Housing, Physical Planning and Environment, Department for Information and International Relations, P.O. Box 20951, 2500 EZ The Hague, The Netherlands.

borne out in the use of enforceable agreements—"covenants"—between industry and the government to achieve environmentally desirable ends, rather than the more adversarial legislative process, whenever possible.

The plans establish a number of "themes" (environmental perturbations such as climate change or acidification) that are then addressed through identification of appropriate management tools and/or policy actions. A critical focus has been the identification of appropriate metrics to determine progress toward sustainability: The initial plan notes that "Making sustainable development measurable . . . is not an easy task, but the results of research on this topic are necessary to enable feedback at the source."

The approach taken by The Netherlands is unique, pathbreaking, and exemplary of industrial ecology. The implicit acceptance of the need for fundamental change in existing economic structures is both commendable and still all too rare at the national government level. We strongly suggest a deeper study of the approach of The Netherlands to all those interested in government policy and law in industry–environment interactions.

6.2 INTERJURISDICTIONAL CONSIDERATIONS

The nation-state is the jurisdiction with which most people are familiar. In general, nation-states have authority over any activities occurring within their borders, and they are often the primary jurisdiction within which remediation and compliance activities are regulated. Many environmental issues, including virtually all of the difficult ones, are beyond the borders of any single nation-state, however: acid precipitation, watershed pollution, ozone depletion, global climate change, and so on. There is thus a dichotomy in scale between the political bodies with the most authority and legitimacy and the environmental perturbations with which they must deal. The global scope of many of the anthropogenically perturbed natural support systems, in combination with an unwieldy international law system, raises questions about the appropriate venue for addressing environmental issues and adds to increasing devolution of nation-state obligations to international organizations, transnational corporations, and political subunits.

The nation-state is not only a compliance entity, it is also the government level at which international environmental treaties and agreements are approved and enforced. International agreements are becoming increasingly numerous, as shown in Fig. 6.1, and increasingly important. Because of their global scope, these treaties and other covenants are of particular interest to the practitioner of industrial ecology. The most obvious example is the Montreal Protocol under which production and consumption of CFCs and other ozone-depleting chemicals are being phased out. Other less familiar agreements include the Basel Convention, under which transnational shipment of hazardous residues is controlled. More such agreements will undoubtedly occur as the results of the 1992 Earth Summit in Brazil are translated into actionable form over the next decade.

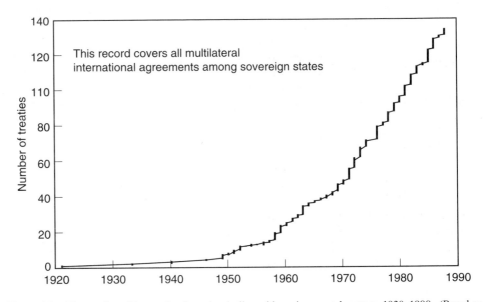

Figure 6.1   The number of international treaties dealing with environmental matters, 1920–1990. (Based on computations by Haas with Sundgren (in press) and derived from UNEP's *International Registry of Environmental Treaties.*) (Reproduced with permission from N. Choucri, The technology frontier: Responses to environmental challenges, in *Global Accord: Environmental Challenges and International Responses*, N. Choucri, ed. Copyright 1993 by the MIT Press.)

Another important aspect of government activity is creating conditions under which trade can occur. Potential conflicts arise because environmental laws generally seek to control the means by which goods are made and to forbid or discriminate against environmentally inappropriate processes, products, or technologies, whereas trade laws in general seek to liberalize the flow of goods among nations. The interaction of trade and environment is thus extremely complex and contentious, involving difficult questions of interpretation of international trade agreements (such as GATT, the General Agreement on Tariffs and Trade, and NAFTA, the North American Free Trade Agreement). Under GATT (which controls the world trading regime), for example, it is permissible for a country to regulate the import of products that are environmentally unacceptable, such as ivory from poached elephants. It is not permissible, however, for a country to attempt to control the manner in which another country produces an otherwise acceptable product. Applying this principle, the United States could not ban the import of Mexican tuna simply because the Mexican fishing fleets killed dolphins in obtaining their tuna, as long as the tuna itself was unobjectionable. Similarly, the creation of free-trade zones carries with it environmental implications: in the case of both the European Union and the North American Free Trade Agreement, considerable concern has been raised by the possibility that the less environmentally conscious participants will degrade the environmental standards of all signatories.

International agreements apply only to signatory countries that specifically agree to be bound by the requirements. In the case of environmental accords, which can impose at least short-term costs on signatories, this limitation can have significant competitive implications. Accordingly, many agreements establish import controls on products made in nonsignatory nations to ensure equality between the signatory parties and other nation-states. For example, when the United States imposed a tax on domestic CFC use, it also imposed a roughly equivalent import duty to ensure that complying domestic manufacturers were not inappropriately penalized for environmentally desirable behavior, at least within U.S. boundaries.

Political subunits of nation-states are increasingly important, because, in many federal systems, considerable power over environmental issues devolves to these subunits. Some of the German Länder, for example, provide funds to encourage ecologically responsible restructuring of small industries, and include environmental reviews in the building permit approval process. In the United States, states led the way in banning packaging foams made using CFCs, in requiring manufacturers to take back batteries and dispose of them properly, in requiring that packaging be free of heavy metals such as lead, in instituting toxic use reduction programs and associated technology diffusion efforts, and in mandating product design changes that improve environmental acceptability of products (such as requiring that products containing nickel–cadmium batteries be designed so that the battery can be removed and replaced).

Political subunit activism cuts both ways, however. It is difficult enough for many corporations to track environmental policy, legislation, and regulation in all the nation-states where they operate; tracking subunits as well is logistically infeasible in many cases. It is unfortunately also the case that many subunits have inadequate technological expertise or advice, and as a result may impose legislation that is impractical, highly uneconomic, or, in some cases, more environmentally damaging than what it replaces. For example, state hazardous waste laws may, by imposing overly burdensome regulation, make it uneconomic to recycle materials that could otherwise be reused. Moreover, the proliferation of differing environmental regulations at the subunit level may impose large transaction costs on international and national trade that are unjustifiable in terms of environmental benefit and drain resources away from more beneficial expenditures.

Local jurisdictions may also impose regulations that pose potential conflicts. For example, local fire law requirements that petroleum storage tanks be buried underground for safety reasons conflict with the environmental benefits of having such tanks above ground, where leaks can readily be detected and repaired. There are often no easy answers to these situations, but seeing them as a whole may produce better resolution than results from piecemeal approaches.

6.3 INDUSTRIAL ECOLOGY AND THE LEGAL SYSTEM

It is relatively straightforward to write legislation to minimize the most egregious environmental consequences of an existing technological process. It is another matter altogether to create legal structures dealing with evolving technology, yet the incentives created by legislation and regulation of corporate and individual behavior are critical to the evolution of our economies toward sustainability. What laws should be implemented to create selective pressures that will encourage the evolution of technologies that serve industrial ecology purposes, when we do not know and cannot tell a priori what those technologies might be? How do we implement boundary conditions that encourage the evolution of environmentally appropriate technologies?

6.3.1   History and Current Status

Environmental laws are not a recent phenomenon. As early as 1306, London adopted an ordinance limiting the burning of coal because of the degradation of local air quality. Such laws became more common as industrialization created substantial point-source emissions. For example, the LeBlanc system for producing soda (sodium carbonate), patented in 1791, resulted in substantial emissions of gaseous HCl to the air in the environs of the facilities. As a result, the English Parliament passed the Alkali Act of 1863, requiring manufacturers to absorb the acid in special towers designed by one William Gossage.

As in these historical examples, much environmental law to this point has reflected the perception of environmental problems as localized in time, space, and media (i.e., air, water, soil). For example, it was not uncommon in the recent past for groundwater contamination by organic solvents to be eliminated by "air stripping," or simply releasing the solvent to the air, where it contributed in many cases to the formation of tropospheric ozone. Similarly, many hazardous waste sites in the United States have been "cleaned" by simply shipping the contaminated dirt somewhere else, which not only does not solve the problem, but creates the danger of incidents during the removal and transportation process. Environmental regulation has thus traditionally focused on specific phenomena and adopted the so-called "command-and-control" approach, in which restrictive and highly specific legislation and regulation are implemented by centralized authorities and used to achieve narrowly defined ends. Such regulations generally prescribe very rigid standards, often mandate the use of specific emission-control technologies, and generally define compliance in terms of "end-of-pipe" requirements. Examples in the United States include the Clean Water Act (by U.S. EPA interpretation applied only to surface waters), the Clean Air Act, and the Comprehensive Environmental Response, Compensation and Liability Act (Superfund), applied to specific landfill sites. Many environmental laws have been predicated on assumptions appropriate to linear flow of materials to waste, rather than the internal cycling characteristic of a sustainable economy. For example, the U.S. Resource Conservation and Recovery Act defines almost any by-product of a linear manufacturing process as a hazardous waste and subjects it to burdensome regulation, thereby limiting the incentives to recycle or reuse the material. The regulation thereby tends to institutionalize the linear manufacturing paradigm, rather than guide industry toward sustainability.

We pointed out earlier that biodiversity would tend to be maintained if new industrial facilities were preferentially built on already altered land, such as urban or industrial areas. It is of interest that such actions might have positive social implications as well. Nonetheless, such actions are discouraged by the common legal practice of making new owners responsible for all existing contamination associated with a site even if they had nothing to do with generating it. Partly as a result of such disincentives, industry continues to develop new sites, degrade land, and add to transportation costs. The legal structure, enacted with noble purposes in mind, thus condemns industrial sites in urban areas to perpetual inactivity.

If properly implemented, command-and-control can nonetheless be effective in addressing specific environmental insults. For example, rivers such as the Potomac and Hudson in the United States are much cleaner as a result of the Clean Water Act. Moreover, where applied against particular substances, such as the ban on tetraethyl lead in gasoline in the United States, the command-and-control approach has clearly worked well. Such regulation, however, is characterized by a great burgeoning of mandatory requirements (see Fig. 6.2), a relative unconcern with economic efficiency, a focus only on the manufacturing stage of industrial activity rather than the life cycle of materials or products, a tendency toward adversarial relations between the regulated community and regulators (which varies significantly by country, however), and, because specific technologies are prescribed, a strong bias against technological innovation. Moreover, in practice, it has proven very difficult to modify such regulations to reflect advances in scientific understanding.

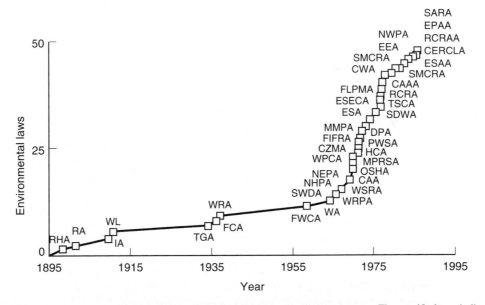

Figure 6.2   The number of United States environmental laws over the past century. The specific laws, indicated by acronymns, are identified in the original reference. (K. E. Yeager and S. B. Baruch, Environmental issues affecting coal technology: A perspective on U.S. trends, *Annual Review of Energy, 12* (1987): 471–502.

6.3.2   Legal Principles Appropriate to Industrial Ecology

From the viewpoint of industrial ecology, command-and-control approaches are clearly a necessary component of an environmental management structure—it should always be illegal to dump untreated electroplating residues containing heavy metals in waterways, for example—but are generally incapable of addressing complex, systemic environmental perturbations or of encouraging innovative approaches. Command-and-control structures were not unreasonable tools to have devised in the first place. However, as legal restrictions on the movement toward sustainability are identified, the tools should be reevaluated to determine how their purposes and the need to achieve sustainability can be balanced. The new approach to environmental regulation that has been evolving recognizes that attempts to micromanage a complex system from a single, centralized node are doomed to failure; dispersed control mechanisms and feedback loops are required. Policies based on this understanding must be implemented if the global economy is to be migrated toward sustainability. Table 6.1 summarizes the differences between the regulatory climates of yesterday and today, which focused on remediation and compliance, and the regulatory environmental management systems that will be required in the future.

The concept of pollution prevention, generally requiring process and product changes to reduce the generation of residues in industrial operations, is increasingly embodied in new legislation in developed countries around the world. Japan, Germany, and other European countries are establishing regulations requiring companies to take back their packaging or products after the consumer is through with them, and recycle or dispose of them properly. Voluntary agreements, such as "covenants" between industry and the government in The Netherlands, or Green Lights (lighting efficiency) or Energy Star (energy-efficient computers) in the United States, are proliferating. Regulations

Table 6.1   Three Stages of Environmental Regulation

Time	Activity	Focus	Geographical/ Temporal Scale	Regulation	Leader
Past	Remediation	Waste substances	Local, Immediate	Command-and-control	USA
Present	Compliance	Emitted substances	Point source, Immediate	Command-and-control, end of pipe	More developed countries
Future	IE/DFE	Products and services over the life cycle	Regional and global systems, all time scales	Establish boundary conditions	European Union, especially The Netherlands and Germany

requiring the reporting of emissions data, without any associated requirements to reduce them, have been implemented in the United States under the Superfund Amendments and Reauthorization Act (SARA), and have proven to be a powerful incentive for manufacturing firms to voluntarily reduce emissions.

Although the new generation of environmental regulation—which might be termed "environmental management"—involves a number of different regulatory tools, it can be characterized in several important ways. It moves beyond end-of-pipe approaches and begins to focus more on operational industrial technologies. It also begins to rely more on incentives to appropriate behavior and less on prescription of technologies, in recognition of the basic fact that industry rather than regulators generally possesses the more detailed technological competency in a modern capitalist economy. This does not mean, however, a naive reliance on industry to "do the right thing"; rather, it involves the development of closer relationships between industry and regulators to work toward common goals and to use meaningful metrics to measure progress toward such goals.

6.3.3   Legal Policy Issues Raised by Industrial Ecology

A number of difficult policy implications arise from the recognition that mitigation of environmental perturbations will require increased reliance on technology. This situation has the effect of empowering private industry, the repository of technology and civilian research and development resources, and reducing the role of environmental regulators and environmental groups, whose technological competency is generally less. Politically, however, it is doubtful that any structure that completely cedes control of environmental progress to industry will be successful. Indeed, polling data show a virtually universal distrust of industry in nations around the world as regards the environment. Accordingly, developing an inclusive, consensual, cooperative process for development of technologically sophisticated environmental management tools would appear to be a necessary element of industrial ecology.

Environmental legal systems reflect not only technological information, but also the culture within which the legal system is contained. Andrews has characterized a number of countries in terms of the openness to stakeholders and the consensual nature of their legal processes (Fig. 6.3). Among those he identifies are the United States, where legal processes are very open but are adversarial rather than consensual; the Netherlands, where legal processes are both open and consensual; Japan, where legal processes take the middle ground on both openness and consensuality; and the United Kingdom, where legal processes are consensual but not open.

Human cultures have evolved during a period of roughly exponential industrial growth, and human intellectual tools thus reflect assumptions and biases appropriate to such continued growth, rather than to sustainability. This inherent bias means that many existing legal concepts and structures that affect the implementation of industrial ecology are not sensitive to newly developed perspectives. Accordingly, the recognition of regional and global environmental perturbations with potentially serious impacts far into the future raises some difficult policy and legal issues. Although they have yet to be

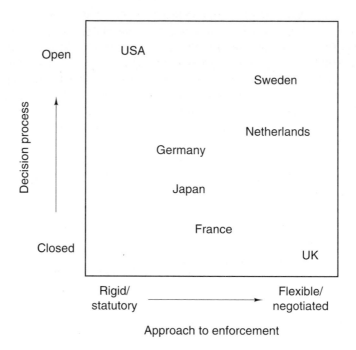

Figure 6.3   Degrees of openness and consensual approaches to environmental legal processes. (C. J. Andrews, Preprint: Policies to encourage clean technology, to be published in *Industrial Ecology and Global Change*, R. Socolow, C. Andrews, F. Berkhout, and V. Thomas, eds. Cambridge, UK: Cambridge University Press, 1994.)

resolved, they must be recognized by the practitioner, as they arise in the context of very real problems.

Conflicts between environmental mitigation efforts and other goals embodied in, for example, antitrust, consumer protection, intellectual property, and trade secret laws are becoming more apparent. Such conflicts are especially likely where cutting-edge competitive technologies or material reuse techniques are involved, precisely the techno-logical systems that environmental mitigation will call forth. Some consumer protection statutes, for example, require an entire product to be labeled as "used" or "second-hand" if it contains any refurbished parts. Such a label can substantially damage a corporation's trademark or reputation for quality; accordingly, such requirements act as a substantial disincentive to implementing recycling of subassemblies and components. Similarly, government specifications and standards often unnecessarily discriminate in favor of virgin material or environmentally unfavorable processes.

Antitrust laws tend to disfavor the establishment of vertical combines that limit "free trade". Such combines, of course, may be a necessary component of product take-back and material management systems, and for ensuring quality control of operations across the life cycle of products and materials.

It is likely that the implementation of industrial ecology will require increased information flow from corporations, accompanied by feedback from regulators and the community. The release of such information, frequently regarded by the corporation as competitively important, has already been a contentious issue with regard to the Community Right-to-Know provisions of the U.S. Superfund Amendments and Reauthorization Act, which required corporations to provide information to the U.S. Environmental Protection Agency on their releases of specified materials to the environment. The trend, however, is clear: It is probable that in the future industrial activities will become a cooperative enterprise between corporations and the communities within which they operate. This change will require that substantial quantities of potentially sensitive information be made public by corporations. Thus, trade secret and proprietary information protection for products and process may be eroded considerably, although intellectual property rights granted by statute, such as patents, will continue to remain viable.

There remain fundamental difficult legal issues that are not yet resolved. A seemingly trivial but vitally important example is the precise definition of waste, which we defined in the first chapter as a residue that our economy had not yet figured out how to use effectively. Among the definitions of the same word used by different legal entities are the following:

(a) Wastes are materials coming from a manufacturing process that are not directly used in another process.

(b) Wastes are materials coming from a manufacturing process for which no further use within the corporation is foreseen.

(c) Wastes are materials coming from a manufacturing process that are marked for disposal.

(d) Wastes are materials coming from a manufacturing process that are released into the environment.

(e) There are no wastes, only residues that should be designed so that an economic use can be found for them.

The situation for a corporation is that whatever is defined as waste is often restricted in its transport, sale, and reuse. If, for example, definition (a) or (b) is used, it will be very difficult for a corporation to transfer the material within its own manufacturing facilities or to sell it to another corporation, even if those actions result in reuse. Conversely, if definition (c) or (d) is applicable, true waste will be properly handled and reusable material can move to its next location of use without overly restrictive and expensive requirements being placed upon it. Finally, if definition (e) is adopted, corporations can attempt to evolve their operations so as to recycle and reuse as much as possible without significant constraints of any kind provided normal safety and environmental impact standards are met.

Another legal (and social) issue that requires resolution is the split between the more developed and less developed countries regarding distribution of resources and technology sharing. This topic is a long-standing and bitter one that has gained new prominence as the global nature of environmental perturbations has been recognized. Less developed countries are more aggressively demanding wealth transfers as the quid pro quo of environmental mitigation efforts on their part, noting, not unreasonably, that the more developed world has used a disproportionate share of the world's resources and has more wealth and expertise available for mitigation purposes.

Analogous to this geographical equity issue is the intergenerational equity problem. Simply put, this is the question of how the interests of future generations in the resource allocation and consumption decisions being made today may be defined and protected by the legal system.

Finally, it is necessary for the legal system to address seriously the issue of how to develop legislative and regulatory structures that can be easily modified as scientific knowledge evolves. A classic example of difficulty in this area is the U.S. Superfund program, which is now generally conceded to have mitigated few risks at enormous expense, in part because the risk assessments made at the time have been largely negated by further studies. Advocates of another approach suggest that legislation must include the possibility of iteration as new information becomes available. Policymakers and their legal advisors are only beginning to devise suitable ways to achieve this goal.

6.4 ECONOMICS AND INDUSTRIAL ECOLOGY

6.4.1 General Principles

The interaction between the discipline of economics and the field of industrial ecology is just now being explored. It is likely that in the future each field will be significantly changed by insights derived from the other: Indeed, economic insights are critical to successful implementation of industrial ecology. Concomitantly, the study of economics without an understanding of industrial ecology will grow increasingly sterile and implausible. For example, the assumption of traditional economics that continuing the rapid growth of population and capital stock into the indefinite future is not only possible but desirable is clearly questionable.

Standard economic analysis strongly holds that money today is worth more than the same amount tomorrow, based on the time value of money and inflation. In other words, future returns are *discounted* compared to today's potential returns. The discounting equation gives the *present value A* of an amount *V* that will be available *t* years from now:

$$A = V \, (1 + i)^{-t} \tag{6.1}$$

where i is the discount rate (cost of capital). In the continuous case, with r equal to the rate of continuous compounding,

$$A = V \, e^{-rt} \tag{6.2}$$

For example, at an annual discount rate of 10%, $100 is worth $90.91 a year from now, $62.11 five years from now, and $38.61 ten years from now.

The use of discount rates to value resources in business and government planning is ubiquitous. Obviously, this approach provides a strong incentive to use resources today rather than save them for the future. The theory is that if money earned today is properly invested, the future will in fact benefit from the most growth-oriented use of existing resources. In general, there is no consideration of the financial impact of eventual resource limitations when discounting methodologies are utilized, and the discounting concept as implemented in practice conflicts with the fundamental industrial ecology principle that the economy should function so as to be indefinitely sustainable.

A second assumption common to standard economic analysis is substitutability among resources and inputs based on dollar value, which in turn reflects scarcity. Using up a nonrenewable resource is acceptable, under this view, as long as the resulting returns are invested in assets useful to the future. The future will have the benefits of the investment, as well as substitutes at whatever prices are efficient. Critics argue that this position is sophistic given that returns are virtually never invested in such a way as to benefit future generations, and conceals the very real problem that the substitutability assumption is questionable. For example, if the electronics industry were to substitute indium alloys for lead alloys in soldering, it might draw down world reserves of indium considerably. Should indium then turn out to be a critical component of the only viable room-temperature superconductor—a necessary complement to existing resources, with no substitutes—the future would indeed have been deprived, regardless of whether the returns related to indium's use were invested or not.

Another difficulty arises because modern economics tends to be highly quantitative. Although such an approach simplifies analysis, it can mean that factors that cannot be quantified are, in practice, simply not included in the analysis. (Although a quantitative analysis is often supplemented in principle by considerations of nonquantifiable impacts, the latter are seldom given the weight of the former.) For example, it is virtually impossible to unambiguously assign a dollar value to any given biological species (methodologies that might do so, such as asking people how much they would pay to preserve a given species, are all both contentious and anthropocentric). Accordingly, economic analyses of development projects that might cause biological disruption tend not to consider any species extinctions that may result. More sophisticated economic analyses might include the potential market value of the species (as a source of new drugs, for example), but will still not include nonquantifiable elements (the value of a species for its own sake).

The failure to integrate nonquantifiable phenomena into economic analyses is of particular concern with environmental perturbations, where quantification is frequently impossible because of fundamental uncertainty and limitations on data. It is impossible to know, for example, what the costs of global climate change will be until it actually occurs: the system is simply so complex, nonlinear, and potentially discontinuous that quantification of global, regional, and local impacts, and associated costs, is for all practical purposes unachievable. This has led some economic analysts to ignore potential discontinuities and to treat such costs as minor, a conclusion that many regard as unrealistic.

Another issue that arises frequently in economic analyses of environmental concerns is that of "externalities". Externalities are costs (or, very occasionally, benefits) that accrue to society or the world at large, but are not captured in prices, and thus not reflected in private economic decisions. For example, when a factory pollutes a river and kills the fish downstream, the costs associated with that activity, and the loss to fishermen and others who formerly enjoyed the river, are, absent legal redress, not borne by the factory. They are externalities.

6.4.2   Evaluation of Costs and Benefits

In considering potential courses of action, a common methodological approach is to attempt to quantify, then compare, the costs and benefits of the activity. This, in the broadest sense, is accounting. In some cases, this may be a relatively formal and rigorous procedure termed a cost/benefit analysis (CBA). In other cases, uncertainty or significant unquantifiable issues, particularly moral or ethical questions, may make less formal procedures appropriate. In any analysis addressing environmental issues, both economic and noneconomic costs and benefits are likely to arise.

Regardless of the format of the cost/benefit analysis, several questions typically must be addressed by the industrial ecologist:

1. What assumptions, implicit and explicit, have been made and are they appropriate? Who is performing the analysis, and what are their interests in the outcome?

2. What are the geographical and temporal distributions of costs and benefits? In particular, what, if any, discount rate is being applied to future costs to justify present benefits?

3. What are the significant uncertainties related to the analysis, and can they be quantified (e.g., a probabilistic estimate of future potential liabilities related to disposal of a given residue)?

4. What is the sensitivity of the conclusion to elements identified under questions 1 through 3?

In answering these questions, the goal should not be to eliminate all use of cost/benefit procedures or resulting courses of action, but to ensure that participants are as well informed as possible.

It is useful to consider questions of cost and benefit at six different levels, each of which raises unique issues: (a) economic costs at the firm level; (b) social costs at the firm level; (c) project costs at the national level; (d) costs and benefits of national environmental regulations; (e) Gross National Product (GNP) and similar national account procedures at the level of the nation-state; and (f) international costs and benefits.

6.4.2.1   Economic Costs at the Firm Level.   Information concerning the economic performance of the firm is generally captured in management accounting systems.

Traditionally, such systems have treated environmental costs—even real, quantifiable environmental costs, such as residue disposal costs—as overhead, and have therefore not broken them out by activity, product, process, material, or technology. The result has been that managers, not having access to environmental cost information concerning their choices, have had neither the incentive nor the data to reduce those costs.

The solution, sometimes called "green accounting," is conceptually simple: develop managerial accounting systems that break out such costs, assign them to the causative activity, and thus permit their rational management. In practice, however, this is a difficult task. For example, in many complex manufacturing operations, developing sensors and systems to provide the physical data on the contributions of different processes and products to a liquid residue stream is a nontrivial task. Moreover, managers tend to resist additional elements of the business process for which they will be made responsible. Also, the assignment of "potential costs", such as estimates of future regulatory liability for present residue-disposal practices, may be resisted for fear of creating unnecessary legal liability (it might be argued that a company that foresaw potential future liabilities thereby was admitting its planned behavior was inappropriate or illegal). Nonetheless, it is clear that development of appropriate managerial accounting systems, and their supporting information subsystems, is critical to completing a necessary feedback loop for environmentally appropriate behavior by corporations.

6.4.2.2 Social Costs at the Firm Level.

There is another type of cost that private firms are increasingly being asked to consider, which may be summarized as a desire for "social costing," or internalization of environmental externalities into the firm's operations and business decisions. This is, after all, the rationale of methodologies such as DFE and LCA: If prices fully captured externalities, such methodologies would be by and large superfluous as firms could simply rely on price to tell them what choices were socially efficient. Because this is not the case, other means are sought to internalize externalities (which, by definition, lie outside the market price). Here, numerous valuation problems make cost/benefit analyses difficult: how to identify and properly quantify social costs; how to treat moral and ethical considerations; how to make decisions when data on impacts are so sparse and uncertain. Moreover, the extent to which social costs can be internalized to the firm absent their implicit internalization in the market (via consumer demand for environmentally preferable options, for example, or government preference for "green procurement") is limited by competitive and economic considerations.

More fundamentally, there is a question as to the extent to which private firms should be encouraged to move independently toward a broader social responsibility for the achievement of a long-term stable carrying capacity, as implied by social costing. Although there are a number of factors encouraging such a trend, such as the critical role of technological evolution if environmental perturbations are to be mitigated, there are also problems with such a scenario. Surveys indicate that private firms have very little credibility with the public on environmental issues, so it is unlikely that people would trust industry with any broader responsibilities in social costing.

6.4.2.3   Project Costs and Benefits at the National Level.   It has long been the case in many developed countries that major projects, especially those undertaken by the government, have required some form of CBA with respect to environmental impacts. In the United States, for example, such requirements are established by the National Environmental Policy Act of 1969, which requires Environmental Impact Statements for major government-supported projects. Increasingly, similar analyses are being prepared for projects supported by national and international lending institutions such as the World Bank.

Although there are clearly benefits to such requirements, which at least mean that environmental issues are considered at some point during project evaluation, the efforts are plagued by the same cost identification and quantification issues raised before. In the case of international projects, these already difficult environmental issues are compounded by cultural and socioeconomic differences that make valuation particularly difficult. A less developed country, for example, may be less inclined to insist on expensive emissions-treatment technology.

6.4.2.4   Costs and Benefits of National Environmental Regulation.   Until recently, the issue of costs and benefits of environmental regulations was not a major policy problem. Although specific industry sectors or firms might face some costs, usually for installing and operating end-of-pipe control technologies or cleaning up polluted sites, the costs of environmental regulation were minimal enough on a national scale that balancing them against benefits was considered superfluous. Whereas this may still be the case in some less developed countries, it is clearly not the case in more developed countries: According to the U.S. EPA, pollution-control activities (only a subset of environmentally related expenditures) for 1972 were $30 billion; for 1987, $98 billion; for 1990 $115 billion; rising to an estimated minimum of $171 billion by year 2000 (all in 1990 U.S. dollars). This amounts to 2.1% of GNP in 1990, rising to at least 2.6% by year 2000. This level of investment in environmental protection is roughly similar to that of other more developed countries.

The question of what benefits have been purchased for this investment is highly contentious. Efforts have been made through, for example, executive orders promulgated by the President of the United States to impose CBA requirements and procedures on the environmental regulatory process. One executive order, for example, requires that agencies "select . . . approaches that maximize net benefits" and ensure that "benefits of the intended regulation justify its costs." Mechanisms and metrics by which these principles can be implemented and measured, however, are partial, incomplete, and hotly debated, with the result that, to date, such requirements have been only partially effective.

The current situation may be summarized as follows: It is apparent that citizens of more developed countries are willing to expend considerable resources for environmental benefits. How these costs compare to the benefits, and how the latter can or should be measured, are still relatively indeterminate.

6.4.2.5 National GNP. The Gross National Product (GNP) is typically defined as the aggregate money demand for all products, including consumer goods, investments, government expenditures, and export spending. It, and similar national account systems, are frequently taken as a measure of individual economic welfare, or, more controversially, as a measure of quality of life. As environmental issues have become more important, such metrics have been increasingly criticized for their failure to depreciate so-called "natural capital" as it is used to produce monetarized assets. Technically, systems based on the United Nations System of National Accounts (SNA), the international standard, recognize land, mineral, and timber resources as assets in a nation's capital stock, but do not recognize them in the income and product accounts. Accordingly, if a natural resource is used, the national income and product accounts show no equivalent depreciation. Thus, it is increasingly argued that if, for example, a forest has been depleted, national income accounts should reflect this reduction in value of a natural asset even as they may reflect money income derived from that depletion. Failure to do so in essence values the existing forest as zero until it is destroyed. Similarly, reductions in the fertility of cropland, mining of groundwater reserves, and other activities that deplete a country's stock of natural resources in order to result in a flow of income are not measured in national income and product accounts. Reflecting this trend, the U.N. Statistical Comission in February 1993 adopted a consensus initiative to establish satellite accounts to the SNA for national environmental accounting. Such accounts are to be maintained separately from the core SNA accounts, but structured so as to be compatible with them.

The failure to account for depletion of natural capital can significantly skew assessment of the real economic costs of policy decisions. For example, Robert Repetto of the World Resources Institute estimated that the depreciation of natural assets in Costa Rica from 1970 to 1989 exceeded 4.1 billion in 1984 dollars—an annual loss of some 5% of gross domestic product (GDP) and one nowhere reflected in the national income accounts. Similarly, a case study of Indonesia indicated that an average annual growth rate of 7.1% in GDP from 1971 to 1984 was reduced to a growth rate of only 4% when resource depletion was considered. Given that these figures represent depletion of only selected natural resources, they are probably significantly underestimated. In countries such as the United States, where resource depletion is minor compared to total economic activity, an environmental accounting system that does not also reflect pollution costs may not significantly affect national accounts.

More fundamental is the question of how an index that reflects quality of life or increases in social welfare rather than growth in income can be constructed. Herman Daly and John Cobb proposed an Index of Sustainable Economic Welfare, but it is safe to say that this issue has not yet been resolved. Until it is, the accounting for the costs and benefits of national economic activity must be regarded from an industrial ecology viewpoint as completely inadequate.

6.4.2.6 International Costs and Benefits. Given the difficulties besetting CBA of national projects and national accounting systems, it can be appreciated that the calculation of international costs and benefits arising from the mitigation of global and regional environmental perturbations and the integration of environment and technology in global economic systems is highly contentious. Issues involving international trade and differing value systems, particularly between less developed and more developed nations, are noteworthy. One additional issue deserving mention is the difficulty of evaluating quantitatively and constructively global environmental perturbations that may have differential geographic impacts. Global climate change, for example, might have the effect of creating new agricultural areas in countries at high latitudes, such as Russia and Canada, and reducing the value of current agricultural land in nations at mid-latitudes, such as in Spain. Low-lying countries, such as Fiji or The Netherlands, would be more impacted by increases in sea level than those with little or no coastline. Little progress has been made in quantifying such differences, or developing an international accounting system that can allocate costs and benefits among nations and regions.

6.4.2.7 Discussion.   We cannot paint a bright picture of current cost/benefit analysis as a useful aspect of industrial ecology, despite its obvious conceptual value. Within a firm, and when restricted to economic rather than social costs, and to the quantifiable portions of the product life cycle, CBA has the potential to play a useful role. Progress on a broader scale will require international collaboration among economists, industrialists, governments, and environmentalists; a usable system is a long way off.

6.5 IMPLICATIONS FOR THE CORPORATION

It is apparent that the increasing complexity of environmental regulation at all levels of government will have significant impacts on the traditional operations and organization of the corporation. Among those that bear emphasizing include the following:

1. The private corporation, as the expert in its technology, needs to become a partner in the development of new regulatory structures. This will require corporations to develop positions that are acceptable to all stakeholders in the society—including regulators and environmentalists—and eschew the adversarial, negative approaches that have too frequently characterized past behavior.

2. New organizations and information flows will have to be developed as corporations internalize environmental issues. In particular, the legal and government affairs organizations within corporations will have to become sophisticated in developing and presenting positions that integrate environment and technology in a responsible manner.

3. Corporations need to participate in efforts to design and implement full-cost accounting, so that environmental costs are automatically incorporated into economic decisions.

4. Corporations need to view society as a whole, and the community within which they operate, as full partners in their activities. They need, in fact, to view communities (in the broadest sense of the word) as customers of their services rather

than just purchasers of their products. Environmental issues will, over time, shift the fundamental raison d'être of private corporations away from a sole focus on shareholders and profits toward a broader social role.

SUGGESTED READING

Aftalion, F. *A History of the International Chemical Industry*, O. T. Benfey, trans. Philadelphia: University of Pennsylvania Press, 1991.

Bhagwati, J. The case for free trade, and H. E. Daly, The perils of free trade, in Debate: Does free trade harm the environment? *Scientific American, 269* (5) (1993): 41–57.

Brimblecombe, P. *The Big Smoke*. London: Methuen, 1987.

Cairncross, F. *Costing the Earth*. Cambridge, MA: Harvard Business School Press, 1992.

Costanza, R., ed. *Ecological Economics: The Science and Management of Sustainability*. New York: Columbia University Press, 1991.

Daly, H. E., and J. B. Cobb, Jr. *For the Common Good*. Boston: Beacon Press, 1989.

Duchin, F. Input-output analysis and industrial ecology, in *The Greening of Industrial Ecosystems*. B. R. Allenby and D. J. Richards, eds. Washington, DC: National Academy Press, 1994.

Environmental Advisory Council. *Eco-Cycles: The Basis of Sustainable Urban Development*, Report Jo 1968. Stockholm: A. Allmänna Förlaget, 1992.

Environmental Protection Agency. *Environmental Investments: The Cost of a Clean Environment*, Report EPA-230-12-90-084. Washington, DC: EPA, 1990.

———. *A Primer for Financial Analysis of Pollution Prevention Projects*, Report EPA/600/R-93/059. Washington, DC: EPA, 1993.

Griefahn, M. Initiatives in Lower Saxony to link ecology to economy, in *Industrial Ecology and Global Change*, R. Socolow, C. Andrews, F. Berkhout, and V. Thomas, eds. Cambridge, UK: Cambridge University Press, 1994.

Kopp, R. J., P. R. Portney, and D. E. DeWitt. *International Comparisons of Environmental Regulations*, Discussion Paper QE90-22-REV. Washington, DC: Resources for the Future, 1990.

Meyer, C.A. *Environmental and Natural Resource Accounting: Where to Begin*. Washington, DC: World Resources Institute, 1993.

Nordhaus, W. D. An optimal transition path for controlling greenhouse gases. *Science, 258* (1992): 1315–1319, 1381–1384.

Popoff, F. P., and D. T. Buzzelli. Full-cost accounting. *Chemical and Engineering News, 71* (2) (1993): 8–10.

Repetto, R. Accounting for environmental assets. *Scientific American, 266*(6), (1992), 94-100.

Todd, R. Zero-loss environmental accounting systems, in *The Greening of Industrial Ecosystems*, B. R. Allenby and D. J. Richards, eds. Washington, DC: National Academy Press, 1994.

United Nations. *Integrated Environmental and Economic Accounting*, Handbook of National Accounting, Series F, No. 61 (interim version). New York: UN, 1993.

Weiss, E. B. *In Fairness to Future Generations: International Law, Common Patrimony, and Intergenerational Equity*. Tokyo: United Nations University, 1989.

EXERCISES

6.1 You are the environmental officer of a chemical firm producing commodity polymers. The government of the country in which most of your production facilities are located has just proposed a broad energy tax, substantially higher than in any other developed country.

(a) List and evaluate the positions your firm can take in response to this public policy initiative.

(b) Which would you choose and why?

(c) What data on your firm's operations would be useful in helping you develop your positions? What organizational elements of the firm should be involved in helping you develop and implement your positions?

6.2 Your firm owns mineral rights currently estimated to be worth $100,000 in a (politically) relatively unstable country.

(a) Assuming a discount rate of 7%, how much will your rights be worth in 5 years? In 10?

(b) What other factors would you consider in making a decision about when to exploit the mineral rights, from the perspective of the following:

(i)   The firm that owns them.
(ii) The government of the country.
(iii) The community near the ore deposit.

6.3 It is frequently said that humans should "plan for the seventh generation"; that is, so actions taken now will not adversely affect the seventh generation of progeny. Assume that a certain product will save 100 lives now, but use up materials that could save substantially more in the future. Assume also a psychological discount rate of 2% for human life, 25 years per generation, and (for simplicity's sake) that all lives saved occur in the year ending the seventh generation. How many lives in the seventh generation must be at risk before a decision not to introduce the product is made now?

PART III: EVALUATING LIFE CYCLES

<table>
<tr><td>
CHAPTER

<big>7</big>
</td><td>

Budgets and Cycles
</td></tr>
</table>

7.1 INDUSTRIAL ECOLOGY: A SYSTEMS DESCRIPTION

In the first chapter, we recalled that traditional biological ecology may be defined as *the scientific study of the interactions that determine the distribution and abundance of organisms*, and remarked that this concept resembled in many interesting and useful ways the interactions of industry and society with natural systems. In developing this idea further, it is instructive to think of the materials cycles associated with a postulated primitive biological system such as might have existed early in Earth's history. At that time, the potentially usable resources were so large and the amount of life so small that the existence of life forms had essentially no impact on available resources. This process might be described as *linear*, that is, as one in which the flow of material from one stage to the next is independent of all other flows. We term this pattern a "Type I" system; schematically, it takes the form of Fig. 7.1(a).

As the early life forms multiplied, external constraints on the unlimited sources and sinks of the Type I system began to develop. These constraints led in turn to the evolution of biotic systems as an alternative to linear materials flows. Feedback and cycling loops were developed as scarcity drove the process of change. In such systems, the flows of material within the proximal domain could have been quite large, but the flows into and out of that domain (i.e., from resources and to waste) eventually were quite small. Schematically, such a Type II system might be expressed as in Fig. 7.1(b).

This latter system is much more efficient than the previous one, but it clearly is not sustainable over the long term because the flows are all in one direction, that is, the system is "running down". To be ultimately sustainable, biological ecosystems have evolved over the long term to be almost completely cyclical in nature, with "resources" and "waste" being undefined, because waste to one component of the system represents

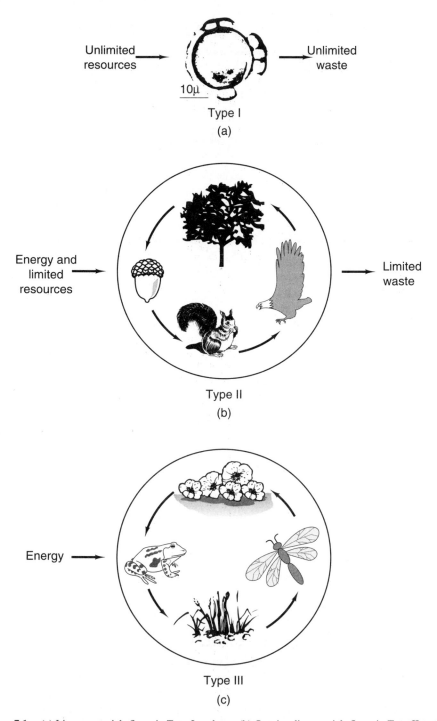

Figure 7.1 (a) Linear materials flows in Type I ecology. (b) Quasicyclic materials flows in Type II ecology. (c) Cyclic materials flows in Type III ecology.

resources to another. This Type III system, in which complete cyclicity has been achieved, may be pictured as in Fig. 7.1 (c). Note that the exception to the cyclicity of the overall system is that energy (in the form of solar radiation) is available as an external resource. It is also important to recognize that the cycles within the system tend to function on widely differing temporal and spatial scales, a behavior that greatly complicates analysis and understanding of the system.

The ideal anthropogenic use of the materials and resources available for industrial processes (broadly defined to include agriculture, the urban infrastructure, etc.) would be one that is similar to the biological model. Many uses of materials have been and continue to be essentially dissipative, however. That is, the materials are degraded, dispersed, and lost to the economic system in the course of a single normal use, mimicking the Type I unconstrained resource diagram. This pattern can be associated with the maturation of the Industrial Revolution of the eighteenth century, which, in concert with exponential increases in human population and agricultural production, took place essentially in a context of global plenty in which copious quantities of energy and raw materials were made available by technological advances. Many present day industrial processes and products still remain inherently dissipative. Examples include lubricants, paints, pesticides, and automobile tires.

On the broadest of product and time scales, many indications demonstrate today that the flows in the ensemble of industrial ecosystems are so large that resource limitations are setting in: the rapid changes in stratospheric ozone, increases in atmospheric carbon dioxide, and the filling of existing waste-disposal sites being examples. Accordingly, industrial systems (and other anthropogenic systems) are and will increasingly be under pressure to evolve from linear (Type I) to semicyclic (Type II) modes of operation. For the past decade or two, industrial organizations have largely been in the position of responding to legislation imposed as a consequence of a real or perceived environmental crisis. Such a mode of operation is essentially unplanned and imposes significant economic costs as a result. Industrial ecology, in its implementation, is intended to accomplish the evolution of manufacturing from Type I to Type II and, ultimately, to Type III, by understanding the interplay of processes and flows and by optimizing the ensemble of considerations that are involved.

The industrial process is conveniently pictured with four central nodes: the materials extractor or grower, the materials processor or manufacturer, the user, and the scavenger (who may also be at least a first stage materials processor and recycler). To the extent that each performs operations within a node in a cyclic manner, or organizes to encourage cyclic flows of materials within the entire industrial metabolic system, they evolve into modes of operation that are more efficient, have less disruptive impact on external support systems, and are more like Type III ecological behavior. The schematic model of such a system is shown in Fig. 7.2. Flows are not necessarily sequential. For example, if excess product is generated by the manufacturer, it may pass directly to the scavenger without having undergone a stage of use.

From the standpoint of an individual node, or from the standpoint of the industrial ecosystem as a whole, waste can be considered to be anything that does not add value. Operationally, one can adopt as goals that every erg of energy used in manufacture

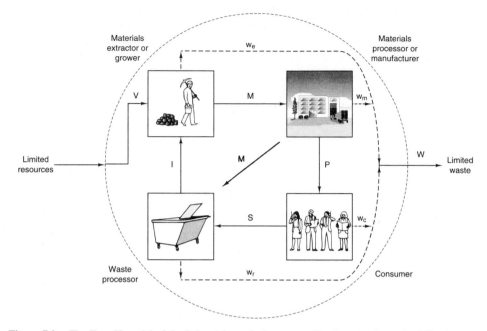

Figure 7.2 The Type II model of the industrial metabolic system. The letters refer to the following mass flows: V, virgin material; M, processed material; P, product; S, salvaged material; I, impure material; and W, waste. The flows are sometimes combined in practice. For example, materials processors and recyclers often produce material ready for transmittal to manufacturers.

should produce a desired material transformation, that every molecule that enters a specific manufacturing process should leave as part of a salable product, and that every product be used to create other useful products at the end of its life. (Material that can be reused in a different product stream within or outside the facility is not properly regarded as waste; it is merely another product.) It is obvious that these goals can never be achieved; it is equally obvious that striving to achieve them will lead toward more environmentally responsible processes.

A perspective on the goals stated before is the realization that the goals require rethinking of the manufacturing process. Indeed, the proper way to look at manufacturing is as an activity in which a spectrum of "products" (traditional end products, scrap material, spent solvents, etc.) are generated. The realization that one is dealing with a product spectrum, with components of the spectrum originating at various points within the manufacturing operation, has profound design implications. For example, one might change a process in such a way that the solvent bath would be a more valuable product even if the end article being produced—today's "product"—remained exactly the same.

7.2 THE BUDGET CONCEPT

Ecology is a concept with cycles at its very heart, and cycles are analyzed by means of materials and energy budgets. Nearly everyone is familiar with the concept of a household or personal financial budget, whether or not being conscientious about making and sticking to it. An approach very similar to that of financial budgeting is used to fashion budgets in industrial ecology. The situation can be appreciated with the aid of the diagram in Fig. 7.3, which shows a tub receiving water from several faucets and having a number of drains of different sizes. When the water is supplied at constant (but probably different) rates by all the faucets and is removed at an equal total rate by drains with (probably) different capacities, the water level remains constant. When the tank is very large, however, and has some wave motion that makes it difficult to tell whether the absolute level is changing, an observer may have difficulty telling whether the system is in balance or not. In that case, the observer may try instead to measure the rate of supply from each of the faucets and the rate of removal in each of the drains over a period of time to see whether the sums are equivalent. A part of this technique involves the determination of the *pool size* (the quantity of water in the tank) and either the rate of supply or the rate of removal. Determination of changes in the pool size then gives information about rates that are difficult to measure. The process of estimating or measuring the input and output flows and checking the overall balance by measuring the amount present in the reservoir constitutes the budget analysis.

Suppose that the input from one of the sources is increased; that is, in our analogy, the flow from one of the faucets increases. Will the water level keep increasing? The answer depends on whether one of the drains can accommodate the additional supply, as can the "trough drain" at the right side of the tank in the figure. If no such drain is

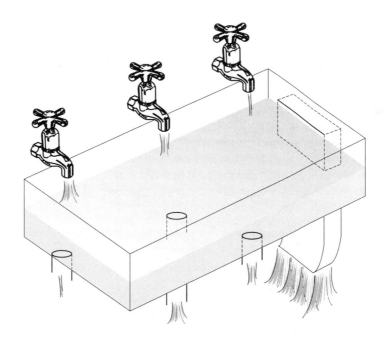

Figure 7.3   A simple conceptual system for budget calculations. The water level in the tub is determined by the water flows in and out, as discussed in the text.

present, then the water level will indeed increase. Conversely, if the flow into a drain is enhanced for some reason, such as the removal of an obstruction, then the water level will decrease in the absence of a corresponding increase in the supply. Such a process will continue in this manner unless the flows through the drains adjust themselves to this new factor or unless the new factor results in other drains changing their functioning.

All of the circumstances mentioned occur in budgets devised for various industrial ecology studies, and all budgets involve the same concepts. One is that of the *reservoir*, in which material is stored. Examples include the shipping department where completed products are prepared for forwarding to customers, or the atmosphere as a whole, where emissions of industrial vapors collect and react. A second concept is that of *flux*, which is the amount of a specific material entering or leaving a reservoir per unit time. Examples include the rate of evaporation of water from a power plant cooling tower or the rate of transfer of ozone from the stratosphere to the troposphere. Third, we have *sources* and *sinks*, which are rates of input and loss of a specific material within a reservoir per unit time. A system of connected reservoirs that transfer and conserve a specific material is called a *cycle*.

Industrial ecology budgets have the same three basic components as those for the tank in Fig. 7.3: determination of the present level (the concentration of a single material or a group of materials), a measurement or estimate of sources, and a measurement or estimate of sinks. A perfect determination of any two of these three components determines the other. Because any material of interest in an industrial facility or in the environment may have several sources and sinks, each source and sink must generally be studied individually.

7.3 QUANTITATIVE MEASURES OF AN ECOLOGY CYCLE

It is instructive to use the materials flows of Fig. 7.2 to define and compute some measures of efficiency so that societal progress in industrial ecology can be assessed. We thus specify the following:

- *Extracting efficiency* ι (a function of good extraction technology, high-quality residue streams, and negotiated manufacturer's specifications):

$$\iota = \frac{M}{V + I} \tag{7.1}$$

  where the principal mass flows are defined in the figure caption. If no virgin material is used and no residue is produced, $\iota = 1$.

- *Manufacturing efficiency* μ (a function of process and product design, and of their implementation):

$$\mu = \frac{P}{P + w_m} \tag{7.2}$$

  where w_m is the manufacturing process-residue stream mass. If no manufacturing residue is generated, $\mu = 1$.

- *Recovery efficiency* ρ (a function of product design, government policies, and recycled materials markets):

$$\rho = \frac{S}{S + w_c} \qquad (7.3)$$

If all material is recycled, $\rho = 1$.

- *Recycling efficiency* λ (a function of "design for disassembly" and government and customer regulations and policies):

$$\lambda = \frac{I}{I + w_r} \qquad (7.4)$$

If no recycling residue is generated, $\lambda = 1$.

An overall *reuse factor* ψ can then be defined as

$$\psi = \frac{I}{M} \qquad (7.5)$$

and *system efficiency* by

$$\sigma = \iota \times \mu \times \rho \times \lambda \qquad (7.6)$$

In a perfect industrial ecology system in which all materials are provided by recycling, $\sigma = 1$.

7.4   TIME SCALES IN BUDGET AND CYCLE ANALYSIS

It is convenient to define a number of different time scales as part of the budget and cycle process. The first is the *turnover time*, τ_0. This parameter is the ratio between the content (β) and the flux of a specific material into (F_i) or out of (F_o) a reservoir that is in steady state:

$$\tau_0 = \frac{\beta}{F_o} = \frac{\beta}{F_i} \qquad (7.7)$$

The turnover time reflects the spatial or temporal variability of a property within a reservoir, with a small variability indicating a long turnover time and a large variability indicating a short turnover time. If material enters or leaves the reservoir by several paths, then the overall turnover time of the reservoir is related to the individual turnover times of the pathways $\tau_{0,i}$ by

$$\tau_0 = \frac{\sum\limits_i F_i}{\beta_i} = \frac{\sum\limits_o F_o}{\beta_i} = \frac{1}{\sum\limits_i \frac{1}{\tau_{0,i}}} \qquad (7.8)$$

A second useful parameter is the *residence time*, τ_r, which is the average time spent in the reservoir by a specific material. If physical rather than chemical processes are involved, then the term *transit time* may be used as an alternative. The average residence time is composed of those of all appropriate molecules, weighted with appropriate probability factors. For example, when one evaluates nitrogen flow into and out of the animal reservoir, one finds that some of the nitrogen rapidly flows through the reservoir, whereas other nitrogen flows much more slowly, a reflection of animal lifetimes and foraging and excretion characteristics. We can define the average residence time in a situation with a variety of residence times as

$$\tau_r = \int_0^\infty \tau\,\psi(\tau)\,d\tau \tag{7.9}$$

where $\psi(\tau)$ indicates the fraction of the constituent having a residence time between τ and $\tau + d\tau$. This probability fraction $\psi(\tau)$ is a function of the reservoir processes. In the case of radioactive decay, for example, it can be shown that $\tau_r = \tau_\alpha$, the time constant for the exponential decay of the radionuclide.

The *age* is the time elapsed since a particle entered a reservoir. The average age of all particles of a specific kind within a reservoir is thus given by

$$\tau_a = \int_0^\infty \tau\,\Psi(\tau)d\tau \tag{7.10}$$

where $\Psi(\tau)$ is the age probability function.

For a reservoir in steady state, the turnover time τ_0 and the average residence time τ_r are equal. They may, however, be significantly different from the average age of the particles in the reservoir, depending on the properties of functions $\psi(\tau)$ and $\Psi(\tau)$. An obvious example showing that $\tau_r \neq \tau_a$ is the human population: In the developed world, the average age is between 35 and 40 years, whereas the average residence time (life expectancy) is more than 70 years.

For a system chosen with a boundary surrounding the entire planet, as with the nitrogen budget, and assuming no significant loss to interplanetary space, the overall quantity of material will not change with time:

$$\sum \beta = constant \tag{7.11}$$

If the entire system is in a steady state, the source and sink fluxes into each reservoir exactly balance, so that in each case,

$$\beta_r = constant \tag{7.12}$$

that is, the contents of each reservoir will not change with time:

$$\Delta_r = \frac{d(\beta_r)}{dt} = 0 \tag{7.13}$$

In a changing system, however, the source and sink fluxes are not equal, and Δ_r is real-valued and expressed as a function of *response time* τ_e, which is the time needed to reduce the effect of a disturbance from equilibrium in a reservoir to $1/e$ of the initial perturbation value. For example, in many chemical applications, the equilibrium of species i in a system is expressed by

$$0 = P_i - \Lambda_i\,\beta_r \tag{7.14}$$

where P_i is the production rate of species i, $\Lambda_i \beta_r$ its loss rate, β_r being the content of species i in reservoir r. If we now introduce a disturbance from equilibrium β_r', the change in β_r with time becomes

$$\Delta = d\,\frac{\beta_r}{dt} = P_i - \Lambda_i(\beta_r + \beta_r') = -\Lambda_i \beta_r \qquad (7.15)$$

so

$$\beta_r'(t) = \beta_r' \exp(-\Lambda_i t) \qquad (7.16)$$

In this case, $\tau_e = \Lambda^{-1}$.

7.5 NATURE'S GLOBAL BUDGETS AND CYCLES

7.5.1 A Survey of Elemental Cycling

Nature provides us with more than 90 natural elements with which to devise and manufacture products. These elements, most of which are present in combined form, occur with great disparities in abundance. A rough measure of their commonality is provided by average elemental abundances in crustal rocks, seen in Table 7.1 for some of the more common elements. Note from the table that a "big eight" of elements—oxygen, silicon,

Table 7.1   Average Elemental Composition of Crustal Rocks

Atomic No.	Symbol	Concentration*
1	H	1,400
6	C	320
8	O	446,000
11	Na	28,300
12	Mg	20,900
13	Al	81,300
14	Si	277,200
15	P	1,180
16	S	520
17	Cl	200
19	K	25,900
20	Ca	36,300
22	Ti	4,400
25	Mn	1,000
26	Fe	50,000
56	Ba	400

*Parts per million by weight.

Source: The data are from B. Mason, *Principles of Geochemistry*, 2nd ed. New York: John Wiley, 1958.

aluminum, iron, calcium, sodium, potassium, and magnesium—make up more than 96% of the weight of crustal rocks. The more than 80 other elements, including almost all those found useful in industry, comprise less than 4% of the accessible mass of Earth. The most common elements are extracted from the most common minerals, the oxides (Table 7.2). Aluminum and iron are the only industrial metals whose oxides appear in the list.

In addition to their abundances in overall Earth system element budgets, relative abundances in Earth's fluid regimes are important, because one can then think of budgets and cycles of more easily accessed portions of the Earth system. Table 7.3 provides selected information for the present composition of the atmosphere. Disregarding highly variable amounts of water vapor, more than 99.9% of the molecules constituting Earth's atmosphere are nitrogen (N_2), oxygen (O_2), and the chemically inert noble gases (largely argon). All of these gases appear to have been present at nearly constant levels during much of the past billion years. The remaining atmospheric constituents, representing less than 0.1% of the atmospheric molecules, are diverse but important because they influence or control a number of crucial processes. For example, carbon dioxide, the chemical feedstock for the photosynthesis of organic matter, is an important factor in

Table 7.2   Average Chemical Composition of Crustal Rocks

Compound	Weight percent	Compound	Weight percent
SiO_2	60.2	FeO	3.9
Al_2O_3	15.6	MgO	3.6
CaO	5.2	K_2O	3.2
Na_2O	3.9	Fe_2O_3	3.1

Table 7.3   Composition of Dry Air at Ground Level in Remote Continental Areas

Constituent	Formula	Concentration (%)
Nitrogen	N_2	78.1
Oxygen	O_2	20.9
Argon	Ar	0.93
Carbon dioxide	CO_2	0.035
Neon	Ne	0.0018
Helium	He	0.0005
Methane	CH_4	0.00017
Krypton	Kr	0.00011
Hydrogen	H_2	0.00005
Ozone	O_3	0.000001–0.000004

Earth's radiation balance. It is chemically unreactive in the troposphere (the lowest portion of the atmosphere), where it is currently present at an average concentration of about 0.035% by volume. The most abundant of the reactive gases is methane, which represents less than 0.0002% of the tropospheric gas. Other reactive species are still less prevalent; the combined concentration of all of the reactive trace constituents in the lower atmosphere seldom totals 0.001%, even in the most polluted environments.

The chemistry of the oceans has some strong and interesting areas of overlap with global budgets and cycles. Conditions typical of the world's oceans, which are of extremely uniform chemical makeup, are given in Table 7.4. As is well known, sodium and chlorine are major constituents, but organic matter from the oceans's rich biological cycling is even more abundant. Ions from water-soluble minerals are common as well. In the surface waters of the ocean, the typical pH is 8.0 ± 0.5 (i.e., slightly alkaline).

Elemental abundances in Earth's reservoirs are a starting point for budgets, but the degree to which elements are available for use is determined by their cycling—how rapidly and in what form they move among reservoirs. That topic can require entire shelves of books for its proper treatment; we limit ourselves here to factors that will influence our perception of the relative impacts of industrial activity on the natural cycles.

From a biological standpoint, the most important cycles are what have been termed the *grand cycles*: those for carbon, nitrogen, sulfur, and phosphorus. The availability of oxygen and hydrogen is crucial as well (although those elements are seldom in short supply). Soil provides the medium in which much of this cycling occurs or through which it is promoted. Without efficient cycling of the elements of the grand cycles, life as we know it on the planet would not be possible.

The halogens (F, Cl, Br, I) form another interesting group of elements, one in which efficient cycling is mediated largely through processes in the oceans. Sea spray and the formation of volatile halogenated organic species by marine organisms are the main processes of interest.

Table 7.4   Concentrations of Major Salt Constituents in Seawater at 25° and Atmospheric Pressure

Constituent	Concentration (mg L^{-1})
DOM[*]	1.0
Na^+	0.48
Mg^{2+}	0.05
Ca^{2+}	0.01
K^+	0.01
Cl^-	0.56
SO_4^{2-}	0.03

[*] DOM = dissolved organic matter.

Source: Data are from M. Whitfield, Activity coefficients in natural waters, in *Activity Coefficients in Electrolyte Solutions*, R. M. Pytkowicz, ed., vol. 2, pp. 153–299. Boca Raton, FL: CRC Press, 1979.

Elements that are abundant in soil dust are distributed by the winds, but enter cyclization only if liberated from their oxides by chemical attack. Of the constituents of Table 7.2, those that turn out to be relevant in this regard are aluminum, iron, and calcium. Those three excepted, the natural cycles of metals and metalloids are slow and inefficient. In a few cases, as with zinc, for example, some modest cycling is important in providing a critical biological chemical in the necessary trace quantities. The essential trace element cycles are generally mediated through lake, river, or ocean processes.

The only one of the rare gases important to this discussion is radon, a product of the radioactive decay of the uranium naturally present in soil and rock. The radon enters the atmosphere from the soil, is itself radioactively transformed to lead in a few days, and the lead lost by incorporation into atmospheric aerosol particles. Its transit among reservoirs is thus anticyclic, because it is destroyed by its internal radioactive transformation prior to any opportunity for recycling.

To summarize, although nature has long-term cycles for all the elements, only between 15 and 20 elements move among reservoirs on time scales rapid enough for their cycles to be readily analyzed. Such elements, being more or less readily available (in a chemical sense) as they cycle, are those that Earth's biota have learned how to utilize in their metabolic processes. Conversely, the elements whose cycling is strongly inhibited by their chemical nature are not familiar to organisms, and their presence in accessible form can be destructive. Similar arguments hold with respect to compounds (combinations of elements) that are not customarily encountered by organisms, and about which another whole level of discussion of budgets and cycles could be undertaken. From the standpoint of biology, therefore, the existence of traditional cycling is crucial, the disruption of those cycles is of concern, and the introduction of new cycles is a strong danger signal.

7.5.2   The Natural Nitrogen Budget

The foregoing concepts can be best illustrated by a specific example, for which we choose the natural global nitrogen budget. A typical ecological way of presenting that budget is shown in Fig. 7.4(a), in which atmospheric nitrogen (in all its forms) is shown to be linked with the pedosphere (the soil-bearing portion of Earth's surface) and the biosphere (the sum of all living things on Earth). The cycle can be regarded as beginning in the box at the upper left with atmospheric dinitrogen (N_2). This nitrogen is *fixed* (made available for use by organisms) through lightning (forming NO and subsequently nitrate (NO_3^-)), and biofixation through bacteria (forming ammonium ion (NH_4^+)). Within the pedosphere and biosphere, nitrogen cycles from the ammonium and nitrate reservoirs through plants and animals, all within a balanced system. Note that some nitrate is given back to the atmosphere; this is material undergoing *denitrification* by specialized soil bacteria. Denitrification depletes the pool of available nitrogen, so the cycle would stop were not nitrogen fixation continually regenerating available nitrogen from atmospheric dinitrogen.

If we cast the nitrogen cycle in the form of the four nodes of Fig. 7.2, we arrive at the budget diagram of Fig. 7.4(b), where we define the mineralizing bacteria (those that

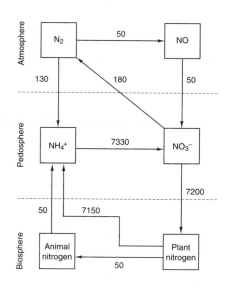

(a)

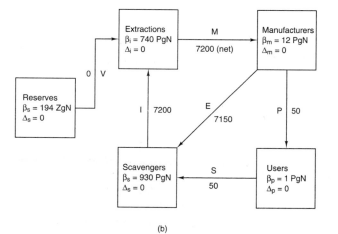

(b)

Figure 7.4   (a) A geochemist's version of the natural global nitrogen budget. (b) An industrial ecologist's version of the natural global nitrogen budget. The units for the reservoir contents are given in each box. The flows are in teragrams (1 Tg = 10^{12} g). The data source for the reservoir contents is T. Rosswall, The nitrogen cycle, in *The Major Biogeochemical Cycles and Their Interactions*, B. Bolin and R. B. Cook, eds., SCOPE 21. Chichester, UK: John Wiley, 1983; that for the fluxes is R. Ayres, W. Schlesinger, and R. Socolow, Human impacts on the carbon and nitrogen cycles, in *Industrial Ecology and Global Change*, R. Socolow, C. Andrews, F. Berkhout, and V. Thomas, eds. Cambridge, UK: Cambridge University Press, 1994.

transform the nitrogen in plants and animals into inorganic form (mostly ammonium ion)) as the *extractors*, the plants as the *manufacturers*, the animals as the *users*, and the soil detriti as the *scavengers*. For completeness, we add a fifth node, the *reserves*, those resources that are exchanged with the rest of the system so slowly that under most circumstances they can be regarded as constant and noninteracting. There are several features of interest in this diagram. First is the vastly different sizes of the reservoirs, that for reserves being nearly a million times those of the extractors and scavengers and even larger when compared with those of the manufacturers and users. Second, over reasonably long time scales all of the reservoir contents are stable, so all the Δ values are zero. Third, most of the manufactured product (nitrate ion) is in excess of the needs of the users, so that portion is eventually scavenged directly rather than cycling through the user reservoir.

Just as shown here for nitrogen, a budget can be devised for any element or compound or group of compounds in nature. Those for carbon, nitrogen, sulfur, and phosphorus, the "grand nutrient cycles" are perhaps the most important from a biological standpoint. Absent major natural disasters and other environmental changes, the natural cycles are in balance over the long term. We might, in fact, define *sustainable development* as the avoidance of serious perturbations to the materials cycles of nature.

SUGGESTED READING

Ayres, R. U. Industrial metabolism, in *Technology and Environment*, J. H. Ausubel and H. E. Sladovich, eds. Washington, DC: National Academy Press, 1989.

Bolin, B. The carbon cycle, in *The Major Biogeochemical Cycles and Their Interactions*, B. Bolin and R. B. Cook, eds. SCOPE 21. Chichester, UK: John Wiley, 1983.

Frosch, R. A. Industrial ecology: A philosophical introduction. *Proc. Nat. Acad. Sci. USA, 89* (1992): 800–803.

Rodhe, H. Modeling biogeochemical cycles, in *Global Biogeochemical Cycles*, S. S. Butcher, R. J. Charlson, G. H. Orians, and G. V. Wolfe, eds., pp. 55–72. San Diego, CA: Academic Press, 1992.

Rosswall, T., The nitrogen cycle, in *The Major Biogeochemical Cycles and Their Interactions*, B. Bolin and R. B. Cook, eds., SCOPE 21. Chichester, UK: John Wiley, 1983.

Schlesinger, W. H., *Biogeochemistry: An Analysis of Global Change*. San Diego, CA: Academic Press, 1991.

EXERCISES

7.1 Choose a common product containing more than one material. Based on your knowledge, identify the product's inputs, life-cycle stages, and disposal fate. Characterize each of the material flows as best you can as Type I, Type II, or Type III. How would you improve the material flows associated with this product?

7.2 For the natural nitrogen cycle of Fig. 7.4, compute μ, ρ, λ, ι, and σ, as well as the turnover time for each node.

7.3 The major reservoir contents (in units of Pg C) and fluxes (in units of Pg C/yr) in the natural global carbon cycle are as shown in the following table.

Reservoir Contents

Atmosphere:	
Carbon dioxide	712
Oceans:	
Dissolved carbon	37,400
Particulate carbon	30
Biota	3
Land:	
Plants	830
Animals, bacteria	3
Soil, peat, litter	1750
Lithosphere:	
Continental crust	1.2×10^8
Oceanic crust	2.1×10^7

Fluxes

Atmosphere–Biosphere:	
Photosynthesis	53
Litter decomposition	50
Fossil fuel combustion	5
Deforestation	3
Atmosphere-sea:	
Air-sea exchange	3

Source: Data are from B. Bolin, The carbon cycle, in *The Major Biogeochemical Cycles and Their Interactions*, B. Bolin and R. B. Cook, eds., SCOPE 21, Chichester, UK: John Wiley, 1983.

Diagram the carbon cycle in both the geochemical and industrial ecology formats and compute μ, ρ, λ, ι, and σ, as well as the turnover time for each node.

CHAPTER
8

An Introduction
to Life-Cycle Assessment

8.1 INDUSTRIAL ECOLOGY COMPARISONS OF INDIVIDUAL DESIGNS AND PROCESSES

A primary thrust of industrial ecology is that manufacturers practice product steward-ship—designing, building, maintaining, and recycling products in such a way that they pose minimal impact to the wider world. Product stewardship should be broadly inter-preted to include services, which should also be performed so as to have minimal impact. The way in which these tasks are addressed in a formal manner is by the process of life-cycle assessment (LCA), a family of methods for looking at materials, services, products, processes, and technologies over their entire life.

The essence of life-cycle assessment is the evaluation of the relevant environmen-tal, economic, and technological implications of a material, process, or product across its life-span from creation to waste or, preferably, to re-creation in the same or another use-ful form. The Society of Environmental Toxicology and Chemistry defines the LCA process as follows:

> The life-cycle assessment is an objective process to evaluate the environmental burdens associated with a product, process, or activity by identifying and quantifying energy and material usage and environmental releases, to assess the impact of those energy and material uses and releases on the environment, and to evaluate and implement opportunities to effect environmental improvements. The assessment includes the entire life cycle of the product, process or activity, encompassing extracting and processing raw materials; manufacturing, transportation, and distribution; use/re-use/maintenance; recycling; and final disposal.

Such an analysis is a large and complex effort, and there are many variations. Nonethe-less, there is preliminary agreement on the formal structure of LCA, which contains three stages: *inventory analysis*, *impact analysis*, and *improvement analysis*. The concept of the life-cycle methodology is pictured in Fig. 8.1. First, the scope of the LCA is defined. An inventory analysis and an impact analysis are then performed, the result being an environmentally responsible product rating (R_{ERP}). This rating guides an analysis of potential improvements (which may feed back to influence the inventory analysis). Finally, the improved product is released for manufacture.

The first component of LCA, inventory analysis, is by far the best developed. It uses quantitative data to establish the levels and types of energy and materials inputs to an industrial system and the environmental releases that result, as shown schematically in Fig. 8.2. Note that the approach is based on the idea of a family of materials budgets, measuring the inputs of energy and resources that are supplied and the resulting products, including those with value and those that are potential liabilities. The assessment is done over the entire life cycle—materials extraction, manufacture, distribution, use, and disposal.

The second stage in LCA, the impact analysis, involves relating the outputs of the system to the impacts on the external world into which those outputs flow. We present aspects of this difficult and potentially contentious stage later in this book. The third stage, the improvement analysis, is the explication of needs and opportunities for reducing environmental impacts as a result of industrial activity being performed or contemplated. It follows directly from the completion of stages one and two, and in implementation is termed "design for environment"; it is the subject of Part IV of this book.

A fundamental concern about present LCA methodologies such as that just described is that they are too complex and detailed to be useful in the real world. An effective LCA methodology must be able to quickly and easily identify, then differentiate between, critical environmental impacts. This will allow designers to concentrate on the most important problems, reserving for later those that produce lesser impacts.

A second limitation of most existing LCA methodologies arises because they were developed for fairly simple products: disposable diapers, drinking cups, consumer personal care goods, and so forth. No satisfactory LCA study has been done on far more complex items such as automobiles, airplanes, or television sets. The importance of dealing with complex systems is that the internal dependencies and linkages involved in designing these products introduce a new level of choices and trade-offs. To take a

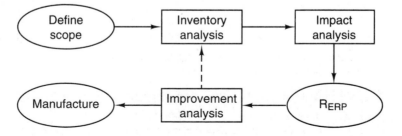

Figure 8.1   Steps in the life-cycle assessment of a product. R_{ERP} is the environmentally responsible product rating.

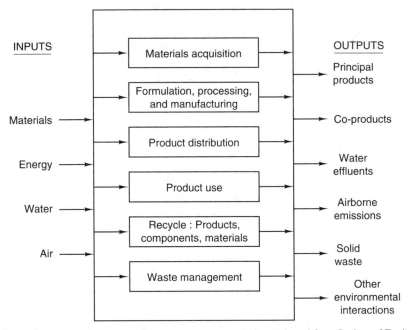

Figure 8.2   The elements of a life-cycle inventory analysis. (Adapted from Society of Environmental Toxicology and Chemistry, *A Technical Framework for Life-Cycle Assessment*. Washington, DC: SETAC, 1991.)

simple example, substituting a bismuth/tin alloy in place of lead solder in electronics manufacture would require a more aggressive antioxident flux (metal cleaner), which would in turn require a chlorinated, not aqueous, cleaning system. Most LCA methodologies, which focus primarily on the environmental impacts of one material, fail to pick up such effects or do so only in an ad hoc manner.

All LCA methodologies are in their infancy. As more experience is gained and more products, processes, and materials are subjected to assessment, the LCA approaches will become more useful and more efficient. It is unlikely that one single methodology will be optimal for all DFE analyses. Packaging, bulk chemical products, consumer care products, food products, complex manufactured items with relatively long lives (airplanes) as opposed to short lives (radios, telephones)—all have quite different characteristics, maintenance needs, life cycles, and environmental impacts. It is reasonable to assume that the tools of LCA evaluation will become more sophisticated with time.

8.2   THE INDUSTRIAL ECOLOGY FLOW CYCLE

To understand better where a material goes once it enters the materials flow system, we need to look at its total budget from the time it is extracted from the ground or harvested until the time it returns. The first steps in that cycle, the industrial production of the materials themselves, are shown in Fig. 8.3. Beginning with virgin materials, the flows proceed through cycles of extraction, separation and/or refining, and physical and chemical preparation to produce finished materials. A typical example might be the extraction from the ground of copper ore and the eventual production of copper wire.

Figure 8.3 shows, in addition to the central flows (M_x), the materials flows that occur away from the central spine. To the right are the wastes consigned to disposal (M_{xD}), typically large fractions in the early stages of the process and smaller fractions at later stages. To the left are flows of recycled material (M_{xR}), such as copper wire recovered from obsolete power distribution systems. Near the center are recycled flows that

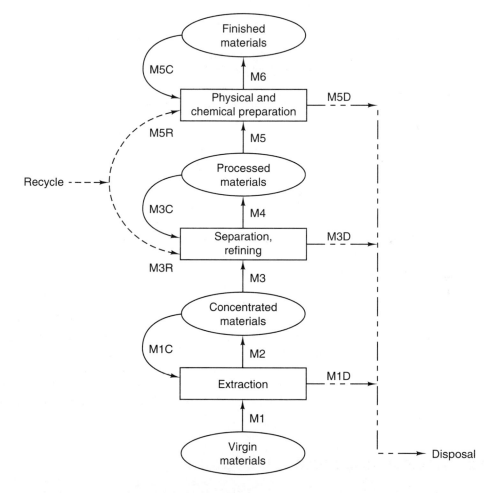

Figure 8.3   Materials flows in virgin materials processing.

occur during the production process itself (M_{xC}), as when scrap from one process stage is reused in the preceding stage. A complementary flow occurs for any chemicals used in processing the material.

The flows of materials and processes as shown in Fig. 8.3 occur within what has traditionally been called the "heavy-industry" community, and the flows are generally under the sole control of the "materials supplier".

The resource flows involved in the manufacture of products from the finished materials produced by the materials supplier are shown in Fig. 8.4. As with Fig. 8.3, waste flows, recycled material flows, and in-process recycle flows are indicated. An important distinction between this figure and Fig. 8.3 is that several finished materials are generally involved, rather than a single one. A typical example is the production of a cable connector from selected metal and plastic.

It is instructive to examine Fig. 8.4 from the perspective of constraints to optimizing the industrial ecology of the process. At the forming step, material may enter from three types of streams: the virgin materials streams (P_{1x}), the process recycle streams (P_{1C}), and the recycled material streams (P_{1R}). Unrecycled waste exits in disposal stream P_{2D}. Optimization involves decreasing or eliminating P_{2D} and utilizing whatever external recycled material is available (P_{1R}), within the constraints of customer preference and existing price structures. The use of any recycled material stream thus

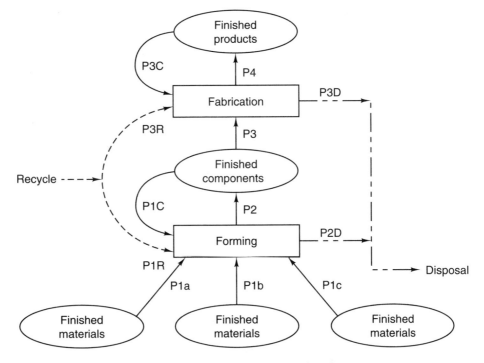

Figure 8.4   Materials flows in manufacturing.

involves a trade-off between ensuring its purity and suitability against the cost of using virgin finished materials. A similar analysis applies to each step of the process.

The flows of materials and processes as shown in Fig. 8.4 are within the "manufacturing" community, and the materials flows are generally under the sole control of the "manufacturer".

The resource flows involved in the customer portion of the industrial ecology cycle are shown in Fig. 8.5. Optimization opportunities here include the need to avoid resource dissipation, especially in the handling of obsolete products. To the extent that customers favor the recycle stream C_{3R} at the expense of the waste-disposal stream C_{3D}, industrial ecology is improved.

Although exceptions exist, the situation under existing economic and governmental constraints is that customer decisions are made independently of either the material supplier (who may also be a material recycler) or the manufacturer.

When Figs. 8.3 to 8.5 are combined, they result in the ensemble industrial ecology cycle shown in Fig. 8.6. One might view this figure as an inverted potential diagram, in which the energy expended to achieve a given flow decreases as one moves upwards in the diagram. That is, it takes much less energy to recycle materials from one of the higher stages to an intermediate stage than to begin at a lower stage. Robert Ayres of the European Institute of Business Administration puts the point differently: that the work done on materials at the expense of energy represents society's battle against thermodynamics, and the energy invested per unit of material decreases as one approaches the top of the materials flow chain. We call the increase in order and usability of processed materials their *embedded utility*, and state as a goal of industrial ecology the preservation of embedded utility in the materials used by industrial processes. One way in which this can be accomplished is to retain in use a large portion of an obsolescent product, recycling and replacing only outdated subsystems and components.

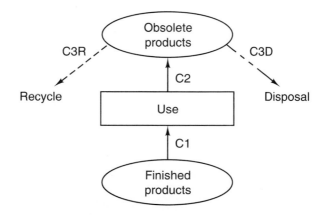

Figure 8.5   Materials flows in customer use.

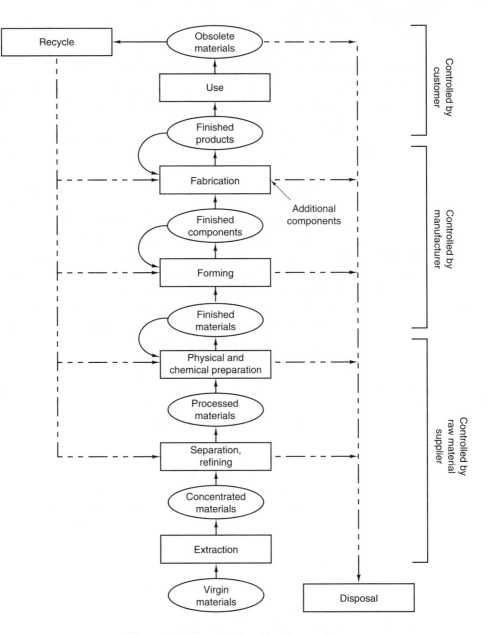

Figure 8.6   The total industrial ecology cycle.

8.3   THE SCOPING PROCESS

There is perhaps no more critical step in beginning an LCA evaluation than to define as precisely as possible the scope of the evaluation: What materials, processes, or products are to be considered, and how broadly will alternatives be defined? Consider, for example, the question of releases of chlorinated solvents during a typical dry-cleaning process. The purpose of the analysis is to reduce emissions. The scope of the analysis, however, must be defined clearly. If it is limited, the scope would comprehend end-of-pipe controls, administrative controls, and process changes. Alternative materials—in this case, solvents—might be considered as well. If, however, the scope is defined broadly, it could include alternative service options: Some data indicate that a substantial number of items are sent to dry-cleaning establishments not for cleaning per se but simply for pressing. Accordingly, offering an independent pressing service could reduce emissions considerably. One could also take a systems view of the problem: Given what we know about polymers and fibers, why are clothing materials and designs still being provided that require the use of chlorinated solvents for cleaning? Among the considerations that would influence the choice of scope in cases such as the preceding are (a) who is performing the analysis, and how much control they can exercise over the implementation of options; (b) what resources are available to conduct the study; and (c) what is the most limited scope of analysis that still provides for adequate consideration of the systems aspects of the problem.

The generality of the options set should also be explicitly defined. In many cases, it is possible to generate numerous potential options, not all of which can be feasibly reviewed. The challenge is to choose representative options that provide valuable guidance for real-world decisions, but are limited and general enough to make an analysis realistic.

Finally, the resources that can be applied to the analysis should also be scoped. Virtually all methodologies considered to date provide the potential for essentially open-ended data collection—and, therefore, virtually unlimited expenditure of resources. As a general rule, the depth of analysis should be keyed to the degrees of freedom available to make meaningful choices among options and the importance of the environmental or technological issues leading to the evaluation. For example, an analysis of using different plastics in the body of a currently marketed, portable, compact disc player would probably not require a complex analysis: The degrees of freedom available to a designer in such a case are already quite limited because of the constraints imposed by the existing design and its market niche. On the other hand, a regulator contemplating limitations on a material used in large amounts in numerous and diverse manufacturing applications would want to conduct a fairly comprehensive analysis, because the degrees of freedom involved in finding substitutes could be quite numerous and the environmental impacts of substitutes implemented widely throughout the economy could be significant.

MOTIVE POWER FOR AUTOMOBILES: A SIMPLE LCA EXAMPLE

Some perspective on LCA analysis is provided by an example comparing the use of fuel to provide power for gasoline and electric automobiles in a particular geographical region. As shown in Fig. 8.7, providing gasoline involves oil drilling, transporting crude oil, refining, transporting and delivering gasoline, and use. For an electrically powered automobile, the energy comes from fossil fuel combustion (mostly of coal and natural gas) at large stationary power-generation facilities.

The LCA inventory analysis considers the emissions resulting from each type of operation. For the U.S. EPA's "criteria pollutants" (those for which air-quality standards have been established), the attributable emissions on a grams per mile basis are shown in Fig. 8.8(a) for three specific gasoline and two electric vehicles. Carbon monoxide and nonmethane hydrocarbon emissions are higher for gasoline engines, sulfur dioxide and particles for electric vehicles. The relative emissions of oxides of nitrogen depend on the particular vehicle and power plant technology that is used.

Two other aspects of environmental impacts of the vehicles have also been investigated. First, the total energy use over the lives of the vehicles, is shown in Fig. 8.8(b); it is apparent that there is little to choose between on that basis. The second, greenhouse gas emissions per mile of operation, is shown in Fig. 8.8(c). Again, the options are nearly equivalent.

It is important to note how much this comparison rests on the way the scope of the analysis is defined. Although electric vehicles are sometimes called "zero-emission vehicles", they also have been labeled "elsewhere-emission vehicles" because their operation requires electricity from pollution-generating power plants. Thus, an important aspect of the comparison is that most emissions attributable to the gasoline-powered vehicle occur where it is being operated, often in congested urban areas, while most emissions attributable to the electric vehicle occur at power plants away from congested areas. Hence, for short-lived pollutants, the population exposure is greater from the emissions of gasoline-powered vehicles. Moreover, control technologies are better applied and monitored at a few point sources than over a diffuse, essentially uncontrolled population of small emission sources. The scope of the analysis in this case determines the conclusion: If emissions from the actual vehicle are all that is considered, the electric vehicle obviously is superior; if emissions of greenhouse gases are of concern from a global-warming standpoint, a different conclusion is reached. We should also point out another aspect of scoping: The inventory analysis refers only to the motive power requirements of the vehicles. A comprehensive LCA would consider in addition the emissions and environmental impacts of the vehicles themselves: their materials, their manufacture, and their eventual recycling. In a broader sense, one must be careful when adding impacts from different stages of a life cycle as well as considering the effects of different scopings, because impacts of different life stages may interact with completely different local and regional environments.

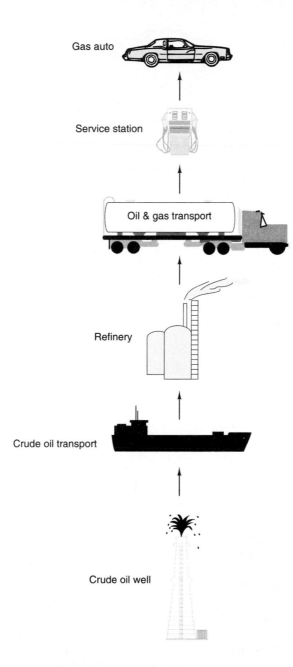

Figure 8.7   Stages of life-cycle inventory analysis for alternative motor vehicle propulsion designs: (a) one using gasoline for motive power, and (b) the other, electricity. Only the provisioning of motive power is diagrammed here, not the life cycles of the materials of the vehicles themselves. (Courtesy of D. H. Moody, National Pollution Prevention Center, University of Michigan.)

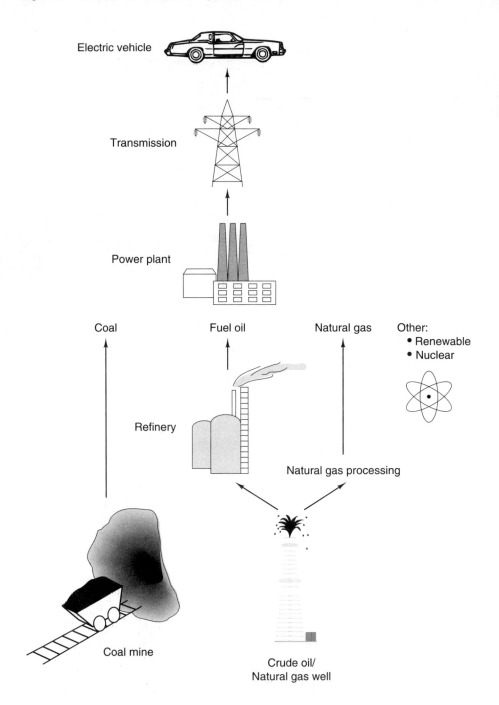

Electric vehicle

Transmission

Power plant

Coal   Fuel oil   Natural gas   Other:
                                                       • Renewable
                                                       • Nuclear

Refinery

Natural gas processing

Coal mine

Crude oil/
Natural gas well

Figure 8.7 (continued)

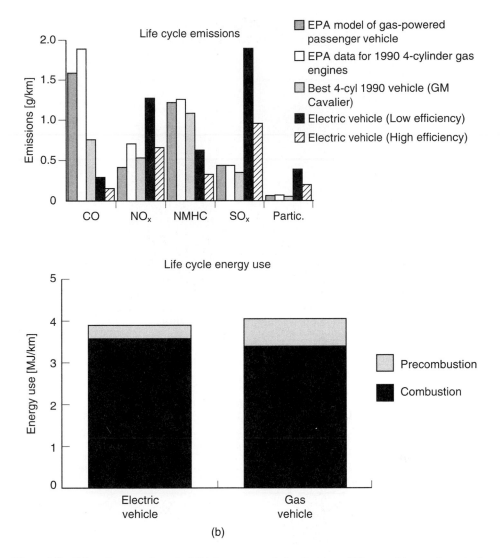

Figure 8.8   Selected outputs from the LCA inventory analysis of motor vehicle propulsion as shown in the Figure 8.7. (a) Attributable emissions (grams per kilometer) of species for which U.S. air-quality standards have been set: carbon monoxide (CO), oxides of nitrogen (NO_x), nonmethane hydrocarbons (NMHC), sulfur dioxide (SO_2), and particulate matter. This plot is based on data from the U.S. EPA, *Compilation of Air Pollutant Emission Factors*, Report AP-42, Washington, DC, 1991; and Q. Wang, M. DeLuici, and D. Sperling, Emission impacts of electric vehicles, *Journal of the Air and Waste Management Association, 40* (1990): 1278. (b) Energy use by typical gasoline and electric vehicles throughout their life cycle. This plot is based on a fuel cycle from Franklin Associates, Prairie Village, KS. The fuel efficiency of the electric vehicle is assumed to be 0.89 km/MJ (2 mi/kwhr), that of the gasoline vehicle 12.7 km/l (30.0 mi/gal). (c) Attributable emissions (grams per kilometer of species with radiative impact equivalent to that of carbon dioxide) of greenhouse gases from typical gasoline and electric vehicles. This plot is based on data from M. DeLuici, R. Johnston, and D. Sperling, Transportation fuels and the greenhouse effect, *Transportation Research Record, 1175* (1992): 33–44. The fuel efficiency of the electric vehicle is assumed to be 0.89 km/MJ (2 mi/kwhr), that of the gasoline vehicle 12.7 km/l (30.0 mi/gal). (Courtesy of D. H. Moody, National Pollution Prevention Center, University of Michigan.)

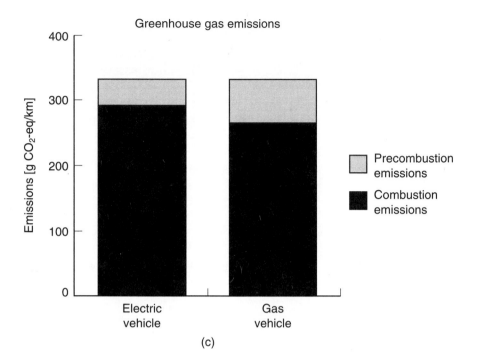

Figure 8.8 (continued)

Another environmental consideration not yet taken into account is that of the relative importance of the different emissions. The presumption from Fig. 8.8(a) is that the emission of a gram of CO is exactly as unwelcome as the emission of a gram of SO$_2$, yet no basis for such a conclusion has yet been presented, as would occur at the impact analysis stage. The third LCA stage, improvement analysis, has not been addressed either.

8.4   STAGES OF LIFE-CYCLE ASSESSMENT

The stages of life-cycle assessment are the subjects of the chapters in Parts III and IV of the book and constitute a substantial part of industrial ecology. We indicate the stages in the flow diagram of Fig. 8.9. The first stage is the scoping of the goals and characteristics of a specific industrial process or product, as discussed earlier in this chapter. The second stage is data collection and data presentation. This is presented in the following chapter, together with assessment tools for one or more design options.

Once each of the possible options is assessed, they must be compared and prioritized. The prioritization depends strongly on the degree to which a given material

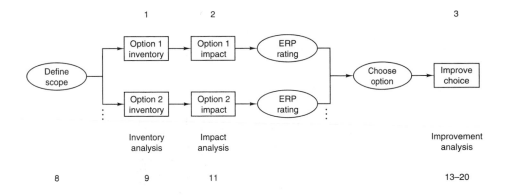

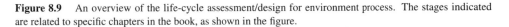

Figure 8.9   An overview of the life-cycle assessment/design for environment process. The stages indicated are related to specific chapters in the book, as shown in the figure.

perturbs the natural Earth system budget, a topic discussed in Chapter 10. Chapter 11 describes various approaches for prioritization, both qualitative and quantitative. Finally, it is instructive to study an example of the prioritization process. We present such an example in Chapter 12, the topic being the suitability of alternatives to lead-containing solder for the attachment of electronic components to printed circuit boards.

SUGGESTED READING

Keoleian, G. A., and D. Menerey. *Life Cycle Design Guidance Manual: Environmental Require-ments and the Product System*, EPA/600/R-92/226. Washington, DC: U.S. Environmental Protection Agency, 1993.

Leiden University Centre of Environmental Science, Dutch Organization for Applied Scientific Research, and the Netherlands Fuels and Raw Materials Bureau. *Manual for the Environmental Life Cycle Assessment of Products (second interim version)*, 1991.

Netherlands Company for Energy and the Environment (NOVEM), The Netherlands Institute for Public Health and Hygiene (RIVM), and The Netherlands National Research Programme for Recycling of Waste Materials, *Methodology for Environmental Lifecycle Analysis: International Developments*, Contract Number 8283, 1992.

Society of Environmental Toxicology and Chemistry. *A Technical Framework for Life-Cycle Assessment*. Washington, DC: SETAC, 1991.

Vigon, B. W., D. A. Tolle, B. W. Cornaby, H. C. Latham, C. L. Harrision, T. L. Boguski, R. G. Hunt, and J. D. Sellers. *Life-Cycle Assessment: Inventory Guidelines and Principles*, EPA/600/R-92/036. Cincinnati, OH: U.S. Environmental Protection Agency, 1992.

EXERCISES

8.1   Figure 8.7(a) shows the stages of the in-use analysis of gasoline for a motor vehicle. Construct a similar diagram for the steel used in making the vehicle.

8.2   Given the diagram constructed for the problem Ex. 8.1, what are the environmental impacts that should be considered for each stage?

CHAPTER 9

Process and Product Audits: LCA's Inventory Analysis Stage

9.1 APPROACHES TO DATA ACQUISITION

Once the scale of the LCA assessment has been established by defining the scope, the process begins with the acquisition of the necessary data. Some of the information needed is straightforward, such as the amounts of specific materials needed for a given design or the amount of cooling water needed by a particular manufacturing process. In other cases, the information is less traditional but no less useful. Accordingly, the industrial ecologist should be willing to approach data needs with a broad perspective. Any methodology, whether it applies to materials selection, processes, components, or complex products, can often be most effective if qualitative, not quantitative. A qualitative approach is somewhat controversial, especially among engineers and business planners who are biased toward quantitative systems. The latter obviously have advantages: They are universally utilized in high-technology cultures, they offer powerful means of manipulating and ordering data, and they simplify choosing among options. However, the state of information in the environmental sciences may not permit the secure quantification of environmental and social impacts because of fundamental data and methodological deficiencies. The result of inappropriate quantification might be that those concerns that cannot be quantified would simply be ignored—thereby undercutting the systemic approach inherent in the industrial ecology concept.

In order to maximize efficiency and innovation and avoid prejudgment of normative issues, an LCA information system should be nonprescriptive. It should provide information that can be used by individual designers and decision makers given the particular constraints and opportunities they face. For example, use of highly toxic materials might be a legitimate design choice—and an environmentally preferable choice from among the alternatives—where the process designer can adopt appropriate engineering controls. Designing products and processes inherently requires balancing many considerations and constraints, and the necessary trade-offs can only be made on a case-by-case basis during the product realization process. Moreover, in many cases, ethical questions may be raised by the analysis, and it is inappropriate to respond to such issues by imposing procedural and methodological resolutions rather than by substantive consideration of the questions raised. LCA information should provide not only relevant data, but, if possible, also the degree of uncertainty associated with that data. This approach is particularly important in the environmental area, where uncertainty, especially about risks, potential costs, and potential natural system responses to forcing, is endemic.

9.2 PROCESS AND PRODUCT BUDGETS

One of the more straightforward budgets in industrial ecology is the materials budget for a manufacturing process. A schematic example is shown in Fig. 9.1 for a chemical process involving the cleaning of a product or product component with a liquid solvent. The process begins with the addition of new solvent to a solvent reservoir, followed by piping or otherwise moving the solvent to the product line, where the solvent wash occurs. Most of the solvent eventually enters a recycling (and/or disposal) stream, but a portion (known as "dragout") is retained on the product. Some of the dragout material remains on the product, and a fraction is lost to the atmosphere by evaporation.

   Mass-balance equations can be set up around any boundary of the system. For example, it is clear that the rate that the solvent leaves the facility must be equal to the rate at which it enters, that is,

$$A = D + E + H \qquad\qquad (9.1)$$

   Once this diagram is drawn, the industrial ecologist can begin to quantify the budget. The quantity of solvent entering the system is probably known, and the rate of solvent purification may be known as well. The rates of loss may or may not be readily quantified. It is clear, however, that if most of the rates are known, some of the others can be computed. If one does not worry too much about absolute accuracy but is content

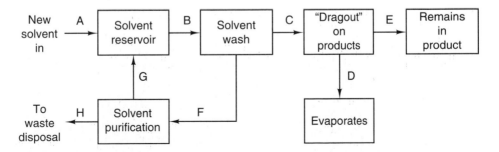

Figure 9.1   Schematic diagram of the flow streams involved in a budget analysis of a chemical solvent in a solvent-washing process. Capital letters indicate the flows of material.

with rough estimates, an approximate budget can often be put together relatively quickly and simply.

Product budgets share many of the characteristics of process budgets, as shown in the schematic example of Fig. 9.2. The first step is to choose a material of interest, generally either one of the predominant materials in the product or one for which significant environmental impact is possible. The next step is to follow that material through the manufacturing process, determining what fraction ends up in the product, what fraction is recycled within the manufacturing facility, and what fraction is lost to disposal.

As with the process budget, a mass-flow budget for a material that is part of a product can be readily put together. For example, the amount of polymer used in the product manufacturing stream must be equal to that leaving the facility as part of the product plus that leaving as part of the residue disposal stream, or

$$A = D + H_M + H_T + H_I = D + J \tag{9.2}$$

Attempting to construct a budget for such a process may demonstrate, for example, that the flow rate of the total residue disposal stream is known, but the individual flows from the manufacturing stages are not. Measurements or estimates then suggest the most important stage upon which to focus if one wants to reduce residue disposal stream J.

The flow diagrams in Figs. 8.3 to 8.5 are strikingly similar to the multistage countercurrent cascades commonly encountered in chemical engineering, and can be analyzed by similar mathematical approaches. The analogy can be appreciated by redrawing Fig. 8.4 so that it takes the form shown in Fig. 9.3. In this diagram, the product inspection steps are treated as process stages that do not receive countercurrent flow. A materials balance can then be set up for each stage or for the entire system, and overall efficiencies can be computed. For stage 1, the materials balance is written as

$$F + R_1 + R_2 = P_1 + D_1 \tag{9.3}$$

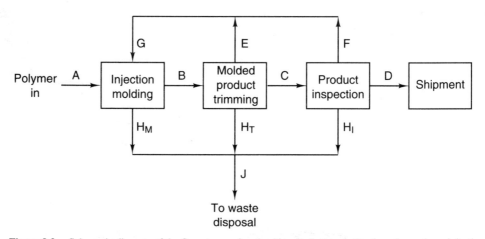

Figure 9.2   Schematic diagram of the flow streams involved in a budget analysis of a polymer in an injection molded product.

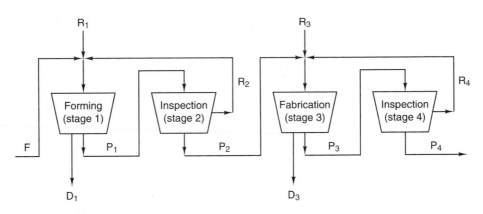

Figure 9.3   Materials flows in manufacturing, treated as a four-stage countercurrent flow system. All symbols on the diagram refer to total mass flows: F, input flow; P, product flows; D, discard flows; R, recycle flows. See text for additional discussion.

and the extraction efficiency ε (for stage 1, for example) is given by the ratio of the output to the input:

$$\varepsilon_1 = \frac{P_1}{P_1 + D_1} \tag{9.4}$$

In the same manner, the mass-balance equations for stages 2 to 4 are

$$P_1 = P_2 + R_2 \tag{9.5}$$

$$P_2 + R_3 + R_4 = P_3 + D_3 \tag{9.6}$$

$$P_3 = P_4 + R_4 \tag{9.7}$$

The reuse efficiency ρ for an inspection stage is given by (for stage 2, for example)

$$\rho_2 = \frac{P_2}{P_2 + R_2} \tag{9.8}$$

and the efficiencies for the entire process are cascaded:

$$\varepsilon = \varepsilon_1 \varepsilon_3 \tag{9.9}$$

$$\rho = \rho_2 \rho_4 \tag{9.10}$$

How the computation of process efficiencies works in practice can be illustrated by an example. Assume for the manufacturing system shown in Fig. 9.3 that the residue stream from stage 3 is 5% of the product stream from that stage, that the internal recycling streams are 3% of the product streams, and that the external recycling streams are 20% of the internal process input streams. Then

$$D_3 = 0.05 P_3 \qquad R_2 = 0.03 P_2 \qquad R_4 = 0.03 P_4$$

$$R_1 = 0.2 F \qquad R_3 = 0.2 P_2$$

Now evaluate the process if the input flow stream to the facility is 100 kg/h and the output flow is 130 kg/h. Performing mass balances around the various stages gives (approximately)

$$P_3 = 133.9 \text{ kg/h} \qquad D_1 = 6.1 \text{ kg/h}$$

$$\varepsilon_1 = 0.95 \qquad \varepsilon = 0.91 \qquad \rho_2 = 0.97 \qquad \rho = 0.94$$

It is straightforward to extend this computation to treat an individual material in streams of mixed materials or to determine efficiencies over the combined product life cycle rather than just the manufacturing stage.

9.3   THE MATRIX CONCEPT FOR MATERIALS AND PROCESS AUDITS

It is standard practice in industrial ecology to compare several responsible designs with each other and determine their relative level of merit. Such comparisons are greatly aided by data-presentation techniques that aid in the comprehension of large amounts of information on widely varying topics. An industrial ecology matrix template is one way to perform and present a materials and process analysis. The typical template consists of two components: (a) a matrix system graphically and qualitatively summarizing the status of a particular design option across the product life cycle; and (b) an accompanying documentation package explaining in detail (and quantifying where possible) the information contained in the matrix's cells.

By using the matrix template, a specific industrial ecology analysis focuses on the options for a particular process. In one implementation of the procedure, four *Primary Matrices* are prepared for each option. Each Primary Matrix has two axes, one comprising the life stages of the product or process and the other consisting of issues categories bearing on the suitability of the option under evaluation. The *Manufacturing Primary Matrix*, shown in Fig. 9.4, focuses on the implications of each option in terms of the manufacturing activity itself. The *Environmental Primary Matrix* looks at the more traditional environmental impacts of technology choices, but does so across the same life-cycle stages of the product and with categories that attempt to minimize the "single-media" fixation of current approaches, that is, the reluctance to consider several alternative impacts in a coherent and structured manner. The *Toxicity/Exposure Primary Matrix* looks at issues that are relatively familiar but does so in a systemic, life-cycle way. Finally, the *Social/Political Primary Matrix* is designed to capture the broader nontechnical aspects of each option; this matrix will pose special challenges as much of the information is virtually ignored under current practices and will be difficult to define and codify with precision. For the most part, the cell designations are self-explanatory. Note that packaging and transportation are listed as life-cycle stages: Such activities, although occurring at a number of points in a material, process, or product life cycle, can have substantial environmental impacts, yet are frequently overlooked in standard analyses. Here they are singled out for special attention in the matrix system.

There are three possible entries for each matrix cell (Fig. 9.5). A straight line through the cell means a category inapplicable to the option under consideration. One or

Life stages

	Initial production	Secondary processing/ manufacturing	Packaging	Transportation	Consumer use	Reuse/ recycle	Disposal	Summary
Process compatibility								
Materials compatibility								
Component compatibility								
Performance								
Energy consumption								
Resource consumption								
Availability								
Cost								
Competitive implications								
Environment of use								

Figure 9.4   The structure of primary matrices designed for evaluating the use of a specific material or the design of a specific manufactured product. Different matrices deal with the following topics: manufacturing, environmental concerns, toxicology and exposure, and social and political concerns. (B. R. Allenby, Design for Environment: Implementing Industrial Ecology, Ph.D. dissertation, Rutgers University, March 1992.)

two plusses ("+" or "++") in a cell indicates positive environmental effects from the design option, and the relative degree of benefit. The third type of entry is an oval with the degree of environmental concern keyed to the geometrical pattern in the oval. Blank would indicate no concern, shading some concern, diagonal lines greater concern, and solid black the highest degree of concern. The degree of uncertainty associated with a particular rating is indicated by the extent to which the pattern fills the oval. Thus, a completely full oval indicates absolute certainty regarding the rating, whereas a 25% filled oval indicates considerable uncertainty regarding the rating. For example, benzene might receive a full, black oval in the Mammalian Acute-Consumer Use cell of the Toxicity/Exposure Primary Matrix because of its status as a known human carcinogen. Methylene chloride, on the other hand, might receive a half-filled, black oval for the same cell, indicating a potentially serious issue (suspected carcinogenicity), but a significant degree of uncertainty associated with that rating.

Upon completion of the Primary Matrices for each of the options being considered, each matrix is given an overall degree of concern/certainty assessment. Alternative methodologies involve summing the assessments of the individual matrix elements in some way. These grouped assessments are then transferred to a Summary Matrix (Fig. 9.6) that displays the results of the assessments for the several alternatives in a form suitable for easy comparison. The choice of the most suitable option is then generally straightforward.

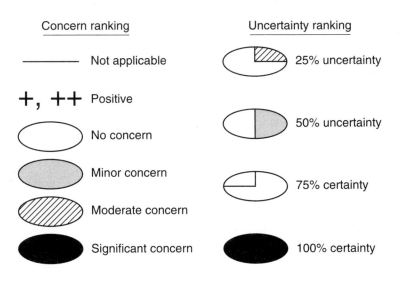

Figure 9.5   Symbols to be used in the matrix cells to indicate degrees of concern and uncertainty. (B. R. Allenby, Design for Environment: Implementing Industrial Ecology, Ph.D. dissertation, Rutgers University, March 1992.)

The matrix assessment concept, described here in overview, has the potential to bring together assessments of all the factors with which industrial ecology deals, and to facilitate the selection of the most desirable of several design options. Many of the evaluations that need to be made as part of the process are difficult, however. It is probable

Summary matrix
options

	Option A	Option B	Option C	Option D
Toxicity/ exposure				
Environmental				
Manufacturing				
Social/ political				

Figure 9.6   The summary matrix designed for evaluating the use of a specific material or the design of a specific manufactured product. (B. R. Allenby, Design for Environment: Implementing Industrial Ecology, Ph.D. dissertation, Rutgers University, March 1992.)

that a comprehensive evaluation would require input from many communities: regulators, the academic community, different industrial sectors, professional engineering and industrial organizations, and so forth. For generic products or processes, such steps as peer review may be desirable. The end product would be designs that, while not optimizing any single factor in industrial ecology, would be environmentally preferable to the designs of today, which frequently reflect fragmented or negligible environmental input.

9.4 AN APPROACH TO THE PRODUCT INVENTORY

One measure of the degree to which industrial ecology is integrated into the fabric of a corporation is whether every product receives an LCA assessment. Such an assessment will reveal whether a design is environmentally responsible and will help the designer or design team to identify changes that will make it more so. Another advantage is that over time, industrial ecology assessment will influence corporate thinking toward the design and manufacture of products with a commonality of responsible choices in materials, processes, and resource conservation across the life cycle of the product.

How should a product inventory be performed? To do so, we retain the matrix concept presented earlier, but adapt it so that it deals not with a single material over all product lines, but with all materials over a single product line. This is the more usual approach to industrial ecology, in which individual products are assessed as they reach the design and development stage.

One axis of the materials audit matrix, that for the Life Stage, is retained. On the other axis, we use Environmental Concern, which we divide into five classifications: materials choice, energy use, solid residues, liquid residues, and gaseous residues. The resulting matrix is shown in Fig. 9.7.

Life stage	Environmental concern				
	Materials choice	Energy use	Solid residues	Liquid residues	Gaseous residues
Resource extraction					
Product manufacture					
Product packaging					
Product use					
Recycling, disposal					

Figure 9.7   Assessment of a product's environmental concerns: The basic matrix.

Designers who have never performed a product audit may wonder about some of the matrix elements in the figure. To aid in perspective, we provide examples in Fig. 9.8 for qualitative entries in each matrix element. The basis for some of these examples is that the industrial process is responsible (implicitly if not explicitly) for the embedded impacts of the processing of raw materials that are used and for the projected impacts as the products are used, recycled, or discarded.

Life stage	Materials choice	Energy use	Solid residues	Liquid residues	Gaseous residues
			Environmental concern		
Resource extraction	Use of only virgin materials	Extraction from ore	Slag production	Mine drainage	SO_2 from smelting
Product manufacture	Use of only virgin materials	Inefficient motors	Sprue, wrapping disposal	Toxic chemicals	CFC use
Product packaging	Toxic printing ink use	Energy-intensive packing materials	Polystyrene packaging	Toxic printing ink use	CFC foams
Product use	Intentionally dissipated metals	Resistive heating	Solid consumables	Liquid consumables	Combustion emissions
Recycling, disposal	Use of toxic organics	Energy-intensive recycling	Non-recyclable solids	Non-recyclable liquids	HCl from incineration

Figure 9.8   Assessment of a product's environmental concerns: Examples of design decisions that would result in an unfavorable audit assessment for each of the matrix elements.

9.5 AUDITING BY LIFE STAGE

Each element in the product audit matrix can be evaluated either as an aspect of the life-stage design or as an aspect of the environmental concern. Consider first the life-stage evaluation. In each life stage, the evaluation can be made with the assistance of the checklists given in Appendix A, as indicated in Fig. 9.9.

Life stage	Materials choice	Energy use	Solid residues	Liquid residues	Gaseous residues
			Environmental concern		
Resource extraction	Checklist 5				
Product manufacture	Checklists 1–4				
Product packaging	Checklist 6				
Product use	Checklist 7–13				
Recycling, disposal	Checklist 14				

Figure 9.9   Assessment of a product's environmental concerns: Checklists applicable to the different life stages.

The auditor begins with the first of the life stages, materials extraction, using checklist 5, Appendix A. For each of the five environmental concerns associated with materials extraction, she or he chooses an ellipse filled as indicated by the following criterion: the product's attributes relevant to the matrix element under consideration should be compared to the designer's best perception of how an ideal product would rank. Comparison to the ideal product is on a four-level scale: excellent, good, fair, and poor. The texture of the ellipse indicates the degree of confidence in the evaluation, black for high confidence, shaded for moderate confidence, and blank for low confidence.

Following the materials-extraction assessment, similar assessments are made for the other life-cycle stages: product manufacture, product packaging, product use, and recycling/disposal. At the completion of the analysis, every matrix element has had a value assigned, and the environmental responsibility of the design is indicated quickly by the density of "blackness" of the matrix.

9.6 AUDITING BY ENVIRONMENTAL CONCERN

Product designers will tend most naturally to perform the product audit in life stages, as indicated before. A complementary audit can be performed by environmental concern, however. In that approach, the audit is performed on the matrix by columns, not rows, as indicated in Fig. 9.10. The general technique is identical to that used for the life-stage evaluation. One begins with the first of the environmental concerns, materials choice, using checklist 5, Appendix A. In each of the five life stages associated with materials choices, the auditor chooses ellipse filling and hatching in the same manner as indicated before.

After both the life-stage audit and the environmental concern audit are performed, there will inevitably be differences in the assessments assigned to the same matrix

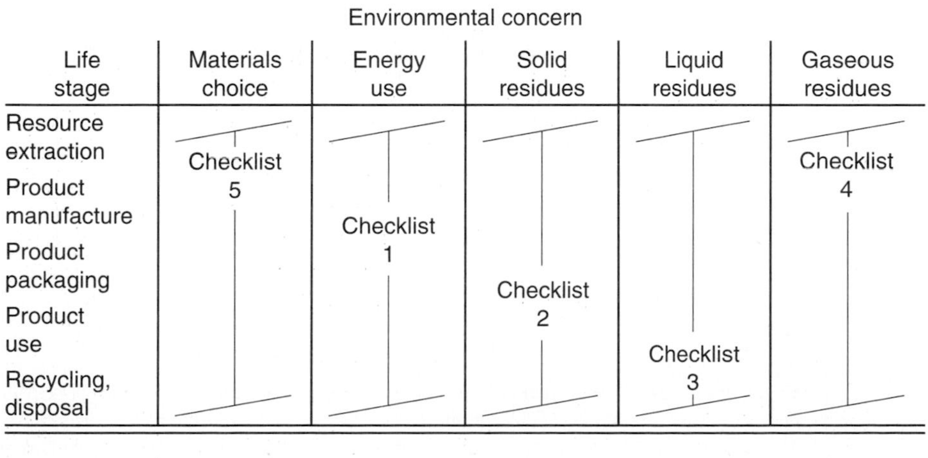

Environmental concern

Life stage	Materials choice	Energy use	Solid residues	Liquid residues	Gaseous residues
Resource extraction	Checklist 5				Checklist 4
Product manufacture		Checklist 1			
Product packaging			Checklist 2		
Product use				Checklist 3	
Recycling, disposal					

Figure 9.10   Assessment of a product's environmental concerns: Checklists applicable to the different environmental concerns.

element by the two evaluations. The resolution of these differences has the potential to be the most rewarding part of the audit, as product and process designers discuss their reasons for the assessments that were made. When resolution is reached, the completed matrix is displayed so that the attributes thereupon indicated can be assimilated and discussed.

DETERGENT-GRADE SURFACTANTS

Surfactants are constituents of detergents that aid in releasing soil from clothing, linens, and other items. Various surfactants can be used, and those surfactants are made from various sources of raw materials. To evaluate whether some options for surfactant sourcing and production were preferable to others, Procter & Gamble Company, in cooperation with Franklin Associates, performed an extensive life-cycle inventory study.

To begin the study, production flow diagrams from raw materials to products must be developed. An example, using palm fruit as the principal raw material, is

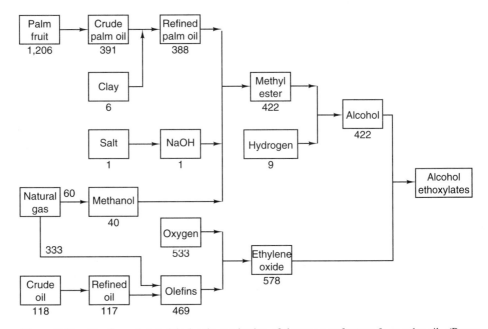

Figure 9.11 The flow of materials for the production of detergent surfactants from palm oil. (Reproduced with permission from C. A. Pittinger, J. S. Sellers, D. C. Janzen, D. G. Koch, T. M. Rothgeb, and M. L. Hunnicutt, Environmental life-cycle inventory of detergent-grade surfactant sourcing and production, *Journal of the American Oil Chemists' Society, 70,* 1–15. Copyright 1993 by the American Oil Chemists' Society.)

shown in Fig. 9.11. Crude oil and natural gas are also needed to synthesize the desired product, a family of alcohol ethoxylates (AE). For each flow of materials, the mass requirements are measured, and emissions (not shown on the diagram) are determined. Similar assessments are made for the production of several different surfactants from petrochemicals, palm kernel oil, and tallow, as well as palm oil. A selection of the results is shown in Table 9.1.

The results show more similarities than differences, but the differences are worth comment. One is the much larger water use of tallow, attributable to the crop irrigation required to provide feed for beef cattle (beef cattle are the primary source of tallow). The energy requirements for petrochemical recovery are higher than those for palm oil or tallow. Particulates and land-applied emissions from the petrochemical feed stock are significantly lower than for the other options. The conclusion of the study was that benefits in one direction appeared offset by liabilities in another, and environmental concerns did not support any fundamental shifts in the worldwide mix of feedstocks used for surfactant manufacture.

Table 9.1 Raw Material Requirements, Energy Requirements, and Net Life-Cycle Emissions for AE Production

Flow Stream	Feedstock		
	Petrochemical	Palm Oil	Tallow
Organic raw materials*	935	899	891
Water use*	40	49	415
Energy use[†]			
Raw materials	50	26	27
Transport	3	5	7
Processing	39	37	40
Atmospheric emissions[‡]			
Particulates	2.2	9.0	8.0
Hydrocarbons	39.2	33.2	34.8
Aqueous emissions[‡]			
Dissolved solids	5.6	5.3	4.3
Oil	0.065	0.12	0.30
Land-applied emissions[‡]	76	111	139

[*] kg per 1000 kg of raw materials.
[†] GJ of energy per 1000 kg of raw materials.
[‡] kg of emissions per 1000 kg of surfactant.
Source: Abstracted from C. A. Pittinger, J. S. Sellers, D. C. Janzen, D. G. Koch, T. M. Rothgeb, and M. L. Hunnicutt, Environmental life-cycle inventory of detergent-grade surfactant sourcing and production, *Journal of the American Oil Chemists' Society, 70* (1993): 1–15.

9.7 SUMMARY

The displays shown and discussed in this chapter are of substantial utility, not only in improving the environmental responsibility of a single product design, but also in comparing the attributes of one design with those of another. The comparison can be done qualitatively, with marked success, by constructing primary matrices for each design and examining them together. More quantitative approaches are just beginning to be developed; ultimately, they may be preferable to the qualitative assessments described here. For the foreseeable future, however, the qualitative assessments will be at least as useful and much easier to generate than the more detailed and quantitative approaches.

SUGGESTED READING

Allenby, B. R. *Design for environment: Implementing industrial ecology*, Ph.D. dissertation, Rutgers University, March 1992.

Graedel, T. E. Regional and global impacts on the biosphere: A methodology for assessment and prediction, in *Energy: Production, Consumption, and Consequences*, J. L. Helm, ed., pp. 85–110, Washington, DC: National Academy Press, 1990.

Portney, P. R. The price is right: Making use of life cycle analyses, *Issues in Science and Technology, 10* (1993–1994): 69–75.

EXERCISES

9.1 Choose one of the following products: a bar of soap, a bicycle, a car wash, or an ocean cargo ship. Use checklists 1 to 5 (Appendix A) to do a qualitative survey of design items of possible concern. Suggest appropriate actions by the design engineer for each of those items.

9.2 Devise techniques for grouping the assessments of Fig. 9.7 by rows and by columns.

9.3 Product inventories have been prepared for two different designs of a high-speed widget. The matrices are reproduced in the following table. The figure on the left side of each matrix element referring to design 1, that to the right to design 2. Using your answer to Ex. 9.2, select the better one from an industrial ecology viewpoint. What features of each design would you address if improvement were needed?

Product Inventory Results

Life Stage	MC	EU	SR	LR	GR
RE	1/1	4/3	4/3	2/2	3/2
PM	2/1	1/2	1/2	2/1	2/4
PP	3/2	1/1	2/3	1/1	1/1
PU	1/2	1/2	1/3	1/1	1/3
RD	2/1	2/2	2/1	1/2	1/2

CHAPTER

10 Perturbing Nature's Budgets and Cycles

10.1 ENSEMBLE INDUSTRIAL MATERIALS FLOWS

Applied industrial ecology is, as we have seen, the study of the driving factors influencing the flows of selected materials among economic processes. When the anthropogenic flows are totaled over the global sweep of industrial activity, their magnitudes turn out to be truly enormous. Table 10.1 presents a selection of those flows on an annual basis, together with some global per capita averages.

The flow of water is the dominant figure in the table. It is listed because about 23% of all water use is for industrial activities and 69% is for agriculture (the numbers are all global averages; the percentage used for agriculture varies greatly with latitude and agricultural crop selection). As with other materials, water is a commodity in restricted supply, especially in some parts of the world, and its measurement and conservation in industrial activities are important components of industrial ecology.

Of the metal flows indicated in the table, that for steel is clearly the largest, aluminum next. The other metals, although crucial in many applications, have much lower total flows. Their natural flows are generally even smaller, however, and thus the perturbation of their natural cycles may be large, even if the absolute amounts appear minimal. Of the fossil fuels, coal and oil consumption are of the same order of magnitude, those for lignite and gas are significantly smaller. In the case of minerals, many are extracted in large quantities, cement being the mineral most widely used. On a global per capita basis, the use of fossil fuels is about equal to that of the combined use of metals and minerals, but is about a factor of 5000 lower than the use of water.

Although Table 10.1 measures materials use, it says nothing about how materials are used and what portions are lost without being productively employed. For example, one might wish to measure the flow of mine residue involved in mineral and metal pro-

Table 10.1 Global Anthropogenic Flows of Selected Materials

Material	Flow (Tg/yr)	Per capita flow [*] (Mg/yr)
Minerals		1.2[†]
Phosphate	120	
Salt	190	
Mica	280	
Cement	890	
Metals		0.3
Al	97	
Cu	8.5	
Pb	3.4	
Ni	0.8	
Sn	0.2	
Zn	7.0	
Steel	780	
Fossil fuels		1.6
Coal	3200	
Lignite	1200	
Oil	2800	
Gas	920	
Water	41,000,000	8200

[*] Per capita figures are based on a population of five billion people and include materials in addition to those highlighted in this table.

[†] Does not include the amount of overburden and mine residue involved in mineral production; neglects sand, gravel, and similar material (but includes cement).

Source: Data are from *UN Statistical Yearbooks* (various years), *Minerals Yearbook* (U.S. Dept. of the Interior, 1985), and *World Resources, 1990–91* (World Resources Institute, 1990).

duction. Platinum group metals, for example, are present in ore at concentrations of about 7 parts per million. Because of this low occurrence, the 143 Mg of purified metal produced each year require the extraction and processing of some 20 Tg of ore. A general picture is provided for the United States by the U.S. Bureau of Mines in the graph of Fig. 10.1, which shows that for some commodity groups, as much as a third of the consumption is released into the environment in a short period of time.

The situation with construction materials used and discarded is particularly worthy of comment. Partly because construction tends to be performed in unstructured environments and to involve many participants, the construction industry as a whole has done little to optimize its flows of materials. As a result, perhaps 15–20% of all materials brought to construction sites is lost, burned, illegally removed, or otherwise not used in construction. One consequence is that construction and demolition debris is a substantial part of most solid-waste streams, and a part that has received little attention. The residue

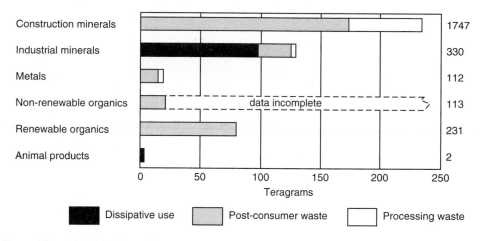

Figure 10.1   Materials released into the environment in the United States in 1990, expressed by commodity group. The numbers at the right margin indicate the total annual consumption for each group. (For nonrenewable organics, dissipative use and processing waste data are incomplete.) (Adapted from *Minerals Today*, p. 10, Washington, DC: U.S. Bureau of Mines, April 1993.)

generally does not pose toxicity problems, but it does pose disposal problems because of its volume, and represents wasted embedded energy as well. Many of the design for environment topics that will be treated in Part IV of this book are highly applicable to the construction industry: packaging takeback, enhanced supplier quality, planning for ultimate demolition, and so forth. The opportunity for societal gains is great.

10.2 INDUSTRY BUDGETS

To what extent can we know where a given material used in industry and commerce goes, and what uses are made of it? The answers require diligent research, partly because the degree of use varies as a function of both the properties and prices of the materials. A materials cycle example, for lead in the world economy, is shown in Fig. 10.2. Lead is particularly interesting in that a large part of its consumption goes to lead-acid batteries, especially for automobiles, and most of that lead is eventually recycled. In many of its other uses, however, lead is dissipated. Hardly any of the lead used for ammunition, pigments, fishing weights, and solder is ever recovered, and these flows, together with nonrecycled battery lead, require annual lead extraction of some 3.3 Tg. (The quantity of ore that must be processed to recover this lead is, of course, very much greater.)

An aspect of national and regional budgets not applicable to the global budget of Fig. 10.2 is that imports and exports of lead in various products traversing the geographical boundaries of the budget region must be considered. Different materials cycles would show a wide variety of behavior with respect to flows of materials across national boundaries, but it is important to recognize that industrial ecology applies to local and

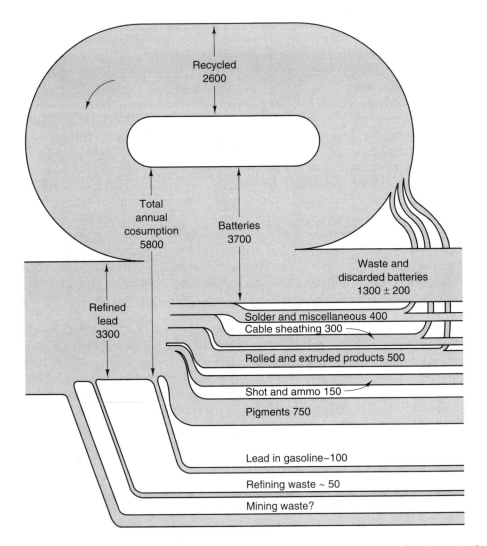

Figure 10.2   Fluxes of lead in the world economy. The data are in units of Gg (thousands of metric tons) and are for the year 1988. (V. Thomas, and T. G. Spiro, Preprint: Emissions and exposure to metals: Cadmium and lead, to be published in *Industrial Ecology and Global Change*, R. Socolow, C. Andrews, F. Berkhout, and V. Thomas, eds. Cambridge, UK: Cambridge University Press, 1994.)

regional cycles as well as global ones. Lead happens to be a particularly good example, because a major component of U.S. use a decade or two ago was as the gasoline additive tetraethyl lead. This use is no longer significant in the United States, but is an important part of the lead cycle in a number of developing countries that have not yet chosen to restrict this use. Much of the tetraethyl lead that is so used is manufactured elsewhere and crosses international boundaries.

10.3 CRADLE-TO-GRAVE BUDGETS AND CYCLES

Materials budgets are often very restricted in scope, to a production line, say, or a manufacturing facility. The results can be put into perspective only when they are viewed in relation to comparable flows in the local area, the regional area, or the planet, depending on the type of emittant. Earth system budgets have traditionally been the province of environmental scientists, using data on both natural and anthropogenic emittants. Once the industrial component becomes significant, however, the industrial engineer has an equal role, and defining cycles and constructing budgets becomes a shared operation.

As an example of a regional materials cycle assessing the interaction of natural and industrial systems, Figure 10.3 treats cadmium in the Rhine River Basin of western Europe. Cadmium is an interesting contrast to lead in that it is not the principal target of a materials extraction operation. Rather, it is a minor constituent of the ores that contain

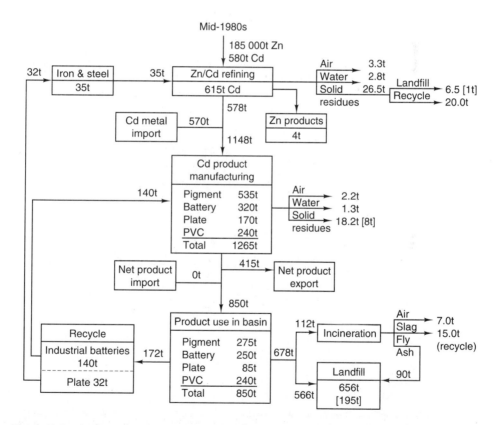

Figure 10.3 Cadmium product use and disposal (in metric tons per year) in the Rhine River Basin in the mid-1980s. Numbers in square brackets indicate the annual availability of cadmium from landfills. (Reprinted by permission from W. M. Stigliani and S. Anderberg, Preprint: Industrial metabolism at the regional level: The Rhine Basin, in *Industrial Metabolism—Restructuring for Sustainable Development*, R. U. Ayres and U. E. Simonis, eds. Tokyo: United Nations University, 1993.)

zinc, and its extraction occurs as a by-product of zinc refining. Mixed metal extraction is a common occurrence, leading to a complication for ameliorating heavy metal use: Should uses of cadmium be decreased while uses of zinc remain at the same level, the cadmium recovered as a consequence of zinc production will still have to be used in some way or disposed of in some manner.

Figure 10.3 is similar to one earlier presented for lead with one important addition: It specifies the environmental reservoirs that receive outflows of material. In the case of the Rhine River Basin, the river interactions are of special interest; potential pathways for these flows are shown in Fig. 10.4. Some of the flows are the result of point sources such as industrial process solution flows containing cadmium, perhaps directed to municipal water-treatment facilities, perhaps not. Others are entirely dissipative, as with the corrosion of building materials or the airborne dispersion of smelting and refining emissions.

The cadmium flows to the river are given quantitative values in Fig. 10.5. Point sources of cadmium from industrial operations are seen to be the most significant. The

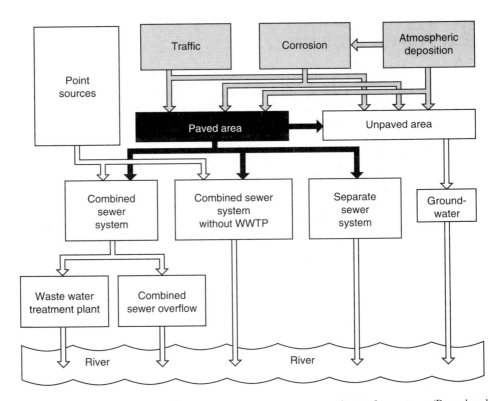

Figure 10.4   Pathways by which pollutants in urban areas are transported to surface waters. (Reproduced with permission from H. Behrendt and M. Boehme, *Point and Diffuse Loads of Selected Pollutants in the River Rhine and Its Main Tributaries*, Research Report. Copyright 1992 by International Institute for Applied Systems Analysis, Laxenburg, Austria.)

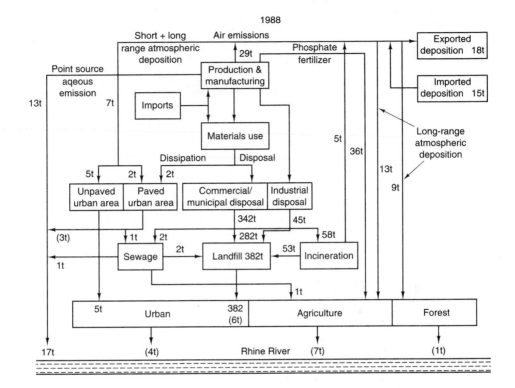

Figure 10.5   Flows of cadmium (in metric tons per year) to the Rhine River, 1988. Numbers in parentheses indicate diffuse sources. (Reproduced with permission from W. M. Stigliani and S. Anderberg, Preprint: Industrial metabolism at the regional level: The Rhine Basin, in *Industrial Metabolism—Restructuring for Sustainable Development*, R. U. Ayres and U. E. Simonis, eds. Tokyo: United Nations University, 1993.)

production of phosphate fertilizer appears in the analysis, because cadmium is a constituent of phosphate ore. Agricultural use of phosphate fertilizer is also a significant source. Other sources worth noting are dissipative corrosion of zinc in galvanized steel (cadmium is an impurity in the zinc, especially that produced by older refining processes) and the mobilization of cadmium from landfills in which cadmium-containing products were interred. The diagram reveals that even if the production and manufacturing sources of cadmium were completely prevented by emissions controls or materials substitutions, about half of the cadmium flow to the Rhine River would continue for many years.

A final consideration with cadmium budgeting is that the figures imply that all flows are instantaneous, whereas in actuality their time scales are very different. Once cadmium emissions to the air are stopped, for example, the air rapidly becomes cadmium-free. In the case of soil, however, the residence times are several orders of magnitude longer. This disparity is demonstrated by data showing that, although cadmium emissions throughout the Rhine River Basin have decreased substantially in the last 20 years or so, cadmium concentrations in the soil have remained little changed.

10.4 INDUSTRIAL PERTURBATIONS TO NATURAL BUDGETS

The industrial budgets presented before are of substantial interest in themselves, but they do not address the overriding question of significance with respect to natural budgets. Clearly, if an industrial cadmium budget furnishes 29 Mg Cd/yr to the Rhine River Basin, if the natural cycle were to furnish a million times that, and if the biological availability of the cadmium from all sources were roughly equivalent, the industrial portion of the Rhine River Basin cadmium budget is unlikely to be important. Conversely, if the natural flux of cadmium is very small compared with the industrial flux, the industrial perturbation is worth noticing. A similar approach can be taken to budgets on any scale: a watershed, a country, a continent, the entire planet. Because some perturbations may have greater impacts than others, industrial ecologists look to environmental scientists first to detect perturbations and then to provide some measure of their relative importance.

Perturbations are generally found by constructing global budgets and cycles that incorporate inputs from both natural and industrial activities. As an example, take the global nitrogen budget, the natural portion of which was presented in Chap. 7. The combined budget is shown in traditional biological form in Fig. 10.6(a), in which the industrial portions are indicated. Although budgets such as these are notoriously difficult to construct, the budget entries are at least of the right order of magnitude, if not completely accurate. The striking characteristic of the figure is that industrial activity is now fixing almost exactly as much nitrogen as is being done by nature: 130 Tg N/yr by chemical conversion of atmospheric nitrogen to inorganic chemicals, mostly ammonia fertilizer, and 40 Tg N/yr by combustion of fossil fuels. The driving portion of the nitrogen cycle is thus being strongly influenced by industrial activities.

The node diagram presentation of the global nitrogen budget appears in Fig. 10.6(b). The inclusion of the reservoir change terms now makes the industrial influence more apparent. Some 70 Tg N/yr are being removed from the atmosphere. The total global reserves are some 194 Zg N, of which 4 Zg N are in the atmosphere, so a loss of 70 Tg N/yr is a trifling amount unworthy of concern. Where does the nitrogen go, however? Some 50 Tg N/yr of it is going to increase the nitrogen contained in plants; this is an annual increase of about 0.4%, and demonstrates that fertilization is having the desired effect of producing more crops. The other 20 Tg N/yr is going to increase the nitrogen reservoir in animals, primarily beef cattle and other domestic stocks. This increase is about 2% of the current value, and is unlikely to be sustainable over very long periods of time.

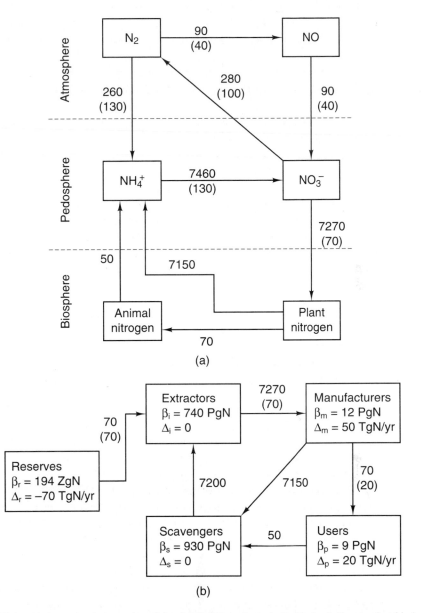

Figure 10.6   (a) A geochemist's version of the global nitrogen budget considering both natural and industrial influences. (b) An industrial ecologist's version of the total global nitrogen budget. The numbers in parentheses indicate the industrial contribution to each of the fluxes. The units for the reservoir contents are given in each box. The flows are in Teragrams (1 $Tg = 10^{12}$g) per year. The data sources are the same as for Fig. 7.4.

THE ATMOSPHERIC CHLORINE BUDGET

Although materials budgets are often very restricted in scope, to a production line, say, or a manufacturing facility, the results can be put into a different perspective when viewed in relation to comparable flows in the local area, the regional area, or the planet. This topic will not receive detailed treatment in this book, but because it is ultimately related to materials balances in industrial ecology, we will discuss it briefly, using an example of particular industrial relevance.

The atmosphere contains substantial amounts of chlorine and smaller amounts of fluorine, bromine, and iodine. In the lower atmosphere, chemical reactions involving the halogens appear to be unimportant except perhaps for corrosion reactions on metal surfaces. In the stratosphere, however, chlorine and bromine compounds are vital reactants in catalytic cycles that lead to lower concentrations of ozone. Particularly for this latter reason, the fluxes involved in the atmospheric halogen cycles are of interest and importance.

A budget for atmospheric chlorine is shown in Fig. 10.7. By far the largest flux indicated in the figure is that from seasalt spray over the oceans. This flux is very uncertain, as is the percentage of seasalt chlorine that is vaporized and the percentage returned to the surface by sedimentation and precipitation. Because these numbers dominate the tropospheric chlorine cycle, their uncertainty renders the entire cycle (though not the subcycles) of only academic interest and of no use in assessing

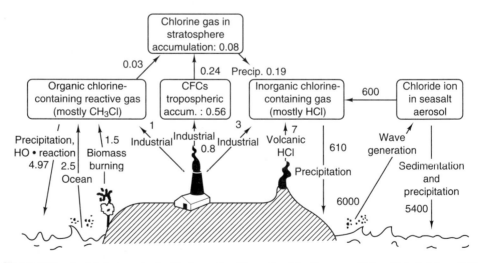

Figure 10.7   A global atmospheric chlorine cycle. The units of the fluxes are Tg/yr. (Adapted from T. E. Graedel and P. J. Crutzen, *Atmospheric Change: An Earth System Perspective.* New York: W. H. Freeman, 1993.)

trends for total atmospheric chlorine. Moreover, seasalt and HCl are efficiently returned to Earth's water and land surfaces, minimizing their long-range transport through the atmosphere and confining them to the troposphere.

Industrial processes of various kinds (including the combustion of coal, which has a modest chlorine content) produce HCl, organic reactive chlorine gases, and the chlorofluorocarbons, which are, depending on the species, slightly reactive or unreactive in the troposphere. The reactive chlorine gases are quite soluble or are converted into soluble compounds, and most of the supply is lost to precipitation. The entire emissions of the CFCs, currently about 0.8 Tg Cl/yr, will ultimately reach the stratosphere. However, that process takes about a decade for the average CFC molecule, and in the interim about 0.56 Tg Cl/yr is added to the tropospheric load and 0.24 Tg Cl/yr is transferred to the stratosphere. Of this latter amount, about 0.16 Tg Cl/yr is broken down by photochemical processes to yield inorganic Cl, while about 0.08 Tg Cl/yr adds to the stratospheric accumulation. The natural supply of chlorine to the stratosphere (as CH_3Cl) is only about 0.03 Tg Cl/yr. Some 0.2 Tg Cl/yr are removed from the stratosphere to the troposphere, mostly as soluble, readily removable HCl.

The main natural source of organic chlorine is the generation of CH_3Cl by kelp. Chlorine from this source does not accumulate in the stratosphere, however, because it is balanced by a roughly equivalent return flow in precipitation and air exchange.

The relative magnitudes of the fluxes in Fig. 10.7 are striking, the natural inorganic flux being some 2000–3000 times that of industrial components or of the natural organic chlorine fluxes. A marked distinction between reactivity, solubility, and (hence) lifetime makes all the difference, however. It is quite impressive to note how the CFCs, a relatively minor branch of the chlorine cycle, can have such a prominent influence on stratospheric ozone and thereby on atmospheric chemistry and on the biosphere. This situation demonstrates that budget studies in various regimes involving anthropogenic and natural sources and sinks are of great value in gaining a proper perspective on the contribution of industrial activity to environmental changes, and that perturbations that appear minor when compared to total volume can nonetheless cause very substantial impacts.

Studies similar to those just described, sometimes very difficult ones to perform, have revealed industrial alterations in each of the four grand cycle elements or in one or more of their subcycles. In the case of carbon, the rapid increase in the concentration of atmospheric carbon dioxide (Fig. 1.2) reflects the high rates of fossil fuel combustion during the last century. The case for nitrogen was described before. For sulfur, the combustion of fuels is again the culprit, the burning of sulfur-rich coal and petroleum releasing sulfur dioxide and producing acid rain. The phosphorus cycle in surface and groundwater has been perturbed on local and regional scales by the mining of phosphate rock and the use of phosphates in detergents (a practice now largely halted). Atmospheric oxygen has also been determined to be decreasing, by about 0.02% per year, largely as a consequence of fossil fuel combustion and the resultant consumption of oxygen in the formation of CO_2.

Perturbations to natural halogen cycles are also easy to identify. The case for chlorine was described in the previous Case Study, where the overall cycle is seen to be unperturbed but the crucial stratospheric subcycle is strongly affected. The same pattern plays out with bromine and fluorine, which are also important participants in the chemistry of the upper atmosphere following their release as insoluble gases for a variety of industrial uses. Another aspect of halogen chemistry is the degree to which it has been purposely used for aggressive interactions with the natural world, as in the use of chlorinated pesticides. Although such applications may be beneficial in some circumstances, they are examples of approaches that need to be regarded with extreme caution.

For virtually all metals and metalloids except iron, silicon, and calcium, the global cycles (though not the budgets) are dominated by industrial activity. Such dominance may not be a problem if the form in which the elements are used is similar or identical to one of nature's forms, calcium sulfate or aluminum oxide being examples. If the form is an alien one from the standpoint of the natural environment, however, care is needed to ensure that it can be used and disposed of with safety. As described before, the use of cadmium as a plating material is one such activity that is problematical from an industrial ecology standpoint.

10.5   THE EVOLUTION OF INDUSTRIAL MATERIALS USE

A well-established historic relationship is that an improved standard of living brings with it an increase in the use of materials. That history for the United States in the twentieth century is shown in Fig. 10.8. The population during the century increased by about a factor of 3, whereas the total materials use increased by about a factor of 10. Most of the materials groups showed reasonably steady gains during that period except for agriculture, which has been a relatively stable user of materials since about 1940.

We will note later that the tendency of modern products to perform equivalent functions while using smaller amounts of materials may act as a control against increased materials use. A powerful counterforce, however, is the large proportion of the world's population that desires improvement in its standard of living and thus can be expected to increase its rate of materials use. Table 10.2 speaks to that situation by listing the 1990 world production and U.S. consumption of a number of different commodities. Also shown in the table are the levels of production that would be needed were the world's population to use materials at the same per capita level as do Americans. In many cases, increases in production of factors of 4–6 would be required. Should such levels of production occur, they would obviously exacerbate any current perturbations of natural budgets. Perhaps, especially for phosphate rock, a perturber of the natural waterborne nitrogen cycle, and copper, a metal that can be toxic to organisms if liberated in a biologically accessible form, a future based on present American levels of consumption appears unrealistic for the planet on a long-term basis. For sheer mass, cement, iron and steel, and salt are also perhaps worth special attention. There can be no simpler expression of how difficult it will be to achieve sustainable development than Table 10.2.

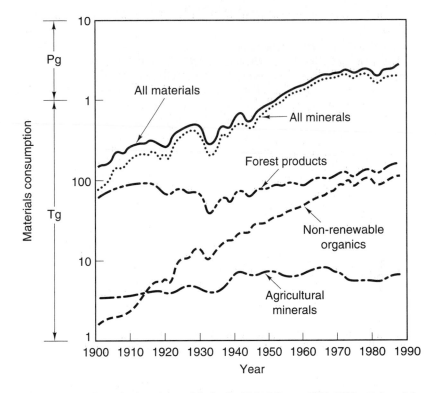

Figure 10.8   Consumption of selected materials in the United States, 1900–1989. (Adapted from D. G. Rogich and staff, Paper presented at ASM International Conference on Materials and Global Environment, Washington, DC, Sept. 13–15, 1993.)

Table 10.2   Materials Consumption Rates and Possible Projections

Commodity	1990 U.S. Consumption	1990 World Production	Conceptual World Consumption Need	Factor of Increase
Plastic	25.0	78.3	530.0	6.8
Synthetic fibers	3.9	13.2	82.7	6.3
Aluminum	5.3	17.8	111.5	6.3
Phosphate rock	4.4	15.7	93.3	5.9
Copper	2.2	8.8	46.0	5.2
Salt	40.6	202.3	860.7	4.3
Potash	5.5	28.3	115.5	4.1
Sand and gravel	24.8	133.1	525.3	4.0
Iron and steel	99.9	593.7	2117.9	3.6
Nitrogen	18.0	107.9	381.0	3.5
Cement	81.3	1251.1	1723.1	1.4

Source: Adapted from *Minerals Today*, p. 17, Washington, DC: U.S. Bureau of Mines, April 1993. All consumption rates are expressed in Tg/yr.

SUGGESTED READING

Ayres, R. U. Industrial metabolism, in *Technology and Environment*, J. H. Ausubel and H. E. Sladovich, eds., pp. 23–49. Washington, DC: National Academy Press, 1989.

Ayres, R. U., W. Schlesinger, and R. Socolow. Human impacts on the carbon and nitrogen cycles, in *Industrial Ecology and Global Change*, R. Socolow, C. Andrews, F. Berkhout, and V. Thomas, eds., in press. Cambridge, UK: Cambridge University Press, 1994.

Keeling, R. F., and S. R. Shertz, Seasonal and interannual variations in atmospheric oxygen and implications for the global carbon cycle. *Nature, 358* (1992): 723–727.

Schlesinger, W. H. *Biogeochemistry: An Analysis of Global Change.* San Diego: Academic Press, 1991.

Turner, II, B. L., W. C. Clark, R. W. Kates, J. F. Richards, J. T. Matthews, and W. B. Meyer, eds. *The Earth as Transformed by Human Action.* Cambridge, UK: Cambridge University Press, 1990.

EXERCISES

10.1 Using Fig. 10.2, diagram the world lead cycle as best you can. What additional data are needed to do a complete job? How might such data be acquired?

10.2 Typical copper ore contains about 1% copper by weight. How much ore is mined each year in the world to produce the copper of Table 10.2? How much ore would be mined if the conceptual world consumption need of that table were to be met? How much would be needed to meet the same goal for an anticipated 2025 world population of 8.5 billion people? Assume an approximate density for the ore and express this last answer in volume units. If spread equally over Earth's land surface, how deep would the residue be?

10.3 For the perturbed nitrogen cycle of Fig. 10.6, compute μ, ρ, λ, ι, and σ, as well as the turnover time for each node.

CHAPTER

11

Prioritizing Options:
LCA's Impact-Assessment Stage

11.1 INTRODUCTION

One of the "ecofacts" that is common knowledge is that a paper cup is more environmentally friendly than one made from polystyrene foam. In a study that amounted to an LCA inventory analysis of these two alternatives, Martin Hocking of the University of British Columbia investigated the relative merit of paper versus polystyrene foam for hot-drink containers in fast-food or other single-use applications. He points out that the paper product requires much more raw material (Table 11.1) and, in addition, consumes as much petroleum as the polystyrene product in the energy needed to go from wood chips to pulp to finished cup. The inorganic chemicals needed in the papermaking process, together with the large relative amounts of water effluent, provide substantial off-site impacts compared with the only significant emittant associated only with polystyrene cup: the pentane used as a blowing agent. The polystyrene cup is easier to recycle and approximately as easy to incinerate. Landfill degradation has been thought to favor the paper cup, but recent studies indicate that even "biodegradable" materials remain undegraded in anaerobic landfills over very long time periods, rendering that advantage relatively unimportant. The conclusion from Hocking's analysis is that "it would appear that polystyrene foam cups should be given a much more even-handed assessment as regards their environmental impact relative to paper cups than they have received during the past few years". Interestingly, a subsequent Dutch study suggests that both of the options are preferable to ceramic, reusable cups unless the latter are washed hundreds of times before being discarded, a conclusion clearly contrary to the popular view.

The implications of studies such as these, still in their infancy, are much larger than whether foam, paper, or reusable ceramic cups are preferable. Indeed, they have

Table 11.1   Raw Material, Utility, and Environmental Summary for Hot-Drink Containers

Item	Paper Cup	Polyfoam Cup
Per cup		
Raw materials		
Wood and bark (g)	33	0
Petroleum (g)	4.1	3.2
Finished weight (g)	10	11.5
Cost	2.5x	x
Per Mg of material		
Utilities		
Steam (kg)	9000–12000	5000
Power (GJ)	3.5	0.4–0.6
Cooling water (m^3)	50	154
Water effluent		
Volume (m^3)	50–190	0.5–2
Suspended solids (kg)	35–60	Trace
BOD (kg)	30–50	0.07
Organochlorides (kg)	5–7	0
Metal salts (kg)	1–20	20
Air emissions		
Chlorine (kg)	0.5	0
Sulfides (kg)	2.0	0
Particulates (kg)	5–15	0.1
Pentane (kg)	0	35–50
Recycle potential		
Primary user	Possible	Easy
After use	Low	High
Ultimate disposal		
Heat recovery (MJ/kg)	20	40
Mass to landfill (g)	10.1	1.5
Biodegradable	Yes	No

potential application to all industrial operations, both those dealing with components entering the manufacturing operation and with products leaving the manufacturing cycle. However, Hocking's assessment has not faced the really crucial issue in implementing LCA in an industrial setting: assigning relative impact values to each of the comparisons made in his analysis. It is with that issue that cultural, ethical, regulatory, legal, environmental, and corporate factors come into play. Nonetheless, Hocking's study and the Dutch study demonstrate that comparative analyses of competitive processes can produce results that may be contrary to the "conventional wisdom".

11.2 LCA IMPACT ASSESSMENT

The previous chapter discussed the component of LCA called inventory analysis. Quantitative information is acquired at that stage in some cases, qualitative information in others. In the data presentations in the previous chapter, it was straightforward to see that some aspects of life-cycle stages are more problematical than others, but the approach begged the question of priorities. One could easily foresee a situation where alternative designs for a product or process each had similar ratings on the ellipse plot, but in different locations. How does the industrial ecologist make a rational, defendable decision among such alternatives? The answer is that (1) the influence of the activities revealed by the LCA inventory analysis on specific environmental properties must be accurately assessed, and (2) the relative seriousness of changes in the environmental properties must be given a priority ranking. Together, these steps constitute LCA's *impact assessment.*

Assessing environmental influences is a complicated procedure, but it can, in principle at least, be performed by employing *stressors*, which are relationships linking factors such as changes in resource consumption or residue generation with resulting changes in environmental properties. Stressors are developed by the environmental science community, and are not always available with the degree of detail and precision needed. Ideally, however, the needed stressors will be established and available for use as part of a standard engineering analysis. By combining LCA inventory results with stressors, a manufacturing process might be found, for example, to have a minimal impact on local water quality, a modest impact on regional smog, and a substantial impact on stratospheric ozone depletion.

The second step in LCA impact assessment implies the ordering of the impacts that have been determined. In effect, this step constitutes the application of risk analysis, the subject of Chap. 4, to the full spectrum of environmental concerns, and requires the close collaboration of industrial ecologists with environmental scientists. Moreover, this collaboration must take place across national boundaries, because multinational corporations cannot rationally design products for countries that have widely differing priority rankings of environmental properties. A major complicating factor is that risk prioritization is not only scientific but also reflects the value system of the community performing the prioritization. It thus requires harmonization across cultures as well as agreement on scientific and technological issues.

Although it has been much more common for environmental scientists to study impacts individually rather to attempt to rank them in priority order, several different approaches to prioritization have been made. In this chapter, we summarize those approaches for which sufficient information is available and discuss their advantages and disadvantages. For illustration, we will show some of the rank ordering of environmental problems that emerges from these efforts. The purpose of the present chapter is not to present or defend specific ordering of impacts, however, but to illustrate the process.

11.3   THE EPA SCIENCE ADVISORY COMMITTEE

One of the first useful exercises in setting priorities on environmental impacts was a 1990 effort of the U.S. EPA's Science Advisory Committee. This committee prioritized impacts with the goal of providing advice to EPA on the best use of its resources. Several ranking parameters were used to sort environmental impacts into priority order:

- The spatial scale of the impact (large scales being worse than small).
- The severity of the hazard (more toxic substances being of more concern than less toxic substances).
- The degree of exposure (well-sequestered substances being of less concern than readily mobilized substances).
- The penalty for being wrong (longer remediation times being of more concern than shorter times).

Given this general approach to ranking environmental hazards, the committee produced the following lists:

Relatively High-Risk Problems

- Habitat alteration and destruction
- Species extinction and overall loss of biological diversity
- Stratospheric ozone depletion
- Global climate change

Relatively Medium-Risk Problems

- Herbicides/pesticides
- Toxics, nutrients, biochemical oxygen demand, and turbidity in surface waters
- Acid deposition
- Airborne toxics, including smog-related constituents

Relatively Low-Risk Problems

- Oil spills
- Groundwater pollution
- Radionuclides
- Acid runoff to surface waters
- Thermal pollution

A related relative risk effort was completed in 1992 by groups of agency representatives, scientists, and citizens for the State of Michigan. The results of that activity were quite similar to those of the EPA Science Advisory Committee, except that the Michigan effort included several societally oriented issues such as incomplete land-use planning and lack of environmental awareness. As far as the scientific aspects are concerned, the two studies together indicate broad acceptance by knowledgeable individuals from varied backgrounds of the ranking parameters advanced by the EPA/SAC effort and of the qualitative results that are obtained therefrom.

11.4 QUALITATIVE LINKING OF SOURCES AND IMPACTS

An important consideration that is often overlooked but which plays a role in the prioritization of impacts is that a single impact generally has many sources and a single source many impacts. As a consequence, the industry–environment interaction generally embodies a summation of impacts of different magnitudes and different spatial and temporal scales. The result is that a choice between two design options may involve choosing between having a modest influence on a single impact of high importance or smaller influences on several impacts of lower importance. A recent example involved the substitution in the electronics industry of organic solvents for some of the CFC cleaning operations; this action constituted the substitution of impacts on photochemical smog, increased energy consumption, aqueous residue streams, and air toxics in place of an impact on stratospheric ozone.

One approach for overall impact assessment is a series of matrix displays, the axes of the matrices being "Sources" (specific activities or industries) and "Critical Properties" (specific impacts). A matrix of this type for atmospheric concerns is shown in Fig. 11.1. To construct this figure, a critical atmospheric component like "precipitation acidification" and its direct and indirect chemical causes are linked with the sources

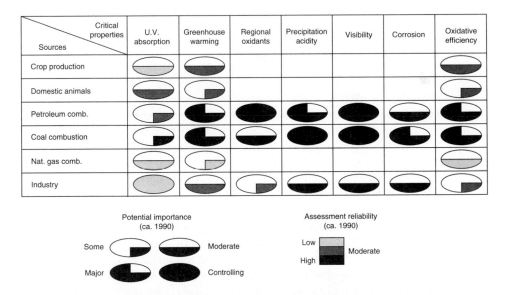

Figure 11.1   An initial ensemble assessment of impacts on the global atmosphere. Critical atmospheric properties are listed as the column headings of the matrix. The sources of disturbances to these properties are listed as row headings. Cell entries assess the relative impact of each source on each component and the relative scientific certainty of the assessment.

responsible for initiating those interactions. The result is a matrix that shows the impact of each potential source of atmospheric change on each critical atmospheric component. The assessment is qualitative, reflecting the present state of knowledge. It also includes estimates of the reliability of that knowledge, an important component of such an assessment effort. The matrix is constructed with separate rows for industrial operations and for the major energy-generating processes. This approach makes it easy to distinguish the impacts of specific types of sources, but does not address the allocation question, that is, whether emissions from energy production should be allocated to the end user rather than to the generating process itself. As will be seen, these secondary links of industry and environment should be regarded as natural components of industrial ecology studies.

The simplest impact assessments of Fig. 11.1 involve only a single cell of the matrix. A typical example is the study of the impacts of a single source, such as a new coal-fired power station, on a single critical atmospheric component, such as precipitation acidification. More complex assessments have addressed the question of aggregate impacts across different kinds of sources. A contemporary example is the study of the net impact on Earth's thermal radiation budget caused by chemical perturbations due to fossil fuel combustion, biomass burning, land-use changes, and industrialization. An alternative approach, especially useful for the purposes of policy and management, is to assess the impacts of a single source on several critical environmental properties. The coal combustion study noted before would fall into this category if the impacts were assessed not only on acidification, but also on photochemical oxidant production, materials corrosion, visibility degradation, heavy metals emissions, and so on. If desired, the columns could be summed in some way to give the net impact of the ensemble of sources on each critical property. Similarly, the rows could be summed in some way to give the net effect of each source on the ensemble of properties.

Figure 11.1 shows that the sources of most general concern for atmospheric impacts are fossil fuel combustion (especially coal and petroleum) and industrial processes. Emissions from agricultural activities, especially CH_4 from rice paddies and cattle and N_2O from fertilizer application, also have significant effects on climate and environment. When contemplating these source-impact assessments it is useful also to consider some of the differing attributes of the sources. Food production through the growing of crops and the raising of animals is potentially sustainable, but current operations require very large resource inputs. Sustainability would thus demand substantial and far-reaching changes in approach. The combustion of fossil fuels, necessarily preceded by extraction and purification, involves little transformation and is purely dissipative in nature. In contrast, industrial operations exist for the purpose of transforming raw materials, and are potentially nondissipative if sustainable energy supplies are assumed.

Among the most troublesome interactions between development and environment are those that involve cumulative impacts. In general, cumulative impacts become important when sources of perturbation to the environment are grouped sufficiently closely in space or time that they exceed the natural system's ability to remove or dissipate the resultant disturbance. For atmospheric perturbations, for example, the basic data required to structure such assessments are the characteristic time and space scales of the atmospheric constituents and development activities. We note that perturbations to gases with very long lifetimes accumulate over decades to centuries around the world as a whole. Today's perturbations to those gases will still be affecting the atmosphere decades or centuries hence, and perturbations occurring anywhere in the world will affect the atmosphere everywhere in the world. Long-lived emittants tend to be radiatively active, thus giving the "greenhouse" syndrome its long-term, global-scale character. At the other extreme, heavy hydrocarbons and coarse particles, being short-lived, drop out of the atmosphere in a matter of hours, normally traveling a few hundred kilometers or less from their sources. The atmospheric properties of visibility reduction and photochemical oxidant formation associated with these chemicals thus take on their acute, relatively local or regional character. Species with moderate atmospheric lifetimes include gases associated with the acidification of precipitation and fine particles, all with characteristic scales of a few days and a couple of thousand kilometers.

This scale diversity is reflected in Fig. 11.2, which reflects source impacts on water, soil, and other environmental properties. The figure contains several messages. One is that industrial activities play at least a small role in all the impacts, unlike other sources. For toxicity, industry plays the major role. For impacts on species diversity, industrial activities are minor compared with crop production (and consequent forest clearing) as significant contributors. The same is true of soil productivity loss. In the case of groundwater quality, industrial activity and food production each can be important, the dominant influence varying from location to location. The industrial system

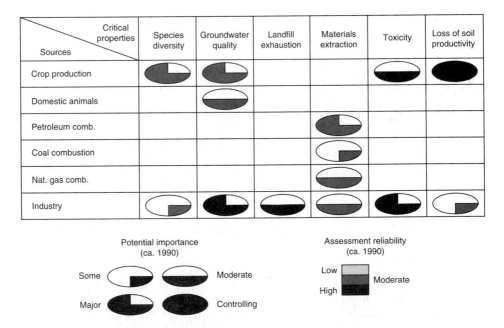

Figure 11.2 An initial ensemble assessment of impacts on global groundwater and soil. The format is the same as that of Fig. 11.1. Although not prescriptive because of their qualitative nature, the displays nonetheless permit ready comparison of the multidimensional nature of the environmental interactions of different industrial actions.

(defined in the very broad sense to include consumers) can be held predominantly responsible for the depletion of landfill capacity.

Industry is a major factor in the case of materials extraction, but not as large as the removal of fossil fuel resources for the production of energy. The profligate use of energy thus bears two burdens: one as a producer of carbon dioxide, the principal anthropogenic greenhouse gas, the other as a major negative factor on environment through the impacts that accompany resource extraction.

11.5 QUANTITATIVE PRIORITIZATION FROM THE ENVIRONMENTAL PERSPECTIVE

The second approach that we discuss is a method developed by Vicki Norberg-Bohm and co-workers at Harvard University. The potential users are international institutions, nongovernmental organizations, and governments that are involved in setting national environmental agendas or developing environmental programs that require international coordination. It was developed and evaluated through its application to four country case studies: India, Kenya, The Netherlands, and the United States.

The method begins with a simple linear model of hazard causation. For each stage in this model, a number of key indicators are defined and used to characterize environmental concerns (see Fig. 11.3). The indicators are designed to reflect both the causes

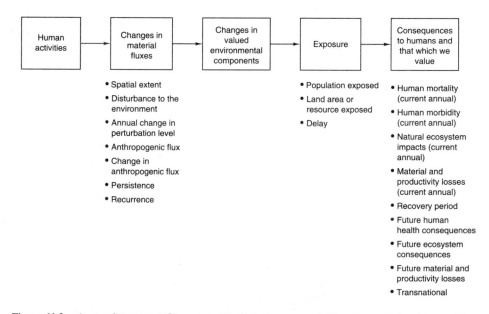

Figure 11.3 A causal taxonomy for comparative hazard assessment of environmental problems. (Reproduced with permission from V. Norberg-Bohm, W. C. Clark, B. Bakshi, J. Berkenkamp, S. A. Bishko, M. D. Koehler, J. A. Marrs, C. P. Nielsen, and A. Sagar, Preprint: International comparisons of environmental hazards: Development and evaluation of a method for linking environmental data with the strategic debate on management priorities, published in *Risk Assessment for Global Environmental Change*, R. Kasperson and J. Kasperson, eds. New York: United Nations Press, 1992.)

available data. For a given environmental problem in a given country, each of these indicators is scored on a scale of 1 to 9. The research team also defined 28 environmental hazards (see Table 11.2), a list designed to be understandable and meaningful to policy makers and other nonspecialists, comprehensive (including virtually all of the and consequences of environmental problems, and also to pose realistic demands on

Table 11.2   The Environmental Hazards Defined by Norberg-Bohm et al. (1992)

Pollution-based hazards

Water quality
1. Freshwater—biological contamination
2. Freshwater—metals and toxic contamination
3. Freshwater—eutrophication
4. Freshwater—sedimentation
5. Ocean water

Atmospheric quality
6. Stratospheric ozone depletion
7. Climate change
8. Acidification
9. Ground-level ozone formation
10. Hazardous and toxic air pollution

Quality of the human environment
11. Indoor air pollutants—radon
12. Indoor air pollutants—nonradon
13. Radiation—nonradon
14. Chemicals in the workplace
15. Accidental chemical releases
16. Food contaminants

Resource-depletion hazards

Land/water resources
17. Agricultural land—salinization, alkalinization, waterlogging
18. Agricultural land—soil erosion
19. Agricultural land—urbanization
20. Groundwater

Biological stocks
21. Wildlife
22. Fish
23. Forests

Natural hazards
24. Floods
25. Droughts
26. Cyclones
27. Earthquakes
28. Pest epidemics

---

potentially significant environmental problems faced by any nation), and at a level of resolution that is neither too general nor too complex to be practical.

By analyzing the data in various ways, the Norberg-Bohm method provides one source of input for hazard prioritization. Whereas it gives a systematic way of ordering key data for a range of environmental problems, it does not specify a manner for interpreting or aggregating these data. Rather, in recognizing the central role of values in creating hazard prioritizations, it can be used to illuminate the implications of different preferences. The use of the method can be illustrated with two examples. Figure 11.4 compares those environmental problems that are thought to pose the greatest current hazards with those problems that are expected to become significantly worse in the next 25 years. Problems that fall in the upper right-hand corner of the graph are those that are

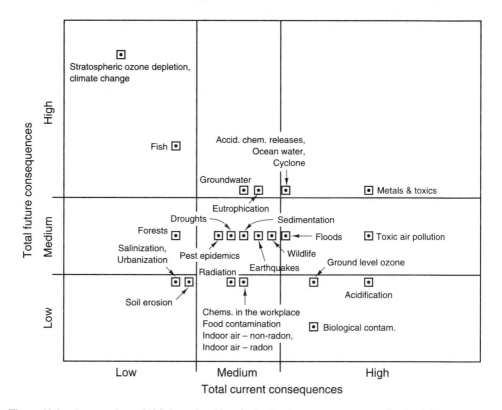

Figure 11.4   A comparison of U.S. hazard rankings in the "total current consequences"—"total future consequences" plane. "Total Current Consequences" is an aggregation of the scores for human mortality, human morbidity, natural ecosystem impacts, and material and productivity losses. "Total Future Consequences" is an aggregation of future human health consequences, future ecosystem consequences, and future material and productivity losses. (Reproduced with permission from V. Norberg-Bohm, W. C. Clark, B. Bakshi, J. Berkenkamp, S. A. Bishko, M. D. Koehler, J. A. Marrs, C. P. Nielsen, and A. Sagar, Preprint: International comparisons of environmental hazards: Development and evaluation of a method for linking environmental data with the strategic debate on management priorities, published in *Risk Assessment for Global Environmental Change*, R. Kasperson and J. Kasperson, eds. New York: United Nations Press, 1995.)

among the worst current problems and are also expected to become significantly more severe in the future. For the U.S., this list includes freshwater metals and toxics, accidental chemical releases, ocean water, and cyclones (a natural hazard included for perspective). Table 11.3 compares the problems that ranked as most severe in terms of total consequences for the four countries of the study. The table demonstrates that although some envionmental problems are shared, we can expect that different countries will have different environmental priorities.

Table 11.3   A Comparison for The Netherlands, United States, India, and Kenya of Hazards Ranked as Most Severe in Terms of Total Consequences†

Environmental Problems	Netherlands	United States	India	Kenya
2. Freshwater—metal and toxic contamination	*	*	*	
5. Ocean water	*	*	*	
10. Toxic air pollution	*	*	*	
6. Stratospheric ozone depletion	*	*		
7. Climate change	*	*		
9. Ground-level ozone formation	*			*
18. Agricultural land—soil erosion			*	*
23. Forests			*	*
1. Freshwater—biological contamination			*	
3. Freshwater—eutrophication	*			
15. Accidental chemical releases		*		
21. Wildlife				*
22. Fish				*
24. Floods			*	
25. Droughts				*
26. Cyclones		*		
28. Pest epidemics				*
4. Freshwater—sedimentation				
8. Acidification				
11. Indoor air—radon				
12. Indoor air—nonradon				
13. Radiation—nonradon				
14. Chemicals in the workplace				
16. Food contamination				

Table 11.3 (continued)

Environmental Problems	Netherlands	United States	India	Kenya
17. Agricultural land—salinization, alkalinization, waterlogging				
19. Agricultural land—urbanization				
20. Groundwater				
27. Earthquakes				

†The total consequences aggregation consists of the descriptors (11) human mortality, (12) human morbidity, (13) natural ecosystem impacts, (14) material and productivity losses, (16) future human health consequences, (17) future ecosystem consequences, and (18) future material and productivity losses.
* Indicates a high-ranking hazard.
Source: Reproduced with permission from V. Norberg-Bohm, W. C. Clark, B. Bakshi, J. Berkenkamp, S. A. Bishko, M. D. Koehler, J. A. Marrs, C. P. Nielsen, and A. Sagar, Preprint: International comparisons of environmental hazards: Development and evaluation of a method for linking environmental data with the strategic debate on management priorities, published in *Risk Assessment for Global Environmental Change*, R. Kasperson and J. Kasperson, eds. New York: United Nations Press, 1992.

11.6   INDUSTRIAL PRIORITIZATION: THE NETHERLANDS VNCI SYSTEM

Using the goal of the management of individual substances in industry as a focus, the Dutch government and the Association of the Dutch Chemical Industry (Vereniging van de Nederlandse Chemische Industrie, VNCI) have developed their own assessment methodology for studying reductions of environmental impacts of a variety of chemical substances. The methodology proceeds in stages from generating options to prioritizing options to planning actions. The second stage, the prioritization of options, will be summarized here.

Prioritization begins with the development of an environmental profile for each option. This profile first identifies the "environmental themes", where implementing the option will have an effect—global warming or disposal of waste, for example—and then assigns an index to the benefit or cost that will result. The next step is to construct an economic profile of each option. Elements typically captured at this stage include net changes in capital and operating costs, both positive and negative. Finally, the two profiles for each option are combined in a prioritization diagram, where options can be compared either individually or in combination.

The inclusion of economic factors is an important and often overlooked aspect of LCA. In the case study in the previous chapter, it appeared that tallow required so much water that it perhaps should be rejected as a surfactant feedstock. Such a decision would overlook the fact, however, that beef cattle are primarily raised for meat, not for tallow. Utilizing the tallow is an efficient use of a low-value co-product that otherwise would require disposal. Hence, a broad economic perspective is often vital in producing a well-reasoned LCA result.

The Dutch procedure is best illustrated by example, for which we use a Dutch study aimed at reducing emissions of HCFC-22 (CHF_2Cl), a gas used in refrigeration and in the manufacture of fluoroplastics. Nine options for doing so were identified, and two were selected for detailed study: establishing a recycling option and improving equipment maintenance. The environmental profile for the first option is shown in Fig. 11.5. The profile shows that beneficial impacts on Dutch contributions to global warming and ozone depletion would result. A slight negative impact is also forecast, due to the SO_2-related acidification that is expected from the incineration of the residue from the HCFC-22 recycling stream.

Figure 11.6 shows the economic profile for the HCFC-22 recycling option. Costs will be incurred in purchasing the equipment necessary for the recycling program and for the operating costs of performing the recycling. However, these costs are offset by reduced HCFC-22 purchases.

The environmental and economic profiles are combined on the prioritization diagram of Fig. 11.7, where the ratings for the increased maintenance option are also given. The quantitative approach of the first two diagrams is retained in constructing this figure, but for convenience in assessment, each axis is divided into low, medium, and high regions, each division indicating a factor-of-10 difference from the adjacent one. The result of the study is that both options were judged to have a significant environmental yield in exchange for a slightly negative (Option B) or slightly positive (Option A) eco-

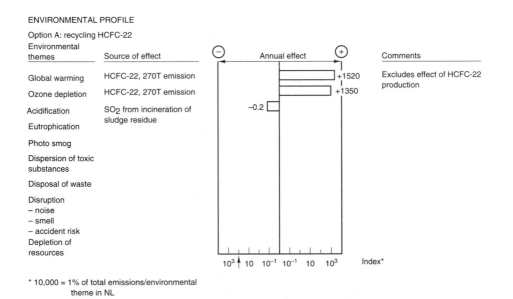

Figure 11.5   The environmental profile diagram for a proposed Netherlands program to recycle HCFC-22. The index is logarithmic and is based on 1% of total Netherlands HCFC-22 emissions being set to a value of 10,000. (Ministry of Housing, Physical Planning, and Environment. Integrated substance chain management, Pub. VROM 91387/b/4-92, 's-Gravenhage, The Netherlands, 1991.)

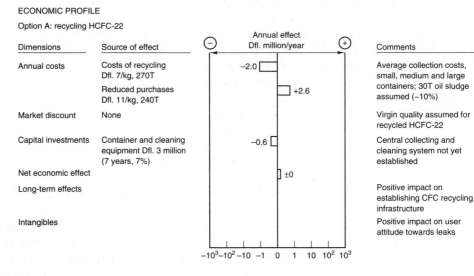

ECONOMIC PROFILE

Option A: recycling HCFC-22

Dimensions	Source of effect	Annual effect Dfl. million/year	Comments
Annual costs	Costs of recycling Dfl. 7/kg, 270T	−2.0	Average collection costs, small, medium and large containers; 30T oil sludge assumed (~10%)
	Reduced purchases Dfl. 11/kg, 240T	+2.6	
Market discount	None		Virgin quality assumed for recycled HCFC-22
Capital investments	Container and cleaning equipment Dfl. 3 million (7 years, 7%)	−0.6	Central collecting and cleaning system not yet established
Net economic effect		±0	
Long-term effects			Positive impact on establishing CFC recycling infrastructure
Intangibles			Positive impact on user attitude towards leaks

Figure 11.6   The economic profile diagram for a proposed Netherlands program to recycle HCFC-22. The index is monetary and logarithmic. (Ministry of Housing, Physical Planning, and Environment. Integrated substance chain management, Pub. VROM 91387/b/4-92, 's-Gravenhage, The Netherlands, 1991.)

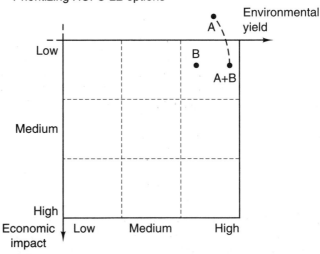

Figure 11.7   The prioritization diagram for a proposed Netherlands program to recycle HCFC-22. (Ministry of Housing, Physical Planning, and Environment, Integrated substance chain management, Pub. VROM 91387/b/4-92, 's-Gravenhage, The Netherlands, 1991.)

nomic impact. The possible choice of implementing both options results in a slightly larger environmental yield, still with only a slight negative economic impact.

The VNCI system is relatively new and not yet refined. Indeed, one can readily discern potential improvements. A significant disadvantage is that the system, as structured, rates all environmental themes equally (although the supporting documents hold out the potential for applying weighting factors). Another potential problem is the logarithmic scale in the economic profile and prioritization diagram. Such an approach would suggest that options differing significantly in cost might fall in the same "medium impact" region; it is doubtful that industrial manufacturing facility managers would regard an option costing three times that of another option as having essentially the same financial impact. Nonetheless, the VNCI system's inclusion of economic costs as an integral part of the analysis is a major favorable attribute. It is reasonable to anticipate that some version of this system could prove very useful for industrial ecology prioritization.

11.7 INDUSTRIAL PRIORITIZATION: THE IVL/VOLVO EPS SYSTEM

An initial attempt to establish a basis for assessing the environmental responsibility of an individual manufacturing process was that of Roger Sheldon of the Delft (Netherlands) Institute of Technology, who proposed in the context of the synthesis of organic chemicals the *atom utilization (AU) concept*, calculated by dividing the molecular weight of the desired product by that of the sum total of all substances produced. Enlarging on this concept, and realizing that the nature of the residue is important as well as its amount, he proposed the *environmental quotient* (EQ), given by

$$EQ = AU \times U \tag{11.1}$$

where U is the "unfriendlyness quotient", a measure of toxicity.

Sheldon offers no advice on how to assign unfriendliness quotients, but this difficult problem has not gone unaddressed. To begin to address this topic in a formal way, the Swedish Environmental Institute (IVL) and the Volvo Car Corporation have developed an analytic tool designated the Environment Priority Strategies for Product Design (EPS) system. The goal of the EPS system is to allow product designers to select components and subassemblies that minimize environmental impact. Analytically, the EPS system is quite straightforward, though detailed. An environmental index is assigned to each type of material used in automobile manufacture. Different components of the index account for the environmental impact of this material during product manufacture, use, and disposal. The three life-cycle-stage components are summed to obtain the overall index for a material in "environmental load units" (ELUs) per kilogram (ELU/kg) of material used. The units may vary. For example, the index for a paint used on the car's exterior would be expressed in ELU/m^2.

When calculating the components of the environmental index, the following factors are included:

- Scope: A measure of the general environmental impact.
- Distribution: The size or composition of the affected area.
- Frequency or Intensity: Extent of the impact in the affected area.
- Durability: Persistence of the impact.
- Contribution: Significance of impact from 1 kg of material in relation to the total effect.
- Remediability: Cost to remediate impact from 1 kg of material.

These factors are calculated by a team of environmental scientists, ecologists, and materials specialists to obtain environmental indices for every applicable raw material and energy source (with their associated pollutant emissions). A selection of the results is given in Table 11.4. A few features of this table are of particular interest. One is the very high values for platinum and rhenium in the raw materials listings. These result from the extreme scarcity of these two metals. The use of CFC-11 is given a high environmental index because of its effects on stratospheric ozone and global warming. Finally, the assumption is made that the metals are emitted in a mobilizable form. To the extent that an inert form is emitted, the environmental index may need to be revised.

Given an agreed set of environmental indices, they are multiplied by the materials uses and process parameters (in the appropriate proportions) to obtain environmental load units for processes and finished products. The entire procedure is schematized in Fig. 11.8. Note that it includes the effects of materials extraction, emissions, and the impacts of manufacturing, and follows the product in its various aspects through the entire life cycle.

Table 11.4   A Selection of Environmental Indices

Raw Materials		Emissions—Air		Emissions—Water	
Co	76	CO_2	0.09	Nitrogen	0.1
Cr	8.8	CO	0.27	Phosphorus	0.3
Fe	0.09	NO_x	0.22		
Mn	0.97	N_2O	7.0		
Mo	1.5E3	SO_x	0.10		
Ni	24.3	CFC-11	300		
Pb	180	CH_4	1.0		
Pt	3.5E5				
Rh	1.8E6				
Sn	1.2E3				
V	12				

*Units: ELU/kg.
Source: Steen, B., and S. Ryding, *The EPS Enviro-Accounting Method: An Application of Environmental Accounting Principles for Evaluation and Valuation of Environmental Impact in Product Design*, Stockholm: Swedish Environmental Research Institute (IVL), 1992.

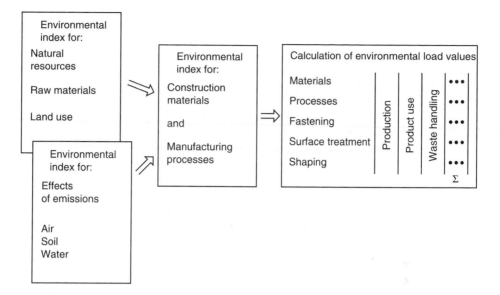

Figure 11.8   An overview of the EPS system, showing how the calculation of summed environmental load values proceeds. (Reproduced with permission from B. Steen and S. Ryding, *The EPS Enviro-Accounting Method: An Application of Environmental Accounting Principles for Evaluation and Valuation of Environmental Impact in Product Design.* Stockholm: Swedish Environmental Research Institute (IVL), 1992.)

As an example of the use of the EPS system, consider the problem of choosing the more environmentally responsible material to use in fabricating the front end of an automobile. As shown in Fig. 11.9, two options are available: galvanized steel and polymer composite (GMT). They are assumed to be of comparable durability, though differing durabilities could potentially be incorporated into EPS.

Based on the amount of each material required, the environmental indices are used to calculate the environmental load values at each stage of the product life cycle. Table 11.5 illustrates the total life-cycle ELUs for the two front ends. All environmental impacts, from the energy required to produce a material to the energy recovered from incineration or reuse at the end of product life, are incorporated into the ELU calculation. To put the table in the LCA perspective, the kg columns are LCA stage one, the ELU/kg columns are the environmental indices, and the ELU columns are LCA stage two. There are several features of interest in the results. One is that the steel unit has a larger materials impact during manufacturing, but is so conveniently recyclable that its overall ELU on a materials basis is lower than that of the composite. However, it is much heavier than the composite unit, and that factor results in much higher environmental loads during product use. The overall result is one that was not intuitively obvious: that the polymer composite front end is the better choice in terms of environmental impacts during manufacture, the steel unit the better choice in terms of recyclability, and the polymer composite unit the better overall choice because of lower impacts during product use. Attempting to make the decision on the basis of an analysis of only part

**Which front end is more
environmentally sound?**

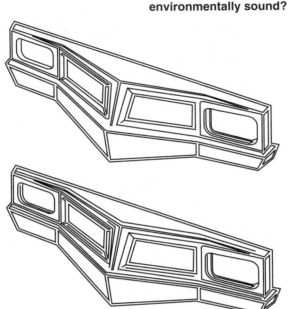

GMT – composite
Material consumption: 4.0 kg
(0.3 kg scrap)
Component weight: 3.7 kg

Galvanized steel
Material consumption: 9.0 kg
(3.0 kg scrap)
Component weight: 6.0 kg
Painted area: 0.6 m^2

Figure 11.9 Design options for automotive front end pieces. (Courtesy of I. Horkeby, Volvo Car Corporation.)

of the product life cycle would result in an incompletely guided and potentially incorrect decision.

The EPS system is currently being refined and implemented by several organizations affiliated with IVL. Corporations in Sweden and elsewhere have expressed great interest in developing EPS systems that are specific to their products and manufacturing procedures. The EPS's greatest strength is its flexibility; raw materials, processes, and energy uses can be added easily. If a manufacturing process becomes more efficient, all products that utilize the process will automatically possess upgraded ELUs. Perhaps its greatest weakness is its need to quantify uncertain data and compare unlike risks, in the process making assumptions that gloss over serious value and equity issues.

11.8 ABRIDGED AND EXPERT SYSTEMS IMPACT ASSESSMENTS

In the ideal world, comprehensive impact assessments would be performed for all products and processes. Experience has shown, however, that it is seldom feasible to undertake such a task on a general basis. One reason is that the staff time and expense needed to complete a comprehensive assessment (often greater than U.S. $100,000 and longer than 6 months) is often prohibitive. A second is that much of the data are difficult or impossible to acquire—factory energy use may not have been allocated to specific

Table 11.5   Calculation of Environmental Load Units for Automobile Front Ends

Materials & Processes	Production			Product Use*			Waste Incineration			Waste Reuse			Total ELU
	ELU/kg	kg	ELU	ELU/kg	kg	ELU	ELU/kg	kg	ELU	ELU/kg	kg	ELU	
GMT composite													
Production:													
GMT material	0.58	4.0	2.32										2.32
Reused production scrap	−0.58	0.3	−0.17										−0.17
Compression molding	0.03	4.0	0.12										0.12
Product Use:													
Petrol				0.82	29.6	24.27							24.27
Recycling:													
GMT material							−0.21	3.7	−0.78				−0.78
Total sum			2.27			24.27			−0.78				25.76
Galvanized steel													
Production:													
Steel material	0.98	9.0	8.82										8.82
Steel stamping	0.06	9.0	0.54										0.54
Reused production scrap	−0.92	3.0	−2.76										−2.76
Spot welding (spots)	0.004	48	0.19										0.19
Painting (m²)	0.01	0.6	0.02										0.02
Product Use:													
Petrol				0.82	48.0	39.36							39.36
Recycling:													
Steel material										−0.92	6.0	−5.52	−5.52
Total sum			6.81			39.36						−5.52	40.65

*The ELU/kg figure is based on 1 year of use. For the automobile, an 8-year life is assumed; hence, the second product use entry is eight times the actual weight.

Source: S. Ryding, B. Steen, A. Wenblad, and R. Karlson, *The EPS System—A Life Cycle Assessment Concept for Cleaner Technology and Product Development Strategies, and Design for the Environment*, Paper presented at EPA Workshop on Identifying a Framework for Human Health and Environmental Risk Ranking, Washington, D.C., June 30–July 1, 1993.

products, residue streams may have been merged, and so forth. Third, with product life cycles and design-to-market intervals of several months to 2 or 3 years on many products, an assessment taking several months to perform simply makes no sense. Consequently, most corporations and many other organizations performing LCAs have adopted the philosophy but have implemented it in a practical form more suitable to their needs and constraints. We call the various forms of this approach "abridged LCA" (ALCA).

In ALCA, as in LCA, the assessment begins with a matrix of life stages and materials inputs and outputs. (A typical matrix is that of the European Community Ecolabel, Table 11.6.) Unlike LCA, where all flows and influences in each matrix element are to be determined, an ALCA approach considers only those elements deemed most important, and then only the constituents of those elements thought to be most significant. For example, noise can be disregarded for many stages of many products and potential water contamination may only occur at certain life stages. If 80% of a product represents a single raw material, all others might be disregarded unless toxic. In these ways, the data and analytic requirements are abridged, often severely.

In principle, an expert is not needed to complete an LCA provided that an impact-assessment methodology is specified. As soon as one moves from LCA to ALCA, however, one moves into the realm of expert analysis, where precise information is combined with heuristics (the "rules of thumb" of knowledgeable practitioners) in order to make decisions. The ALCA abridgement strategy is the first decision point. The second is generally the determination of the values of the matrix elements that remain. Often, the values are still not capable of being specified with numerical precision, and may instead be assigned qualitatively, along the lines discussed for inventory analysis in Chap. 9. The matrix is then summed to arrive at the overall result. If two competing product or process designs are being evaluated, a 1 value can be assigned to each matrix element for which the desirable characteristics of one product exceed those of the other in the opinion of the expert or experts. The preferred alternative is then the one with the greater number of 1 values.

The process of assigning a quantized value to a matrix element is one that could be potentially suitable for *expert systems analysis*, in which a computer-based system is designed to assimilate the knowledge of human experts and make that knowledge available to others. Expert systems is a component of artificial intelligence (AI), and has been implemented successfully on such diverse problems as medical diagnosis and railroad locomotive maintenance. Among the advantages of expert systems are several that are particularly relevant to industrial ecology: It permits the sharing of expertise, it pools expertise from multiple sources, and it frees experts from routine analyses to concentrate on unusual situations. An example of a possible dialog by a nonexpert practitioner of ALCA with an expert system is shown in Fig. 11.10.

Expert system implementation is feasible if the problems presented to the system are such that an expert could be expected to promptly solve them. That level of performance has not yet been achieved in a consistent way by experts, however, much less expert systems, largely due to the considerable uncertainty still remaining in impact assessment. Hence, expert systems is a tool on the ALCA horizon rather than in the ALCA toolbox. Nonetheless, we anticipate that expert systems will have a significant role to play in ALCA, and can be expected to be implemented over time when the general level of expertise by life-cycle analysts becomes adequate to permit it.

SYSTEM	What is the material?
USER	Steel.
SYSTEM	What is the life stage being assessed?
USER	Disposal or recycling.
SYSTEM	Is the steel welded to other metal?
USER	No.
SYSTEM	Is it painted or plated?
USER	Plated.
SYSTEM	What is the plating material?
USER	Chromium.
SYSTEM	How is the plating applied?
USER	In a hexavalent chromium plating bath.
SYSTEM	The ALCA rating is 3.
USER	Why?
SYSTEM	Steel is eminantly recyclable, but the chromium plating material will very likely be lost. Other options would not dissipate material. This materials choice is superior, however, to those whose only fate would be disposal because they cannot be reused.

Figure 11.10   A possible dialog with an expert system programmed to assign a rating of 0, 1, 2, 3, or 4 to each element of the impact assessment matrix.

11.9 ALCA MATRIX MANIPULATION

The LCA process can be represented mathematically as an exercise in matrix manipulation. To demonstrate, consider the matrix of Table 11.6. If the matrix elements $f_{m,n}$ are filled with inventory analysis data, we have the LCA first-stage matrix, which we call $\mathbf{F}$. If the matrix elements $s_{m,n}$ are filled with impact-assessment data of the type of Table 11.4 (a one-time operation except for revisions), we have the LCA second-stage matrix, called $\mathbf{S}$. The LCA assessment for a single constituent n is then given by

$$L_n = \sum_1^m f_{m,n} \times s_{m,n} \tag{11.2}$$

Similarly, the LCA assessment for a single stage m is given by

$$L_m = \sum_1^n f_{m,n} \times s_{m,n} \tag{11.3}$$

The overall assessment is given by

$$\mathbf{L} = \sum_1^m \sum_1^n f_{m,n} \times s_{m,n} \tag{11.4}$$

As with any matrix, some of the $\mathbf{F}$ matrix elements may contain zeros. This situation will occur under either of two situations: if a null inventory value obtains, as might

Table 11.6   The European Community Ecolabel Matrix

Critical Property	Product Life-Cycle Stage				
	Materials	Production	Distribution	Utilization	Disposal
Solid waste					
Soil pollution and degradation					
Water contamination					
Air contamination					
Noise					
Consumption of energy					
Consumption of natural resources					
Effects on ecosystems					

be the case for anticipated soil pollution and degradation during the distribution of a product, or where the inventory value is deemed unimportant, as will often happen in ALCA. Zeros will occur in the **S** matrix if no impact is foreseen from a product or process. For wire fencing, for example, no consumption of energy occurs during use no matter how the fencing is designed or from what materials it is constructed.

For expert systems LCA, the **L** matrix elements are often quantized: They may be either binary (as in problem/no-problem decision systems) or one of several digits (as in a 1–5 severity ranking system). In any of these cases, the straightforward matrix-algebra approach outlined before is appropriate for use. Its important characteristic is that it generates a numerical rating for a design or for alternative designs, thereby providing a target for improvement. As has been said, "What gets measured gets managed". The result of the ALCA matrix-algebra computation fulfills the requirement for such measurements.

11.10 IMPLEMENTING IMPACT PRIORITIZATION IN INDUSTRIAL ECOLOGY

Although none of the techniques summarized before are ideal solutions to the need for producing priority rankings of environmental impacts, they provide useful guidance toward improved solutions. Without making a judgment on the appropriateness of specific rankings, the methodology exemplified by the Volvo/Swedish Environmental

Research Institute/Federation of Swedish Industries collaboration seems the most suitable of those currently available for implementing quantitative impact prioritization in industrial ecology, especially if an economic impact analysis along the lines of the Netherlands VNCI approach can be simultaneously incorporated. The rankings themselves, destined always to be in a state of flux, will be aided by the insights of the other efforts discussed before. Both the methodology and the degree of confidence that can be placed in the prioritization itself, as well as the capability of easily integrating new data, are important components of a usable system.

From the standpoint of both regulated and regulating sectors, it is important that any LCA impact-assessment system be regarded as unbiased. For this purpose, it is desirable that a competent third party, rather than an individual corporation, perform the actual prioritization, as was done in Sweden. With the increasing importance of multinational corporations, it would be even more helpful if such a system were generated by an international organization such as a committee of the International Geosphere-Biosphere Program (IGBP). Regardless of the organization performing the prioritization, provisions for regular updates are crucial in view of the evolving nature of environmental science.

Finally, it is worth stating that it is the prioritization activity that provides the natural and necessary link between the activities of industrial product and process design engineers, environmental scientists, and society as a whole.

SUGGESTED READING

Edmunds, R.A. *The Prentice Hall Guide to Expert Systems.* Englewood Cliffs, NJ: Prentice Hall, 1988.

Environmental Protection Agency Science Advisory Committee. Reducing risk: Setting priorities and strategies for environmental protection, Report SAB-EC-90-021 and 021A. Washington, DC: EPA, 1990.

Graedel, T. E. Industrial ecology: Definition and implementation, in *Industrial Ecology and Global Change*, R. Socolow, C. Andrews, F. Berkhout, and V. Thomas, eds. Cambridge, UK: Cambridge University Press, 1994.

Hocking, M. B. Paper versus polystyrene: A complex choice. *Science, 251* (1991): 504–505, and *252* (1991): 1361–1362.

Michigan Relative Risk Analysis Project. *Michigan's Environment and Relative Risk.* Lansing, MI: Department of Natural Resources, 1992.

Ministry of Housing, Physical Planning, and Environment. Integrated substance chain management, Pub. VROM 91387/b/4-92. 's-Gravenhage, the Netherlands, 1991.

Norberg-Bohm, V., W. C. Clark, B. Bakshi, J. Berkenkamp, S. A. Bishko, M. D. Koehler, J. A. Marrs, C. P. Nielsen, and A. Sagar. International comparisons of environmental hazards: Development and evaluation of a method for linking environmental data with the strategic debate on management priorities, in *Risk Assessment for Global Environmental Change*, R. Kasperson and J. Kasperson, eds. New York: United Nations Press, 1992.

Sheldon, R. A. Organic synthesis—Past, present, and future. *Chemistry and Industry*, no volume number, issue 23 (December 7, 1992): 903–906.

Steen, B., and S. Ryding. *The EPA Enviro-Accounting Method: An Application of Environmental Accounting Principles for Evaluation and Valuation of Environmental Impact in Product Design.* Stockholm: Swedish Environmental Research Institute (IVL), 1992.

Van Eijk, J., J. W. Nieuwenhuis, C. W. Post, and J. H. de Zeeuw. *Reusable Versus Disposable: A Comparison of the Environmental Impact of Polystyrene, Paper/Cardboard, and Porcelain Crockery.* Deventer, the Netherlands: Ministry of Housing, Physical Planning and Environment, 1992.

EXERCISES

11.1   Using the U.S. EPA Science Advisory Committee lists of risks (Chap. 4), create your own risk-prioritization list. Explain and defend your choices, in each case differentiating between scientific and technical assessments on the one hand and values and ethical judgments on the other.

11.2   What is the relationship of Figs. 11.1 and 11.2 to Fig. 9.7? If you were doing a survey of the environmental impacts of crop production, how would you use these figures?

11.3   A comparison for different countries of ranked hazards is shown in Table 11.3. Consider four other countries, each with a common border with one of those listed: Belgium, Mexico, Bangladesh, and Ethiopia. For each, predict five problems that will be identified as high hazards. Defend your reasoning.

11.4   (a) Suppose that global warming is thought less likely to occur than had previously been assumed, and that as a result the ELU/kg for product use is lowered to 0.6. What effect does this have on the comparitive ratings of the two front ends of Figure 11.9? (b) A new high-strength honeycomb steel has been developed and is being considered for use in automobile front ends. Rather than the steel front end weighing 6.0 kg, a satisfactory front end weighing only 4.0 kg can be formed from 6.0 kg of the new steel. The new steel's improved properties, which are due to added trace alloying elements, have negligible effects on processing or recycling, the same ELU/kg assessments apply to those stages (Table 11.5). Compute the ELU values for the new front end and compare them with the two options in the table. (c) Assume that the global warming revision of part (a) occurs as well as the availability of the new honeycomb steel front end. What effect do these two changes taken together have on the relative impact results? (d) Suppose that the shortage of petroleum (used as a feedstock for the manufacture of plastic composites) became so great that the ELU/kg of the composite materials was set to 1.90 and that the honeycomb steel front end was available. What effect do these two changes taken together have on the relative impact results? (e) What are the messages to designers implied in the analyses in the earlier parts of this exercise?

CHAPTER 12

The Materials and Process Audit for Electronic Solders and Alternatives: A Detailed Case Study

12.1 INTRODUCTION

Comparative DFE analyses can be quite complex, and it is useful at this point to provide an example of such an assessment to illustrate some of the considerations involved. We choose an important one for modern industry: whether the use of lead solder is the preferable way to fasten electronic components to circuit boards, given that lead is a toxic substance. The example uses the steps laid out in Chap. 8: (1) scoping of options and depth of analysis; (2) capture of relevant data in the DFE matrices; (3) brief discussion of translation of results into design directions; and (4) dissemination of results. Because the purpose of the exercise is to illustrate the procedure, not make soldering experts of the reader, much of the detail of the analysis has been omitted, but further details can be found in Appendix F.

12.2 DFE ANALYSIS OF LEAD SOLDER IN PRINTED WIRING BOARD ASSEMBLY

The principal use of lead solder in electronics manufacturing is in the assembly of printed wiring boards, which form the heart of every electronics device. This use of lead solder is a concern to industry for two reasons. First, lead is unquestionably a toxic heavy metal, so any consumption that might result in human environmental exposure raises some concern. Second, government regulation of lead use is increasing and there is always the possibility that regulatory constraints will be placed on its use. It is important to recognize that the comparative volume of lead involved in this particular use is quite small (0.7% of U.S. domestic consumption). Indium and bismuth, two metals frequently suggested as substitutes for lead in soldering applications, are little used materials. As a consequence, full replacement of lead solders with indium/tin alloy (for example) would result in a jump in domestic demand for indium from approximately 28 metric tons to some 11,000 metric tons; the comparable figures for bismuth are 1400 metric tons of current consumption compared to approximately 12,000 metric tons for replacement of lead in solder. The case with the option using silver as the metallic filler for conductive epoxies is intermediate depending on the technologies chosen.

Moreover, both indium and bismuth occur in extremely low concentrations in their ores. Both, in fact, are produced almost entirely as by-products of mining of other ores: indium from zinc and bismuth from lead. Again, silver occupies an intermediate position: Some silver deposits are relatively rich, but approximately two-thirds of silver is produced as a by-product from mining other ores.

12.2.1   The Scoping Function

12.2.1.1   Option Selection.   As a general rule, the options chosen for analysis should be both limited in number and generic. Although this restriction limits the applicability of the analysis somewhat, it is obviously critical if the analysis is to be at all manageable. In this particular case, there are a myriad of potential substitutes for tin/lead solder, many of which can be rejected because they also contain lead or cadmium. Of those that are usable in the same general temperature range as lead solders, alloys based on indium and bismuth are predominant. Metal-containing polymer matrices are also potentially attractive. Accordingly, a generic tin/indium solder, a tin/bismuth solder, and a nonalloy alternative consisting of an epoxy matrix filled with conducting silver beads were selected for analysis.

12.2.1.2   Depth of Analysis.   Because electronics manufacturing technologies and techniques have coevolved with lead solder, any significant shift to an alternative material could involve substantial redesign of existing processes and considerable capital investment in new equipment. Redesign of existing products and components would also be a likely consequence. An unsuccessful shift to an alternate technology could result in significant competitive penalties and quality degradation. Accordingly, a "cradle-to-reincarnation" analysis of potential alternatives is clearly warranted.

12.2.2   Manufacturing Primary Matrices

From a manufacturing standpoint, potential problems exist with substitution for lead as a result of the scarcity and by-product status of indium and bismuth, and, to a lesser extent, silver. These problems are reflected in the cost and availability categories and, for indium and bismuth, the energy consumption category (these metals occur in such low concentrations that significant energy must be expended to extract and purify them). The indium manufacturing primary matrix is presented as Fig. 12.1 as a sample of these matrices; those for lead, bismuth, and silver epoxy appear in Appendix F.

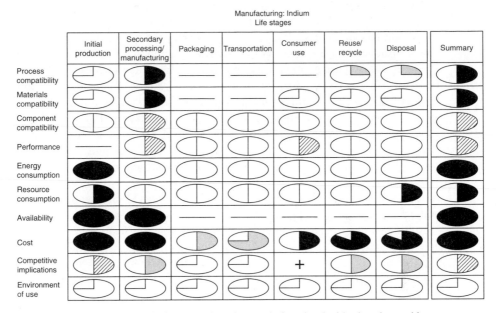

Figure 12.1   The indium manufacturing matrix for printed wiring board assembly.

Among the questions raised by the manufacturing primary matrix analysis is that of inappropriate resource consumption for indium, bismuth, and, to a lesser extent, silver. These relatively rare metals are being taken from ores, where they are recoverable, and put into product, where they will probably be unrecoverable. The standard neo-classical economic response is that alternatives will be found as prices increase, but this assumes that future generations will find other materials that are completely substitutable. If any of these materials turn out not to have substitutes, this assumption fails, and the current generation will have consumed material(s) of unique value to future generations. The current price structure does not reflect this potential cost to future generations—indeed, cannot, given the inherent uncertainty of the future value of existing materials, which in turn depends on future technological developments.

A second methodological issue concerns the costs of energy associated with initial materials production. The question is how to allocate energy consumption from mining and processing more ore to obtain increased supplies of replacement solder. If, for example, increased indium supplies are being produced, leading to concomitant production of more zinc, how should the incremental energy costs (and incremental environmental costs) be allocated and how should the environmental costs associated with zinc produced as a by-product of indium production be treated? This question illustrates the fact that taking a systems-based approach entails the need to identify, and resolve, a set of methodological issues that to date have not even been recognized.

12.2.3 Social/Political Primary Matrices

The indium social/political primary matrix is illustrated in Fig. 12.2. This matrix is the most difficult to evaluate because of data and methodological deficiencies. The unspoken but standard assumption is that the price structure for various options captures these effects: This is, in practice, far from the truth. Especially when operations that may involve the Third World or relatively powerless political classes or localities are involved, most social costs simply remain externalities. Indeed, in many cases, these costs may not even be recognized as costs. Take, for example, the increases in mining and ore-processing activity implied by substantial substitution of indium, bismuth, or silver (epoxy) options for lead solder. Not only will there be greatly increased environmental disruptions (captured in the environmental primary matrices), but the impact of such activities on local communities can be profound. Mining by definition draws down the natural capital of a region. Mining operations also create temporary needs for labor, resulting in temporary communities. This negatively affects not only the communities the labor may be drawn from, but the mining community itself, which will be abandoned in due course. Note that such effects are more likely to be associated with by-product metals than with lead, because lead used in printed wiring board assembly is such a small fraction of existing overall lead demand.

Similarly, the labor impacts of increased mining and processing are mixed. On the one hand, increased employment is generally considered a benefit; on the other hand, the counterargument could be made that the temporary nature of employment associated with mining generates disruption of the labor market as employees and potential employees flock to the site, to be followed by unemployment at some future time as the ore body plays out or market conditions shift. Moreover, activities associated with mining and processing ore may or may not result in transferable skills that can be employed in other sectors, and the hazardous nature of mining has long been recognized—

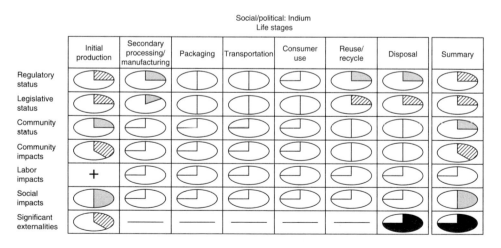

Figure 12.2   The indium social/political matrix for printed wiring board assembly.

especially in developing countries, where worker safety and environmental controls may not be as stringent as desirable.

An additional conundrum is raised where an option presupposes substantial demand increases, as with indium, bismuth, and, to some extent, silver. The demand restructuring will raise prices and reduce material availability for competing uses. Bismuth is heavily used in pharmaceuticals, as a substitute for more toxic lead in a number of metallurgical applications, and in fire-control systems. Indium is used for, among other things, infrared reflecting window coatings, windshield defrosters for airplanes, as a surface coating on engine bearings, and in new-generation solar cells. Although some concern in indicated in the significant externalities category for this reason (as well as the resource-depletion problems noted before), the more general point is that methodologies for evaluating such social welfare externalities do not exist. As is the case with intergenerational issues, the methodologies will be difficult to develop because ethical and value judgments are required.

12.2.4   Toxicity/Exposure Primary Matrices

The indium toxicity/exposure primary matrix is presented in Fig. 12.3. This section of the analysis covers topics that have traditionally been part of environmental, health, and safety practices in government and private industry. Accordingly, they are relatively self-explanatory. The only caution that must be exercised in this category is not to let the available data distort the analysis. This is difficult because, in most instances, even for a substance used as long and as extensively as lead, there will be far more data available on mammalian toxicology than on nonmammalian toxicology. For less well-studied

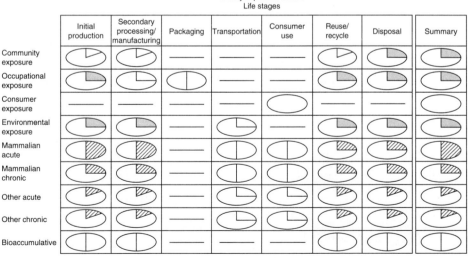

Figure 12.3   The indium toxicity/exposure matrix for printed wiring board assembly.

materials, such as indium and bismuth, the nonmammalian toxicology is essentially unknown, and projections must be made from few data indeed.

In the same sense, reliably predicting exposures may be difficult. This will be especially true where anticipated exposures would change dramatically as a result of the implementation of an option. In the present assessment, one recognizes that there are currently no indium or bismuth recycling operations, so it is difficult to evaluate what exposures would result if substantial substitution of these materials for lead in solder were to occur and recycling operations were to be developed. Projections from current recycling operations can be made, of course, but they must be treated as approximations at best.

The point to be drawn from these matrices is that, even in an area where substantial attention has been paid to date, a systems-based approach reveals considerable data and methodological deficiencies. Nonetheless, qualitative assessments turn out to be both possible and defendable.

12.2.5   Environmental Primary Matrices

The indium environmental primary matrix is presented in Fig. 12.4. What is most notable is that almost all severe impacts occur at life-cycle stages not under the control of the firm using the lead solder to manufacture printed wiring boards: The initial production life-cycle stage is controlled by the mining/processing firm, whereas the reuse/recycle and disposal stages occur after consumer use. Accordingly, were the firm to evaluate only the environmental impacts of its operations in choosing among options,

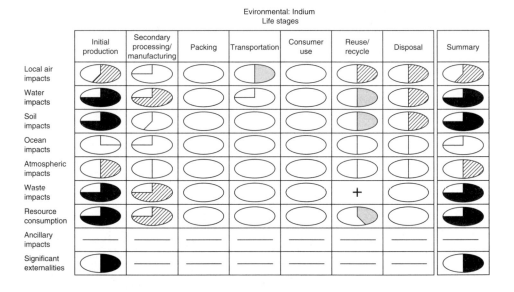

Figure 12.4   The indium environmental matrix for printed wiring board assembly.

it would be ignoring virtually all of the major environmental impacts associated with its choice. The necessity of a life-cycle, systems-based approach is thus clearly demonstrated by these matrices.

In the secondary processing/manufacturing life-cycle stage, several additional considerations arise. First, it is axiomatic that replacement of any existing technology before its useful life is over in some sense "wastes" existing capital stock. All other things equal this is undesirable, as the embodied energy and resources in that stock are therefore not used as extensively as they otherwise might be. This potential loss is particularly likely for the conductive epoxy technology, which would render existing soldering equipment obsolete.

12.2.6 Summarization

The four matrices have been presented as though they were aspects of the same stage of analysis. In actuality, they perform (in a qualitative way) two stages of life-cycle assessment. The first stage, *inventory analysis*, is effectively captured by the manufacturing primary matrix. The other three matrices deal not with emissions, consumption, or other inventory parameters, but with the effects of those activities; they constitute a qualitative *impact assessment*, stage two of LCA. Thus, although the analysis does not involve the level of computational detail of, for example, the Swedish EPS technique, it goes well beyond the one-stage LCA approaches most often found in the industrial ecology literature.

Given this perspective, let us examine the summary matrix for the analysis of electronic solders and alternatives. The matrix (Fig. 12.5) is constructed by recording the most serious levels of concern obtained for the bismuth, indium, lead, and epoxy options

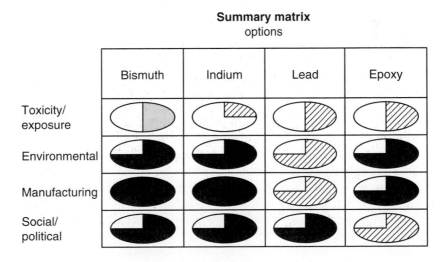

Figure 12.5   The summary matrix for printed wiring board assembly.

in each primary matrix. For the toxicity/exposure primary matrix, however, the exposure and toxicity rankings for each substance, separately displayed in that primary matrix, are combined into one ranking in the summary matrix. This ranking reflects the overall hazard risk posed by the option, and some additional explanation is appropriate. For bismuth, the ranking reflects a low level of concern because exposure potentials are limited and existing data on bismuth toxicity do not raise serious issues. Indium receives a moderate concern ranking for overall risk based on moderate levels of toxicity but limited exposures across the life cycle. Lead poses clear and severe toxicity issues, especially in the mammalian chronic category. However, the amount of exposure associated with this particular use of lead is so low that it is not appropriate to rank that risk as serious. Finally, some of the organic constituents of epoxy pose significant toxicity risks and the silver component exhibits substantial aquatic toxicity for some organisms. Moreover, there is the possibility that some exposure to epoxy precursors may occur at the initial production and secondary processing/manufacturing life-cycle stages. Accordingly, a moderate risk ranking for silver epoxy is appropriate.

The results of the impact assessment can be stated simply: the status quo, lead solder, is preferable to substantial substitution of alloys containing significant amounts of bismuth or tin or by epoxies containing significant amounts of silver. When the relatively minor component of overall lead demand attributable to printed wiring board assembly applications is contrasted with the significantly expanded mining and processing activities the other options would entail, lead-based solders are the least environmentally harmful choice. Thus, a systematic analysis has led to what, for many people, is a counterintuitive result.

12.3 FROM ASSESSMENT TO PLANS

The third stage of LCA, *improvement analysis*, follows directly from the summary matrix. It is clearly appropriate to reduce the use of lead in printed wiring board assembly where feasible. First, it makes sense to investigate low-volume, niche applications for indium, bismuth, and conductive epoxy alternatives. Second, it is advisable to investigate new fastening techniques and new electronic circuit designs requiring fewer interconnections.

The analysis also identified important areas for future research, including the possibility of niche applications of indium and bismuth alloys, and the development of epoxy and polymer systems containing minimal amounts of conductive metal. This research-and-development effort would be guided by the analysis: Most resources would flow toward the generally environmentally preferable alternatives of low metal content polymeric systems, with relatively less expended to evaluate indium and bismuth alternatives.

There are two other noteworthy facets of the analysis. The first is that the environmental impacts that drive the analytical conclusions occur in life-cycle stages that a traditional environmental analysis by the manufacturer, focused on its operations and its customers, would miss. The second is that even this mundane, real-world exercise raises value and ethical issues that have not yet been addressed in a comprehensive manner. Such value questions require cultural and social responses, not responses by private firms responding to short-term profit motives. They can be identified, but are not resolved, by DFE analyses.

12.4 DISSEMINATION OF RESULTS

The results from a DFE analysis usually will be first provided to the firm sponsoring the effort. There are two primary internal customers. The first is the design community that, through the questions incorporated in the DFE module, will now be encouraged to focus on reducing lead solder use, but substituting only where the alternatives are clearly superior. Because of the fundamental nature of the questions addressed and the many basic materials science issues raised by the alternatives, however, there is a second primary customer, the internal research-and-development community. This community will use the assessment results as a guide to desirable long-term technological development.

Although in many cases the results of a DFE analysis will be disseminated only within the firm, there is also a need in an assessment study for what might loosely be called "technology diffusion". It is desirable to inform other stakeholders—electronics manufacturers, suppliers, regulators, the public—of the results, determine whether the data, methodologies used in the analysis, and assumptions are as accurate as possible, and present the value judgments made during the course of the analysis. In the long run, such information dissemination will ensure that more environmentally preferable options are made available to electronics manufacturers.

SUGGESTED READING

Allen, B. M. *Soldering Handbook*. London: Iliffe Books, 1969.

Allenby, B. R., *Design for Environment: Implementing Industrial Ecology*. Ph.D. dissertation, Rutgers University, 1992 (available through University Microfilms, Inc.).

Allenby, B. R., J. P. Ciccerelli, I. Artaki, J. R. Fisher, D. Schoenthaler, T. A. Carroll, D. W. Dahringer, Y. Degani, R. S. Freund, T. E. Graedel, A. M. Lyons, J. T. Plewes, G. Gherman, H. Solomon, C. Melton, G. C. Munie, and N. Socolowski. An assessment of the use of lead in electronic assembly, in *Proceedings of the Technical Program*, pp. 1–28, Surface Mount International, San Jose, CA, 1992.

Buckley, J. D., and B. A. Stein, eds. *Joining Technology for the 1990s*. Park Ridge, NJ: Noyes Data Corp., 1986.

Clark, R. M. *Handbook of Printed Circuit Manufacturing*. New York: Van Nostrand Reinhold, 1985.

Manko, H. M. *Solders and Soldering*. New York: McGraw-Hill, 1964.

National Reuse of Waste Research Programme (CNOH), Conference Reprints from the First NOM European Conference on Design for the Environment, Nunspeet, The Netherlands, September 21–23, 1992.

Noble, P. W. J. *Printed Circuit Board Assembly*. New York: John Wiley, 1989.

EXERCISES

12.1 Evaluate the technique for the assessment of electronic solder alternatives. What portions of the procedure seem most soundly based? What portions seem most problematic? How would you revise the assessment technique if given the opportunity?

12.2 About 5% of total lead consumption in the United States has traditionally been used for ammunition. Birds eat lead shot and can be killed by the ingestion of a few pellets, a situation that has resulted in significant mortality in many species of birds worldwide. Using the information in this chapter and in Appendix F, perform an LCA assessment on this use of lead. Does your assessment suggest alternatives?

12.3 You are asked by your firm, a washing machine manufacturer, to evaluate three materials for use in machine housings:

  • Material A: Steel, the current material, is capable of being recycled with current techniques and infrastructure.

  • Material B: A simple and very light plastic. It provides significantly less impact protection than Materials A and C and can be recycled, although no infrastructure for doing so currently exists.

  • Material C: a layered carbon/epoxy composite. This material is also very light and is much stronger and wears better than the other alternatives. It cannot be recycled using any forseeable technology. It has never been used for washing machines before.

  (a) Qualitatively identify the major advantages and disadvantages of each option. Include environmental, manufacturing, toxicological, and social aspects in your analysis.

  (b) What additional information would be useful in helping you to make a design choice?

12.4 The dry cleaning industry uses and releases significant amounts of chlorinated solvents, primarily perchloroethylene, in the performance of its functions. You are a technical analyst for the U.S. Environmental Protection Agency, and have been asked to assess the situation with a goal of identifying how to reduce air emissions of solvents.

  (a) Identify the options you believe should be analyzed.

  (b) What additional information would be useful in helping you to better define an option set?

12.5 You work for a firm that is considering a product that would create a significant additional demand for copper. The copper would be in a soluble form in your product and would dissipate in the environment during use and when the product is discarded (no takeback program is planned). Substantial sales in developing countries are anticipated. You are asked to do an analysis of the potential environmental impacts of this product.

  (a) Identify potential concerns raised by such a product.

  (b) If there are any concerns, can you suggest possibilities for mitigation that could be explored?

  (c) What additional information would be useful to you in performing this analysis?

PART IV: DESIGN FOR ENVIRONMENT

CHAPTER

13

Industrial Design
of Processes and Products

13.1 INTRODUCTION

Industrial ecology involves both products and processes, and the distinction between the two is important. Products are what is sold by a corporation: paper clips, toothpaste, and airplanes, for example. Processes are the techniques by which those products are made: the production of glass from lime, soda ash, and sand, for example. It is often the case that the people who design processes and the people who design products are different. It is also the case that the interactions of process design with environmental concerns are somewhat different from those of product design. The industry–environment interaction is thus heavily influenced by two rather separate groups of designers, both of whom must be addressed if industrial ecology is to be effective.

Processes are much more universal than products, and a successful process design often has great importance to and great staying power for an entire industry. It is often the case, moreover, that groups of processes, each dependent on the others, will co-evolve. Almost all steel mill products of today are made by either the basic oxygen or thermoelectric process, and the intricate procedures culminating in the integrated circuit are the basis for hosts of products from portable radios to computers. Corporations that make silicon furnaces or high-pressure reaction vessels market them as the products they are, but they are received and used by customers as process tools that enable them in turn to create their own products. Processes are the ways in which feedstock material of one sort or another is transformed into intermediate materials. Thus, processes define much of the flow of solids, liquids, gases, and energy into a manufacturing facility and are responsible for much of the flows of solids, liquids, and gases leaving that facility. Once a process is thoroughly embedded in an industry, it is difficult and expensive to do more than make incremental changes, at least in the short term.

Product designers, in contrast, have considerable flexibility when choosing the processes needed for their products. It is often possible, for example, to use wood, plastic, or metal to accomplish the same design purpose, and the decision may rest on cost or aesthetics as much as on materials properties. If a product arises as a consequence of an assembly process, as occurs with the transformation of fabric into an article of clothing, little processing in the materials sense can occur. Conversely, if a producer incorporates into its operation the cycle from feedstock materials to intermediates to products, a single facility can incorporate many facets of industrial ecology. In any case, product designers must consider industry–environment interactions that are largely outside the province of the process designer: choice of materials, product packaging, environmental impacts during product use, and the optimization of product recycling.

The distinction between process and product thus leads to a difference in expectations concerning the two. Processes are unlikely to change very much except on those rare occasions when a new process proves to be beneficial enough to supplant the old. Product designers, however, often can use combinations of existing or evolving processes to design products with newly minimized and tailored industry–environment interactions. In the ideal situation, processes and products are developed and introduced together, and the opportunity to integrate the entire industrial operation becomes greater. The close linking of research and development promotes concurrent process and product design activity. In this chapter and throughout the book, we will often identify and discuss the differences, similarities, and interrelationships between processes and products.

13.2 MODERN APPROACHES TO INDUSTRIAL PROCESS DESIGN

13.2.1 Products and By-products

Traditional process design emphasizes the optimization of yield for the primary product or products. Two types of process can be distinguished. The first is one in which transformations occurring in a chemical flow stream are the primary focus, such as the production of ethylene glycol from ethylene chlorohydrin and sodium bicarbonate. The second is one in which the chemical flow stream acts on a separate product flow stream, as in the etching of a metal part by an acid bath. In neither case have secondary chemical products been much of a consideration in process design, both because their characteristics are perceived to be of little interest relative to the principal product and because their generation is difficult to predict by customary chemical process design tools. Nonetheless, it is often these secondary products that are of concern when they appear as impurities in primary products or trace constituents of residue streams.

Sheldon Friedlander of the University of California at Los Angeles has pointed out that the concentrations of these by-products are often extremely low relative to the primary product concentrations (Fig. 13.1). Predictions of by-product generation may require new approaches to process design if industrial ecology concerns are to be properly taken into account.

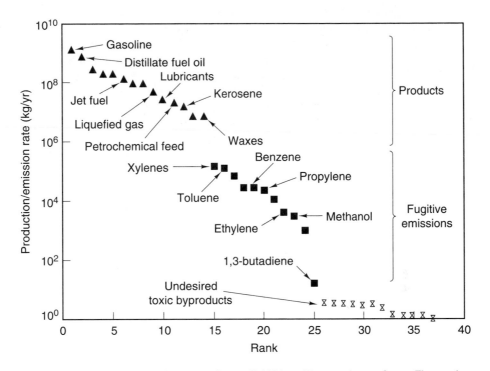

Figure 13.1   Rank/order diagram for outputs from a 50,000-barrel/day petroleum refinery. These estimates are based on composite data from several sources and should be considered illustrative. The generation and possible release of pollutant by-products such as the polycyclic aromatic hydrocarbons indicated by the double triangles in the bottom right of the diagram are usually the most difficult parts of the refinery design to measure or predict, and they are often as much as six orders of magnitude below the principal product-generation rates. (Reproduced with permission from S. Friedlander, Preprint: The two faces of technology: Changing perspectives in design for environment, in *The Greening of Industrial Ecosystems*, B. R. Allenby and D. J. Richards, eds. Washington, DC: National Academy Press, 1994.)

13.2.2   Steps in Process Design

In today's industrial regime, the rates of process change are becoming extremely rapid. Bruce Paton of Hewlett-Packard Corporation uses the example of retrofitting processes to remove CFCs to illustrate the concept that industrial ecology often can be advanced most quickly by focusing on new, rather than existing, processes. He lists several strategies, which we summarize in what follows as important in new process design.

13.2.2.1   Selection of New Process Technology.   New processes can arise from a need to formulate a particular type of material, such as a new plastic that requires specific temperatures and pressures in its manufacture, or from a need to change the characteristics of an existing process, such as the high energy consumption and air-quality impacts of smelting and refining processes. The selection and development of process approaches and technology often establish the general characteristics of a process's suitability from the standpoint of industrial ecology.

13.2.2.2 New Equipment Selection. In implementing new processes, designers often have a choice among equipment alternatives. The act of purchasing equipment freezes industry–environment process interactions for long periods, so considerable thought should be given not only to the ability of equipment to produce the desired product rapidly and efficiently, but also to the embedded environmental impacts.

13.2.2.3 Materials Management. The choice of materials used in developing a process plays a major role in defining a facility's industrial ecology. For example, a process may involve cleaning, rinsing, air drying, solution replenishment, and so forth, all steps that have the potential for residue stream emissions. It is to the process designer, and particularly to the materials management component of that designer's function, that the industrial ecology maxim, "Every molecule entering a manufacturing facility should leave it as a component of a salable product" is particularly directed.

13.2.2.4 Batch vs. Continuous Processes. A process designer sometimes has a choice between a batch process and a continuous process. Where possible, the continuous process is almost always the preferable choice from an environmental standpoint, because the need for recleaning is minimized and heating and cooling cycles are avoided.

13.2.2.5 Summary. The characteristics of most processes are that they have long lead times, that they are often purchased from a vendor rather than developed from within, and that significant changes are likely to be evolutionary rather than revolutionary, at least on the time scales readily influenced by industrial process engineers.

13.3 MODERN APPROACHES TO INDUSTRIAL PRODUCT DESIGN

13.3.1 Design for X

The sequence of events that culminates in a product begins with product definition, which is a statement of the features that a specific product should have when development is completed. These features normally include what the product will be used for, how it will function, what its properties will be, the range of probable cost, and (if appropriate) its aesthetic attributes. That product definition list gives the designer many things to consider simultaneously, but modern designers have a list much longer still, because they need to consider related product attributes that may, in the end, determine the product's success or failure. The paradigm for these latter considerations is called "Design for X" (DFX), where X may be any of the following:

- *Assembly* (A): The consideration of assemblability, including ease of assembly, error-free assembly, common part assembly, and so on.

- *Compliance* (C): Consideration of the regulatory compliance required for manufacturing and field use, and including such topics as electromagnetic compatibility.
- *Environment* (E): This component of DFX, and the philosophy on which it is based, is a principal subject of this book.
- *Manufacturability* (M): The consideration of how well a design can be integrated into factory processes such as fabrication and assembly.
- *Material Logistics and Component Applicability* (MC): The topic focuses on factory and field material movement and management considerations, and the corresponding applicability of components and materials.
- *Orderability* (O): The consideration of how the design impacts the ordering process from the customer perspective, and corresponding manufacturing and distribution considerations.
- *Reliability* (R): The consideration of such topics as electrostatic discharge, corrosion resistance, and operation under variable ambient conditions.
- *Safety and Liability Prevention* (SL): Adherence to safety standards and design to forestall misuse; thus, the prevention of costly legal action.
- *Serviceability* (S): Design to facilitate initial installation, as well as repair and modification of products in the field or at service centers.
- *Testability* (T): Design to facilitate factory and field testing at all levels of system complexity: devices, circuit boards, systems, and so forth.

DFX design practices are already being implemented by leading manufacturing firms. Accordingly, the least difficult way to ensure that environmental principles are internalized into manufacturing activities in the short term is to develop and deploy Design for Environment (DFE) as a module of existing DFX systems. Moreover, the fact that DFE is intended to be part of an existing design process acts as a salutary constraint, requiring that DFE methods and analysis be implementable in the real world.

An increasingly important aspect of the design process is the degree to which it is linked to the computer. Most modern industrial design teams utilize computer-aided design/computer-aided manufacturing (CAD/CAM) software, which can incorporate standard component modules into a design, check a design for spatial clearances, produce lists of materials, and so forth. To the degree that DFE can be integrated into these design tools, it will become automatically a part of the physical design process. DFE in CAD/CAM is just beginning, and diligent effort will be needed to bring it to a high level of development.

13.3.2   Steps in Product Design

13.3.2.1   Product Definition.   Earlier in this chapter, we stated that product definition is an initial and crucial stage in the product development process. It is at that stage where the environmental attributes of a product can be identified and built into the design. It is important to recognize that DFE will require some portion of the designer's effort, and that, like all other aspects of design, thoughtful choices made early in the

design process are by far the most cost-effective. Given both the internal and external impetus for environmentally responsible design, DFE should be automatically a component of the product definition and creation cycle.

13.3.2.2 Materials Management.   Designers generally have at least modest freedom within the product definition guidelines concerning choices of materials, and, once those choices are made, concerning the ways in which those materials are incorporated into products. In a subsequent chapter, we provide guidance about these choices.

13.3.2.3 Detailed Product Design.   Detailed product design is the stage at which DFX considerations, including DFE, are taken fully into account. Designs inevitably involve trade-offs among such attributes as utility, cost, reliability, and so forth. This stage, which we mention here in passing, is the central specialty of physical designers.

13.3.2.4 Product–Process Interactions.   In the case of many new designs, the products cannot be manufactured without concurrent evolution in industrial processes. An example from the electronics industry is the continuing push to design products using finer and finer resolution in integrated circuits in order to increase speed and reduce size. Such designs are reasonable only if the finer resolution can be achieved given the manufacturing tools available. In the same way, DFE goals can be realizable only if process designers work closely with product designers to provide the manufacturing tools needed to make environmentally friendly products.

13.3.2.5 Interactions with Suppliers.   It has become obvious within the past decade that the goals of efficiency, quality, reliability, and cost minimization cannot be achieved in the manufacturing process without the active participation of a corporation's suppliers. The partnership with which the most experience has been gained is probably the "just-in-time" delivery of supplies, a technique that minimizes storage costs and overstocking and maximizes component quality. From an industrial ecology standpoint, these relationships can be used to develop access to sources of recycled material, to create customers for recyclable manufacturing residues, and to create environmentally appropriate standards and specifications for purchased items.

13.3.2.6 Marketing Interactions.   Product designers and product managers can promote industrial ecology goals by working with customers as well as suppliers. Often the customers will be retailers rather than individuals, and this permits the designer to optimize product packaging (including packaging return or other recycling), minimize stocks of unneeded products (thus minimizing reprocessing or disposal), minimize unneccessary transportation, and make information on environmentally related aspects of products and their eventual recycling widely available.

13.3.2.7 Summary.   The characteristics of products are short lead times, designs largely or entirely by the manufacturer, and significant changes from product to product. In each case, these contrast with the characteristics of processes and offer significant opportunity for the rapid implementation of industrial ecology.

SUGGESTED READING

Design for X. Special Issue. *AT&T Technical Journal, 69* (3) (May/June 1990).

Paton, B. Design for environment: A management perspective, in *Industrial Ecology and Global Change*, R. Socolow, C. Andrews, F. Berkhout, and V. Thomas, eds. Cambridge, UK: Cambridge University Press, 1994.

EXERCISES

13.1 Select a wrench or other simple hand tool and, to the degree possible, evaluate it qualitatively for each of its "design for X" features.

13.2 Repeat Exercise 13.1 for a kitchen electrical appliance of your choice.

CHAPTER

14

Designing for Energy Efficiency

14.1 ENERGY AND INDUSTRY

Industry uses substantial amounts of energy and, as a consequence, contributes significantly to energy-related environmental problems. In the United States, for example, manufacturing activities account for some 30% of all energy consumed and much of that energy is very inefficiently employed. Figure 14.1 shows that the use of electricity (mostly generated from fossil fuels) is concentrated in a few industry types, such that six industry groups consume more than 85% of total industrial energy or energy equivalents. This information says nothing about the efficiency with which that energy is used, however. A useful index of industrial energy use, though not a complete one, is *energy intensity*, the energy consumption per dollar of gross domestic product. Individual corporations within the same industry vary widely in energy intensity. In general, industries dealing largely with raw materials rather than finished or semifinished products have higher energy intensities.

Although the focus of this chapter is to examine how energy is used and how to reduce energy use while maintaining industrial operations, it is also of interest to mention the consequences of energy generation, because plant and process engineers sometimes have optional energy sources upon which to draw and can consider the selection of the most environmentally benign source for the energy they need. In this connection, we list in Table 14.1 the chemical species emitted to the air by a variety of energy-generation processes. (The effects of those emissions were discussed in Chap. 3.)

The table illustrates the well-known fact that energy supplied by fossil fuel combustion has the potential to be more environmentally harmful than energy produced by nuclear power or by natural energy sources, at least so far as the atmosphere is concerned. Incineration is intermediate in impact, and its ability to serve as a waste-disposal

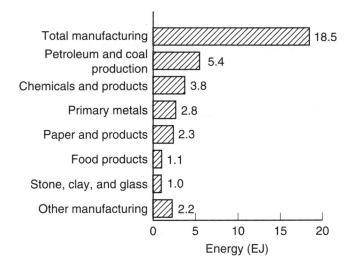

Figure 14.1   Consumption of energy in selected manufacturing industries. (U.S. Dept. of Energy, Energy Information Administration, *Manufacturing Energy Consumption Survey: Changes in Energy Efficiency 1980–1985*, DOA/EIA-0516(85), Washington, DC, 1990.)

Table 14.1   Gaseous Species Emitted by Energy Generation Processes

Process	Gaseous Species						
	CO_2	CH_4	NO_x	SO_2	H_2S	HCl	Particles
Fossil Fuel **Energy Sources:**							
Coal	*	*	*	*			*
Petroleum	*	*	*	*			
Natural gas	*	*	*				
Other Anthropogenic **Energy Sources:**							
Nuclear power							
Refuse incineration	*					*	*
Biomass incineration	*	*	*				*
Natural **Energy Sources:**							
Hydrothermal steam					*		
Solar power							
Hydropower							
Wind power							

alternative as well as an energy source makes it suitable in some industrial applications. To the extent possible, energy obtained for industrial use should be from environmentally benign sources.

Energy efficiency must often be balanced with toxicity concerns. For example, more efficient lighting relies on mercury vapor in lamps, and many catalysts that reduce energy requirements in industrial and consumer applications are either toxic or are scarce, nonrenewable resources. Superconducting materials may offer the possibility of significant energy savings, but are likely to contain toxic materials. Identifying and resolving these "impact-balancing" situations is a difficult but necessary task as products and processes evolve toward sustainability.

14.2 PRIMARY PROCESSING INDUSTRIES

In the chapter on materials flows, we commented that energy intensity was highest during materials extraction processes. This fact received renewed emphasis in Fig. 14.1, showing the industries that are the biggest energy users. These industries are the suppliers of processed materials to the intermediate processing industries, so one cannot plan to decrease industrial energy use by eliminating the extraction industries. Rather, one needs to investigate opportunities within the extraction industries for reductions in energy intensity.

Of the industry groups listed in Fig. 14.1, the one with the largest energy use is petroleum and coal production. Most of this energy use is attributable to the refining of petroleum. The trend toward desulfurization of crude oil, and the production of high-octane gasoline without the use of metal-containing additives, place increased energy demands on the refining operation. Refinery operations are generally subject to careful supervision and continuing engineering effort to improve efficiencies, but increased attention to cogeneration, heat exchange, and leak prevention will offer opportunities for further improvements.

Chemicals and chemical products rank second among the industries on Fig. 14.1, although about a third of the amount represents petroleum and natural gas used as feedstocks for products rather than fuel that is consumed to produce energy. Of the remaining two-thirds, most is used in the generation or removal of process heat as a result of temperature differences between the process streams and the heating and cooling streams. Physical designers should attempt to develop processes that minimize these temperature differences, perhaps by better in-plant use of process heat, by process redesign involving different feedstock materials or improved catalysts, or by capture of process heat for subsequent use or sale. The production of compressed gases is another energy-intensive area; about 70% of the cost of the gases represents electricity costs, and improving the energy efficiency thus has the potential to pay rich benefits for that industry.

Primary metals is the third industry listed in Fig. 14.1. Although the extraction of ore from the ground and its shipment are quite energy-intensive, the bulk of the energy use is in generating the large amounts of process heat needed to extract metal from ore and to produce ingots and other products. Historically, major changes in the use of energy in the primary metals industry have occurred as a result of the introduction of new processes. In the case of steel, for example, the relatively new electric arc furnaces are much more efficient than the older open hearth and basic oxygen processes. Another example is the large historical decrease in electrical energy consumption needed to produce aluminum, shown in Fig. 14.2. It is worth noting that for aluminum and other metals, the practical and thermodynamic limits to the energy needed for processing are beginning to be approached, suggesting that major gains from process changes alone may have already occurred among the more advanced manufacturers.

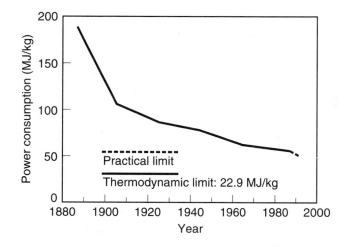

Figure 14.2   The history of electric power consumption in the production of aluminum. (Adapted with permission from P. R. Atkins, D. Willoughby, and H. J. Hittner, in *Energy and the Environment in the 21st Century*, J. W. Tester, D. O. Wood, and N. A. Ferrari, eds., pp. 383–387. Copyright 1991 by The MIT Press.)

14.3 INTERMEDIATE PROCESSING INDUSTRIES

The intermediate processing industries are too diverse to be discussed individually, but several general techniques for improving their energy efficiency can be described. The first, and one that is thus far quite infrequently implemented, is the use of computerized systems for the management of energy use. The overall concept is that energy should be used only when needed, and not because inattention or lack of on-site personnel makes it impractical to exercise control. Thus, equipment should be started and stopped as dictated by time of day or by sensors of product-stream characteristics. Among the types of energy-using equipment that can be controlled in this way are motors, boilers, fans, and lights.

A second technique, previously discussed in connection with the chemical industry, is the utilization by the corporation or by its infrastructure partners of residual heat from process streams, product streams, exhaust streams, and the like. Often these actions will take the form of increased attention to process redesign so that the exchange of heat among material flow streams can be optimized. Alternatively, the heat can serve unrelated processes, as in Nova Corporation's use of residual heat from a natural-gas compressor station in Alberta, Canada, to provide heat for greenhouses producing flowers, plants, and tree seedlings.

Third, increased use can be made of modern-design motors, especially those with variable-speed drives. The gains that can be expected are quite dependent on the application, but 20–50% decreases in energy use have been realized in several test cases.

14.4 ANALYZING ENERGY USE

It is the usual case in industry that the amount of energy required to operate a facility is well monitored, whereas the energy required for each operation or set of operations within a facility is not known. In such cases, an energy audit is advisable to show where the opportunities for gains might lie, as well as to provide data for "green" accounting systems. Figure 14.3 shows such an audit for a facility that uses oil, coal, and electricity to provide energy for three different industrial processes, as well as for lighting and heating. The diagram demonstrates that more than enough energy is available in losses from the process A energy stream to operate processes B and C and heat and light the entire factory in the bargain. The diagram also suggests that boiler losses would be the most logical place to consider for improvement, and that steam losses also constitute a target of opportunity.

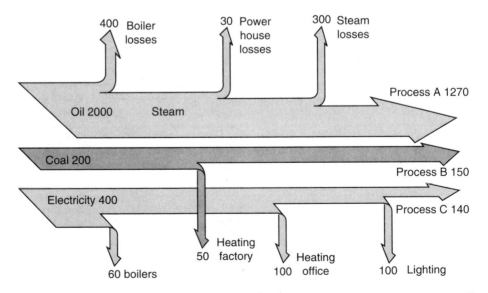

Figure 14.3 A "Sankey diagram" of energy sources, uses, and losses for a typical industrial facility. The units are arbitrary. (Reproduced with permission from *Climate Change and Energy Efficiency in Industry.* Copyright 1992 by International Petroleum Industry Environment Conservation Association.)

For a particular process, one wishes to audit the energy use at each stage of manufacturing. For the production of aluminum cans, for example (Fig. 14.4), the big energy use is in the separation and purification of aluminum contained in the ore. Production of sheet and of cans is also significant, but on a much-reduced scale. The transport of material between stages is a minor contributor to total energy use. With this information as a basis, one might choose to increase the amount of recycled material used to produce metal products rather than extract metals directly from ore. Aluminum is the optimum material for such a strategy, because the use of many different kinds of scrap material results in energy savings of 30% or more.

To examine the energy-use implications of virgin material and recycled material, consider the process sequence shown in Fig. 14.5. Each processing step has associated with it an energy per unit of throughput. For simplicity, we choose the amount of output material to be 1 ton. β is the fraction of throughput that becomes "prompt scrap" rather than output material: rejected material, sprues, runners, lathe turnings, and so forth. The energy consumed per ton of output material is then given by

$$\Phi = E_p + E_f(1 + \beta) + E_m(1 + \beta)$$

$$= E_p + (E_f + E_m)(1 + \beta) \tag{14.1}$$

It is obvious from this equation that manufacturing operations that produce a smaller fraction of scrap will require less energy per unit of output than those where a large fraction of material must be refabricated.

A more relevant case for industrial ecology is a manufacturing sequence that uses both virgin and consumer recycled material. The latter need undergo only secondary

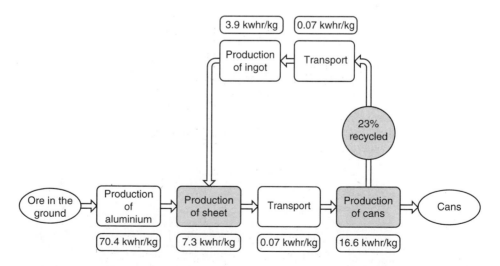

Figure 14.4   A process energy-use diagram for the production of aluminum cans. (Reproduced with permission from *Climate Change and Energy Efficiency in Industry*. Copyright 1992 by International Petroleum Industry Environment Conservation Association.)

Table 14.2   Energy Input (GJ/Mg) Required for the Production of Various Metals

Metal	Primary Production	Secondary Production
Steel	31	9
Copper	91	13
Aluminum	270	17
Zinc	61	24
Lead	39	9
Titanium	430	140

Source: The data are from P. F. Chapman and F. Roberts, *Metal Resources and Energy*, Boston: Butterworths, 1983.

production, which is much less energy-intensive than primary production. The situation is illustrated in Fig. 14.6, where ϕ is the fraction of output material from primary production, Ω is the amount of material entering the process in the ore, and ψ is the amount of the material entering the process as consumer scrap. The energy consumed by this system per ton of output material is given by

$$\Phi = E_p(\phi)(1 + \beta) + E_s(1 - \phi)(1 + \beta) + E_f(1 + \beta) + E_m(1 + \beta)$$

$$= [\phi E_p + (1 - \phi)E_s + E_f + E_m](1 + \beta) \tag{14.2}$$

Because $E_p \gg E_s$, total energy use is minimized by making ϕ and β as low as possible. It should be noted that product designers who specify virgin materials in their products may not be directly paying the high energy cost that results, but the virgin material specification forces the cost to be borne at some point within the industrial ecology system.

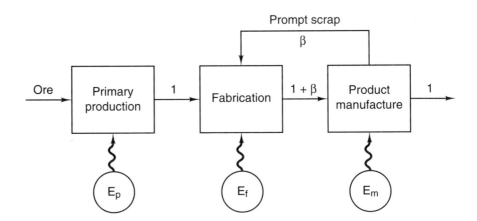

Figure 14.5   Schematic diagram of a metals processing system using only virgin material (after P. F. Chapman and F. Roberts, *Metal Resources and Energy*, Boston: Butterworths, 1983.)

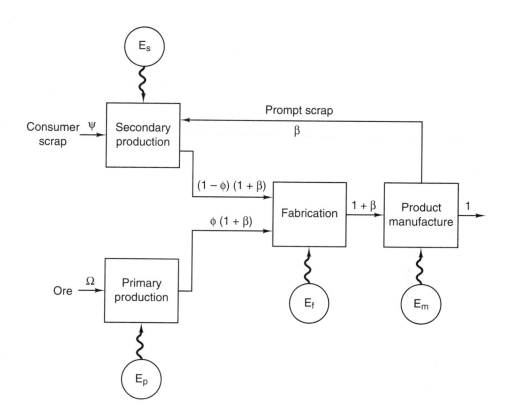

Figure 14.6   Schematic diagram of a metals processing system using both virgin material and consumer scrap. (After P. F. Chapman and F. Roberts, *Metal Resources and Energy*, Boston: Butterworths, 1983.)

14.5   GENERAL APPROACHES TO MINIMIZING ENERGY USE

Former U.S. Senator Everett Dirksen was once quoted on budget matters as saying, "A billion here, a billion there, and pretty soon it adds up to real money". The same is true of energy use. Energy conservation in all its facets is good management, responsible action, and progress toward increased corporate profitability, and every little bit helps. In the previous paragraphs, we have mentioned those aspects of energy conservation dealing with specific types of industries. In this section, we discuss some more general approaches to industry's use of energy that can be applied across the entire industrial sector.

14.5.1   Heating, Ventilating, Air Conditioning (HVAC)

The "lighter" the industry, the greater is the percentage of energy use that tends to be attributable to HVAC. This is not only because light industry is inherently less energy-hungry than heavy industry, but also because its manufacturing operations often involve precision control of the in-plant environment. Substantial energy savings may be available by improving the "shell efficiencies" of buildings, that is, weatherstripping, window treatments, proper planting of trees and shrubs, and the like. Detailed maintenance of HVAC equipment is an often overlooked action that can be very beneficial. Major gains are possible by replacing aging HVAC equipment with modern, computer-controlled varieties, which can use 30–90% less energy depending on the specific application.

14.5.2   Lighting

The provision of adequate lighting generally accounts for 20% or more of industrial energy use. The traditional use of incandescent lights, or of fluorescent lights without high-reflectance fixtures, electronic ballasts, and high-efficiency bulbs can readily be improved upon, often with payback times of 2 years or less. Figure 14.7 shows the energy efficiencies of various light sources, and demonstrates that the same amount of light can be provided at a tenth or fifteenth of the incandescent-energy consumption.

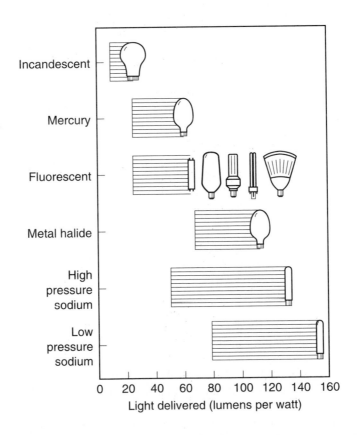

Figure 14.7   Ranges of energy efficiency of various light sources. (U.S. Dept. of Energy, *Energy Efficient Lighting*, DOE/CE-0162, Washington, DC, 1986.)

14.5.3   On-Site Energy Generation

Although decreased use of energy is a primary goal of energy assessment, an alternative beneficial activity is the use of energy that is present within an industrial facility but is not used. A common form for such energy is process heat that is not captured. If not utilized in heat exchangers, it can often become available to generate electric power on site. One way to do so is by using both the heat and power from a single thermodynamic cycle, a practice called *cogeneration*. There are many variations, one of which is illustrated in Fig. 14.8. In these integrated energy systems (IESs), the energy stream is a desired output just as is the product stream. Any excess energy that is generated can be sold back to the electric utility and become a small part of the integrated power grid, or perhaps be captured in chemical compounds for subsequent liberation and use.

In a number of industrial facilities, including steel mills, petrochemical complexes, and oil refineries, it is possible to use a process residual stream as a feedstock for power generation. Depending on the type of process and the availability and cost of commercial energy, designing and constructing an IES facility can be a sound way to utilize available resources for a variety of purposes.

Environmental impacts in IES facilities sometimes can be minimized by careful choice of energy feedstocks. Biomass fuel may be an option in some places, for example, hydropower in others. An unusual but commendable example of energy feedstock switching is that of Monsanto's Sauget, Illinois, facility, which produces energy by burn-

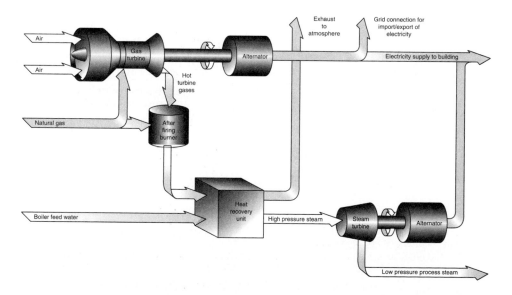

Figure 14.8   A schematic diagram of a cogeneration facility for the simultaneous production and use of heat and power. (Reproduced with permission from *Climate Change and Energy Efficiency in Industry*. Copyright 1992 by International Petroleum Industry Environment Conservation Association.)

ing a mixture of shredded scrap tires and coal. The mixture is cheaper and cleaner-burning than coal alone, and disposes of some one million scrap tires per year.

Sometimes energy feedstock switching among more usual supplies is warranted if the total environmental impact can be lowered. Choices thus can be made among fossil fuel options or other potential sources. A successful example of this principle is that of the AT&T manufacturing facility in Columbus, Ohio, which has contracted to buy methane from a nearby landfill site. This gas, a product of anaerobic biological degradation of landfill materials that would otherwise be vented to the atmosphere and enhance the potential for global warming, is used in place of fossil natural gas.

14.5.4 Energy Housekeeping

Good energy housekeeping involves taking the industrial situation as it exists and devising ways to modify or change it to make it more energy-efficient. Opportunities abound. For example, more efficiently designed computers or, in the interim, better control of their use of energy, can reduce consumption notably. It is important to realize that because all employees use energy in their jobs, they can make useful contributions to energy housekeeping in offices, laboratories, and production facilities.

Perhaps the most successful energy conservation program to date was initiated by the Louisiana Division of the Dow Chemical Company in 1982. Many of the improvements embodied techniques useful industrywide, such as heavy insulation on pipes carrying hot fluids, cleaning heat-exchanger surfaces often to improve heat-transfer efficiency, and employing point-of-use fluid heaters where storage or long pipelines create the potential for heat loss. The company's energy contest results are summarized in Table 14.3.

There are two central messages in the table. One is that all the good projects were not thought about in the first year. Rather, good ideas kept coming. The second point is that the return on investment is substantial and easily demonstratable. Over the 1982–1992 period, the savings to the corporation, computed after subtracting the relatively small costs involved in implementing some of the ideas, were some $170 million!

Table 14.3   Energy Contest Results: Lousiana Division, Dow Chemical

	1982	1984	1986	1988	1990	1992
Winning projects	27	38	60	94	115	109
Average return on investment (%)	173	208	106	182	122	305

Source: K. E. Nelson, Practical techniques for saving energy and reducing waste, in *Industrial Ecology and Global Change*, R. Socolow, C. Andrews, F. Berkhout, and V. Thomas, eds. Cambridge, UK: Cambridge University Press, 1994.

EPA's ENERGY STAR COMPUTERS PROGRAM

The ubiquitous presence of computer, printers, and other electronic office machines exacts a toll on energy consumption. It is estimated that computer systems currently account for about 5% of all the electricity that is used by the commercial business community, and the rate of use is increasing. Much of this total is the power needed by the machines themselves, but because each watt of power consumed also requires about 1.5 watts for the facility containing the equipment, there is an air conditioning load as well. A high percentage of electronic office machines draw power all through the work day, but may be in actual use only a small fraction of the time. To address this problem, the U.S. EPA began its "Energy Star Computers Program" in 1992 (Figure 14.9). The goal was to encourage the manufacture of desktop computers and accessories that would automatically switch to a low energy state when left idle. Such technology, already in operation on most laptop computers, was expected to ultimately yield a 50–75% reduction in power consumption by electronic office machines.

A personal computer and its monitor typical of today's technology draw about 150 watts of power. The Energy Star computers must enter a low-power state in which the computers and the monitors each draw 30 watts or less. Similar requirements apply to computer printers of various types and sizes. A manufacturer meeting these requirements is eligible to use the Energy Star logo in advertising its products.

Corporations representing more than two-thirds of all computer and printer sales in the United States have accepted the Energy Star partnership arrangement, and Energy Star products are becoming widely available as this book is written. Achieving Energy Star status is more than altruism: the U.S. government will only purchase desktop computer equipment meeting Energy Star specifications, provided availability and performance needs are met. Other customers may be expected to impose the same requirement once sufficient equipment is available.

EPA POLLUTION PREVENTER

Figure 14.9   Logo for the U.S. EPA's "Energy Star" program.

14.6 SUMMARY

Energy provides an example of a situation in which the process designer plays an equivalent or larger role than does the product designer. Sample checklists for both are included in Appendix A. As with most situations, collaborative efforts between the two are likely to produce the greatest energy savings while still promoting efficient and effective manufacturing. Perhaps more than most of the other topics discussed in this book, energy minimization is a particularly appropriate arena for incremental change as well as for complete process change. As we write this, the European Union is considering the imposition of a tax on carbon emissions and The Netherlands is considering imposing energy-use requirements on industrial products. How these topics are played out remains to be seen, but there seems little doubt that energy will become increasingly expensive as resources dwindle under the demands of a rapidly growing global population and as environmental concerns inspire more and more legislative activity. Reductions and pattern changes in energy use are thus extremely likely to be good investments in future corporate profitability as well as environmental responsibility.

SUGGESTED READING

Hoffman, J. S. Pollution prevention as a market-enhancing strategy: A storehouse of economical and environmental opportunities. *Proceedings of the National Academy of Science of the U.S. 89* (1992): 832–834.

Ross, M. Improving the efficiency of electricity use in manufacturing. *Science, 244* (1989): 311–317.

Special Issue, Energy for Planet Earth. *Scientific American, 263* (3) (September 1990).

Tester, J. W., D. O. Wood, and N. A. Ferrari, eds. *Energy and the Environment in the 21st Century*. Cambridge, MA: The MIT Press, 1991.

U.S. Congress, Office of Technology Assessment. *Changing by Degrees: Steps to Reduce Greenhouse Gases*, OTA-O-482. Washington, DC: U.S. Government Printing Office, 1991.

EXERCISES

14.1 Assume a materials processing system as shown in Fig. 14.5, with E_p = 31 GJ/t, E_f = 5 GJ/t, E_m = 5 GJ/t, and β = 0.1. Compute Φ.

14.2 To the system of the previous problem, add a secondary production component to reprocess consumer scrap with E_s = 9 GJ/t and ϕ = 0.7. Find ψ, Ω, and Φ.

14.3 In the system of Exercise 14.2, a fraction λ of the material entering the primary production process is irretrievably lost to slag. Reformulate Eq. (14.2) to take this loss into account. If λ = 0.2, compute ψ, Ω, and Φ.

14.4 An office building in your community has 50 offices, each with an average of four desks. Each desk has a desk lamp that can use either a 60-watt incandescent bulb or a 13-watt fluorescent unit. The average use of a lamp is 7 hours per day. How much power is required the building per year for each of the two options? Given your local energy cost, what is the annual cost of each of the two options? If the price of an incandescent bulb is 88 cents and that of a fluorescent unit is 12 dollars, how long will it take to justify the purchase of fluorescent units, assuming everything is newly purchased and no discount rate is assumed?

CHAPTER 15

Industrial Process Residues: Composition and Minimization

15.1 ON EMISSIONS AND EFFLUENTS

Industrial activities release substantial amounts of materials to the environment. One measure of those releases is the U.S. EPA's toxic release inventory. A portion of the information contained in the latest report from that program is presented in Table 15.1 for some of the larger industries. These emissions are direct consequences of the types of manufacturing processes being used, and are, at least in principle, under the control and potential improvement of the process designer. The metals and chemical industries dominate industrial solid-residue production. This dominance basically reflects the fact that the materials in those industries are generally encountered much closer to their extraction stage than is the case with other industrial groups.

A few things need to be said to put the tabular values into perspective. First, they are not presented as a function of the size or value of the respective industries, attributes that would affect the relative rankings if incorporated. Second, the implication of the table is that the weight of the emissions is the only relevant property, neglecting such factors as relative toxicity and the sensitivity and capacity of the media into which the emissions are discharged. Finally, data over a several year period clearly show that emissions rates in all categories are undergoing a steady decrease as new control technologies and new processes are implemented. Notwithstanding these caveats, however, it is clear from the table that plenty of process design opportunities exist for improvement in the industry–environment relationship.

Environmental legislation directed at minimizing the consequences of anthropogenic activity has traditionally treated the different environmental regimes as separate entities unrelated to each other. Indeed, one gets no indication of interaction of the regimes by scanning Table 15.1. In practice, however, the process engineer can often

Table 15.1   Toxic Chemicals Released by U.S. Industrial Groups in 1990 (kg)

Industry	Air Emissions	Surface Water Discharges	Underground Well Injection	Land Releases	Off-Site Transfers
Chemicals	316,000	60,600	300,000	44,700	114,000
Electrical	37,200	190	8.6	1,240	15,900
Furniture	27,300	2	0.03	3,350	1,960
Lumber	17,100	95	0.04	58	3,330
Machinery	24,000	95	0.26	63	6,160
Metals	160,000	5,700	9,100	143,000	165,000
Paper	111,000	17,100	0.03	3,350	8,370
Petroleum	32,400	2,260	17,200	1,410	4,200
Plastics	87,400	210	6.8	91	10,200
Printing	23,600	0.5	0.16	2	1,990
Textiles	15,400	252	0.02	17	1,420
Transportation	87,900	107	0.15	885	17,900

Source: U.S. Environmental Protection Agency, *1990 Toxics Release Inventory: Public Data Release*, Washington, DC: Office of Pollution Prevention and Toxics, 1992.

choose whether the residue from a process goes to air, water, or soil. A process using a solvent, for example, might do equally well with a very volatile material or one not so volatile. The former would be emitted into exhaust air streams, and the latter could be part of a facility's water effluent, provided in each case that the concentrations of the solvent molecule were within legal limits. For low-volatility solvents, it is often possible to collect them in process sludge (a term referring to slush or sediment). Sludge is almost universally landfilled, yet it is often more than half water by weight.

As the solvent example demonstrates, treating environmental regimes as separate entities from a planning or regulatory viewpoint is completely artificial in many cases. For convenience, we discuss air, water, and solid phases separately in this chapter. Nonetheless, we remind the reader that a goal of industrial ecology is to minimize total environmental impacts, not those to a selected regime. For any process residue, the goals are to reduce the amount and to substitute for any toxic materials, not to move the residue from a more highly regulated regime to a less highly regulated one. This chapter provides some ideas on how to accomplish those goals. Specific guidance is provided by the checklists in Appendix A.

15.2   GENERATION OF INDUSTRIAL SOLID RESIDUES

15.2.1   Process Residues

Many manufacturing processes produce solid residues of varying types and to varying degrees. In the case of dry processes, such as ball milling, the solid material that makes the process work has a certain useful lifetime and then must be replaced, often having become a residue of metal powder. Similarly, liquid process solutions eventually become exhausted and must be replaced, often having produced a sludge. Much of this residue generation is a function of process or machine design. If residue is generated, and especially if it is then discarded, it violates one of the principles of good industrial ecology: that to the maximum extent possible, every molecule entering a manufacturing facility should leave that facility as part of a saleable product. In many cases, there is a direct correlation between the concentration of a species in a residue stream and the ability and economic viability of extracting and recycling it. This limitation suggests that process residue streams, like products, should be designed to facilitate recovery and recycling.

15.2.2 Product Residues

Product residues are distinguished from process residues in that the former represents material that is intended to (and in the ideal industrial operation would) be incorporated into saleable products. An example of the generation of solid residues during manufacturing is provided by polymer-molding operations, which require an initial charge of material to prepare the molding machine for product flow; this material is later generally discarded. Another consequence of molding is that products inevitably are produced with "sprues" and "runners", small pieces of excess material. To the extent that the machine or mold designer can reduce or eliminate this residual material, the amount of solid residue generated is minimized. A further step is to design machines, materials, and products in such a way that this manufacturing residue can be reused, either within the facility or by transfer to another facility or another user. In practice, a large percentage of this "prompt scrap" is recycled within the facility where it is generated.

Similar principles apply to metals preparation processes such as casting, trimming, or polishing. In each case, scrap material is produced as a consequence of the preparation. If the metal has toxic properties, minimization of material and proper disposal of the detritus becomes a health and residue-management issue as well as a materials issue. Process redesign and materials substitution need to be considered, and recycling of all extraneous material, either inside or outside the facility, should be the goal of the process designer.

15.2.3 Packaging Residues

Packaging residues come in two varieties: packaging that enters a facility with the components and supplies that are purchased from others and the packaging that a facility puts on the products it makes. The latter topic is discussed in a later chapter. The former is addressed here, from the perspective that any packaging material entering a facility complicates its operations without producing in itself anything that translates into product.

The most important general guide to packaging is that the supplier who brings packaging into the manufacturing facility should take it out again. Enforcing this general principle will lead to the gradual achievement of another: the minimization of packaging. Recycling packaging to a supplier can take several forms. In the case of liquid products that arrive in bottles or drums, the containers can be returned after use, either flattened or, even better, filled with spent chemicals to be recycled by the supplier. For solid packaging, including boxes, wooden pallets, plastic wrapping, foam cushioning, metal strapping tape, and all the other mixture of materials that one commonly sees, it is imperative that manufacturers work with their suppliers to minimize and recycle this material, as it otherwise is a major contributor to landfills (and a major contributor to costs not directly involved in the manufacture of products.) In the case of pallets, substituting other materials for wood may be useful. Corrugated cardboard has been found suitable for one-time use in selected applications, and plastic pallets show good multiple-use properties.

15.2.4   Miscellaneous Solid Residues

Not all solid residues produced within a manufacturing facility are from processes, products, or packaging. An example of the miscellaneous residues that are generated is the dusts accumulated on and in the air filters that must be cleaned or replaced periodically. By its very nature, this material is chemically heterogeneous as well as difficult to recover. In such a circumstance, facility engineers should attempt to clean up the processes generating the material being collected, and then to use filter material and air exchange rates that optimize the capture of the material. For example, segregation of air streams can lead to more homogeneous, easier-to-recycle dusts. To the degree that reusable filter material can be utilized, such an approach is preferable to the use of material that must be discarded.

15.3   DEALING WITH INDUSTRIAL SOLID RESIDUES

15.3.1   Types of Industrial Solid Residues

Industry is the major generator of the solid waste discarded in landfills. Although municipal waste is a (deserved) topic of public concern, that waste in the United States is less than 2% of the amount of industrial waste regulated under the Resource Conservation and Recovery Act (RCRA). Figure 15.1 shows the sources of that waste (as defined by RCRA); some 70% of the weight of these wastes is estimated to be water contained in sludges and aqueous solutions. No matter how defined, the amount of material that is discarded by industry is enormous.

Industrial processes are quite individualistic, yet there is enough commonality to make some useful generalizations concerning solid-residue emissions. For the species cited in earlier discussions on environmental concerns, we list in Table 15.2 some of the common industrial sources responsible for their emission.

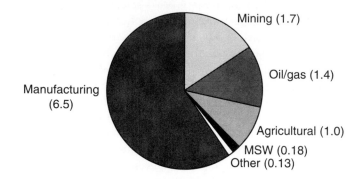

Figure 15.1   Sources of solid residues (in Pg) from industrial activities, as defined by the U.S. Resource Conservation and Recovery Act. Mining residues exclude mineral-processing residues, oil/gas residues exclude waters used for enhanced oil recovery, and the "other" category includes residues from utility coal combustion. MSW is municipal solid waste, shown for perspective. (U.S. Congress, Office of Technology Assessment. *Managing Solid Wastes from Manufacturing, Mining, Oil and Gas Production, and Utility Coal Combustion,* OTA-BP-O-82. Washington, DC: U.S. Government Printing Office, 1992.)

Table 15.2   Solid Residues Associated with Common Industrial Processes and Products

Source Activity	Solid Species							
	Trace Metals	Plastic Residues	Paper Residues	Biological Residues	Radioactive Residues	Sludge	Powders	Mixed Trash
Process residues								
Dry processes (general)	*						*	
Dry processes (radionuclide)					*	+		
Wet processes (general)	+					*		
Wet processes (biological)				*		*		
Product residues								
Casting sprues	•							
Mold sprues		•						
Rejects	+	+						+
HVAC Filters							*	+
Packaging								
Incoming		•	*					*
Outgoing		+	+					+

• = Major source; * = Modest source; + = Minor source.

15.3.2   Trace Metals

This group of compounds, some of which are also called heavy metals, includes antimony, arsenic, cadmium, chromium, lead, mercury, and selenium. Aluminum, copper, nickel, and silver are sometimes included as well because of their biological toxicity when in solution.

Several industrial groups dominate the emissions of trace metals to solid residues, different activities being important for different metals. A summary of current estimates is reproduced in Table 15.3, which shows that soils have deposited on them or in them metals from many different industrial sources. Three stand out as overwhelmingly important: the disposal of fly ash from coal combustion, residues from manufacturing processes involving metals, and the disposal of commercial products. Neither the first nor third of these sources is directly caused by or under the control of the manufacturing engineer. However, by its use of energy, the manufacturing process contributes in a major way to the consumption of coal and the production of fly ash. Similarly, discarded commercial products are a failure of industrial ecology, not in the manufacturing stage, but in the disposal stage. Thus, whereas the direct generation of trace metal residues in manufacturing is of concern because of toxicity problems and disposal costs, it is the indirect generation that plays the major role in the pollution of soils.

Table 15.3   Inputs of Trace Metals to Soil (Tg/yr)

Source category	As	Cd	Cr	Cu	Hg	Ni	Pb	Sb	Se
Agric. residues	0–6.0	0–3.0	4.5–90	3–38	0–1.5	6–45	1.5–27	0–9	0–7.5
Animal residues	1.2–4.4	0.2–1.2	10–60	14–80	0–0.2	3–36	3.2–20	0–0.8	0.4–1.4
Logging residues	0–3.3	0–2.2	2.2–18	3.3–52	0–2.2	2.2–23	6.6–8.2	0–5.5	0–3.3
Urban refuse	0.1–0.7	0.9–7.5	6.6–33	13–40	0–0.3	2.2–10	18–62	0.2–1.3	0–0.6
Metal manufacturing residues	11.7–20	4.3–7.3	0.7–2.4	660–1580	0.6–3.1	57–130	330–780	24–40	0.4–0.6
Coal fly ash	6.7–37	1.5–13	149–446	93–335	0.37–4.8	56–279	45–242	2.6–22	4.1–60
Disposal of commercial products	36–41	0.78–1.6	305–610	395–790	0.55–0.82	6.5–32	195–390	0.8–4.0	0.1–0.2

Source: Abstracted from J. O. Nriagu and J. M. Pacyna, Quantitative assessment of worldwide contamination of air, water and soils by trace metals, *Nature, 333* (1988): 134–149.

15.3.3 Plastic Residues

Thermoplastic residues are often relatively easy to reuse within the facility in which they are generated, as they can be added to incoming streams of virgin polymer material. Thermoset material is much less easy to reuse, and processes incorporating thermosets should therefore strive to minimize residue generation and, if possible, to substitute thermoplastics. The recycling of these plastic materials is treated in more detail in Chap. 19. If the residue is so degraded as to be unsuitable for recycling, it should be incinerated to recover part of its energy content.

15.3.4 Paper Residues

Paper and paperboard residue is often easy to recycle, though usually not within the facility in which it is generated. To the maximum extent possible, paper and paperboard use should be minimized, and any that is used should be of recycled rather than virgin stock. Any packaging that enters the facility with components or materials should be removed by the supplier. Remaining paper and paperboard residue should be sorted as appropriate and recycled.

15.3.5 Biological Residues

Processes that utilize biological organisms to make products (vaccine manufacturing is an example) generate biological residue as well. Traditionally, this concern has resided almost exclusively in the pharmaceutical industry, but the advent of genetic engineering and biomaterials science suggests that the potential for biological residue generation will become more widespread in the future. As with other residue, biological residue is minimized by striving to make manufacturing processes as efficient as possible. No matter what the efficiency, it is important that any biological residue generated be rendered non-infectious prior to disposal, and that the disposal be managed carefully. In this evolving area, process designs that produce by-product biological material that can serve as input to another process will be highly preferable to those involving disposal only.

15.3.6 Radioactive Residues

Small amounts of radionuclides are used industrially to manufacture such devices as medical equipment and smoke detectors. The manufacture of products containing such material inevitably produces radioactive residues. Because the restrictions on manufacture, use, and disposal of products containing significant amounts of radionuclides are so onerous, the use of radionuclides should be avoided unless their specific characteristics make such use necessary. If used, the quantity of materials should be minimized as much as possible and disposal requirements followed rigorously.

15.3.7   Sludge

Many industrial processes produce sludge that is of no use to the firm but should not be consigned to disposal. Minimizing sludge generation is within the province of process designers and of the process engineers who are operating the facility. After such minimization as is possible, efforts then should be made to find a use for the sludge, either internally or externally. Sometimes this can require some additional processing, but the environmental and economic benefits, including avoidance of disposal costs and potential liability, can well justify this extra effort.

15.3.8   Mixed Residues

The principal rule for dealing with mixed residues is to try not to generate any, because segregation of individual materials groups from mixed residues is labor-intensive and expensive. Thus, to the extent that solid residues can be kept subdivided by material, they are easier to recycle within the facility or easier to sell to another user outside the facility. If mixed residues contain even small amounts of materials deemed toxic, special disposal precautions may be needed for the entire residue stream. Hence, should mixed residues be generated unavoidably, efforts should be made to sort them so that reuse can be optimized. The disposal of mixed residues is a clear signal of failure in industrial ecology practice.

15.4   GENERATION OF INDUSTRIAL LIQUID RESIDUES

A major source of water-quality problems is the acidified water that is a consequence of mining operations. Acid mine drainage arises from the metallic sulfides generally present in mineral-rich or coal-rich deposits. Once exposed to the air and moisture, the sulfides are oxidized to sulfates and sulfuric acid, which are transported to streams and rivers by flowing artesian springs and by rainfall. Mine drainage waters are often as acidic as pH 2–3.

A second industry whose activities have substantial potential to affect water quality is agriculture, because the land areas involved are large and the control of runoff of soil or chemicals into streams and rivers is generally difficult. A primary concern is sediment transported to surface waters through runoff and erosion; sediment is surely the most abundant water pollutant on a weight basis. Another concern is animal residues from feedlots and poultry facilities. Still another is phosphorus, nitrogen, trace metals, and organics from land treated with fertilizers, pesticides, and herbicides. Finally, agriculture is the biggest single user of water itself, mostly for irrigation. In some areas, agricultural withdrawal threatens the long-term availability of adequate local and regional water supplies.

A third major source of effluents to surface and ground waters is industrial materials processing and manufacturing. Table 15.1 lists chemicals, metals, paper, and petroleum as the industries generating the largest quantities of discharges of liquids. At least in some geographical regions, however, perhaps the most important industrial activity involving water is its use in physical and chemical processes and in cooling. Nearly all the water is eventually returned to the reservoirs from which it was taken, but it has often been been heated or adulterated in the process. The incorporated effluents can include suspended solids, nutrients, acids, and heavy metals. Nearly all of these constituents receive some on-site treatment prior to their discharge into public water supplies, at least in more developed countries. Especially where toxic chemicals are involved, on-site treatment is both extensive and expensive.

15.5 DEALING WITH INDUSTRIAL LIQUID RESIDUES

15.5.1 Introduction

Industrial processes generate many different types of liquid emissions. For the species cited in earlier discussions on environmental concerns, we list in Table 15.4 some of the common industrial activites responsible for their emission into liquid-residue streams.

For each of the species of Table 15.4, it is helpful to discuss their principal uses in industrial processes, the relative importance of the industrial emissions to the particular environmental concern, and typical remedial measures that can be used to minimize or eliminate the impact. From the standpoint of the designer, these topics are of concern only if his or her design involves the potential use of these species.

Table 15.4 Liquid Species Emitted by Common Industrial Processes and Products

Process	Liquid Species					
	Trace Metals	Nutrients	Solvents, Oils	Organics	Acids	Suspended Solids
Agriculture		•		•		•
Chemical manufacturing	+		*	*		+
Electronics			*	+		
Electroplating	•		*			+
Fertilizer		*				+
Food production		+				
Leathermaking	+			*		
Metal cleaning			*	*		*
Mining, smelting	*				*	•
Pesticides, herbicides				*		

* = Modest influence on local, regional, or global scale; + = minor influence on local, regional, or global scale; • = major influence on local, regional, or global scale.

15.5.2 Trace-Metal Emissions

Several industrial activities (excluding agriculture) are involved in the emission of trace metals to aquatic ecosystems, as shown in Table 15.5. Among the common sources of trace metals to aquatic ecosystems that should be called to the attention of readers are industrial electroplating, tanning of leather, the use of heavy metals in inks (now declining significantly as less toxic alternatives are introduced), and the use of batteries containing high concentrations of heavy metals.

It is of interest to examine in the table the processes that are the primary sources of the trace metals emissions to aquatic ecosystems. Smelting is the principal source for arsenic, nickel, and selenium. For cadmium, both metals processing and chemical manufacture play major roles. Metal manufacture is dominant for emissions of antimony, chromium, copper, lead, and mercury. Sewage sludge is also important in the lead cycle; this reflects the runoff into integrated sewer systems of lead aerosols from gasoline as well as contributions from industrial sources.

A related issue dealing with the industrial use of metals is the associated use of cyanide solutions. The cyanide ion has been widely employed because its strong chelating properties promote the cleaning and stripping of surface adulterants from metals as well as holding high concentrations of metal ions in electroplating solutions. Nonetheless, cyanide's toxic properties have made it a material to be avoided if alternatives are available. For many applications, including metal stripping and gold plating, suitable commercial noncyanide products can readily be obtained.

15.5.3 Nutrients

Most of the problems of nutrient addition to natural waters relate to commercial or agricultural fertilizers or other products undergoing dissipation in use. The manufacture or industrial use of these materials has the potential for local impacts, however. Minimiz-

Table 15.5   Industrial Inputs of Trace Metals to Aquatic Ecosystems (Tg/yr)

Source category	As	Cd	Cr	Cu	Hg	Ni	Pb	Sb	Se
Metal extraction	0–0.75	0–0.3	0–0.7	0.1–9	0–0.15	0.01–0.5	0.25–2.5	0.04–0.35	0.25–1.0
Smelting, refining	1.0–13	0.01–3.6	3–20	2.4–17	0–0.04	2.0–24	2.4–8.8	0.08–7.2	3.0–20
Manufacturing:									
Metals	0.25–1.5	0.5–1.8	15–58	10–38	0–0.75	0.2–7.5	2.5–22	2.8–15	0–5.0
Chemicals	0.6–7.0	0.1–2.5	2.5–24	1.0–18	0.02–1.5	1.0–6.0	0.4–3.0	0.1–0.4	0.02–2.5
Pulp and paper	0.36–4.2		0.01–1.5	0.03–0.39		0–0.12	0.01–0.9	0–0.27	0.01–0.9
Petroleum prods	0–0.06		0–0.21	0–0.06	0–0.02	0–0.06	0–0.12	0–0.03	0–0.09
Sewage sludge	0.4–6.7	0.08–1.3	5.8–32	2.9–22	0.01–0.31	1.3–20	2.9–16	0.18–2.9	0.26–3.8

Source: Abstracted from J. O. Nriagu and J. M. Pacyna, Quantitative assessment of worldwide contamination of air, water and soils by trace metals, *Nature, 333* (1988): 134–149.

ing discharges and recovering materials from process flow streams is generally the most sensible approach.

The removal of phosphates from detergents and thus the disappearance of a major source of nutrients to natural waters is one of the major success stories of industrial ecology. This action, originally undertaken by the Henkel Corporation of Germany, required more than a decade of research and development by the world's detergent manufacturers. Today, detergents are almost universally biodegradable and free from phosphates, and are recognized by consumers as being both efficacious and environmentally less harmful.

15.5.4   Solvents and Oils

A great many industrial processes are dependent upon the use of organic solvents. These solvents can produce toxicity, ozone depletion, or smog formation, however, and it is thus important to assess their use.

Stripping solutions are widely used in industry to remove scale, rust, and other surface contaminants from metals. These solutions are chemically aggresive, and tend to cause difficult disposal problems when exhausted or contaminated. David Benshoof of Best Lock Corporation has shown that attention to minimizing stripping processes can result in significant decreases in the volume of stripping solution used. In that case, metal parts were being cleaned prior to being lacquered or painted, and parts that had been rejected for quality reasons needed to have the coatings removed before being reworked. The stripping solution used was methylene chloride (CH_2Cl_2), a toxic priority pollutant listed under the U.S. Clean Water Act. The operation was modified in the following ways:

- Clean storage areas were provided for parts awaiting surface coating, thus minimizing the cleaning needed.
- Flow analysis of the stripping process permitted substantial reductions in solution degradation and need for subsequent recharge.
- The surface-coating formulations were modified so that less stripper solution was needed to remove them.

The result of these efforts over a period of several years is shown in Fig. 15.2. The amount of methylene chloride use decreased by 94% during that time, thus decreasing the quantity purchased and (more important) the quantity of hazardous residues sent for reprocessing. The only cost incurred was to add clean storage areas for the parts awaiting surface coating. The payback period was approximately 2 months.

SOLVENT SUBSTITUTION

It is often possible to substitute one solvent for another and to reduce total solvent use even if it is not possible to eliminate organic solvent use altogether. The Polaroid Cor-

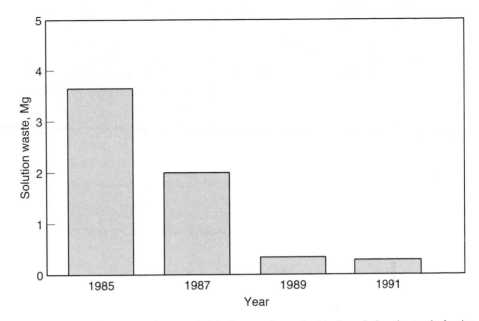

Figure 15.2   Reductions over a 6-year period in the use of organic stripping solutions in metal cleaning. (Courtesy of D. H. Benshoof, Best Lock Corporation, Indianapolis, IN.)

poration has provided an example of such a substitution in one of its dye processes. Rather than a relatively standard process using toluene as the solvent and producing a residue that required incineration, Polaroid scientists developed a process using trimethyl orthoacetate, which could be recovered by distillation. The result, as shown in Table 15.6, was a substantial reduction in residue generation and a substantial cost saving as a consequence.

An important characteristic of solvent substitution to recognize is that a substitute can be a less effective solvent than the material it replaces, and one may need to

Table 15.6   Characteristics of Proposed Polaroid Dye Processes

Measurement Parameter	Toluene Process	TMOA Process
Toluene use per unit of production (kg)	43.1	0
TMOA use per unit of production (kg)	0	43.1
On-site recycle and reuse of TMOA (%)	0	87
Off-site incineration of TMOA (%)	0	13
Off-site incineration of toluene (%)	100	0

balance the concerns for increased solvent use against the advantages of substitution, at least in the short term. In many cases, for example, switching from organic to aqueous solvent systems will require use of additional energy in the cleaning process.

A graphic illustration of this situation is shown in Fig. 15.3, which shows the usage of different solvents in a pharmaceutical production process. The impetus for substitution was that the process as designed used mostly methylene chloride, a suspect carcinogen, as the solvent. Much of that compound was eliminated from the process by the substitution of isopropyl acetate, but the poorer solvent capabilities of the latter compound increased the total solvent use. During the following year, further effort on substitution resulted in complete removal of both methylene chloride and isopropyl acetate, as a variety of other specialty solvents were tested and approved. The final result is a process that requires only about a third of the amount of solvent used 3 years previously, and the solvent constituents have no significant negative toxicological properties.

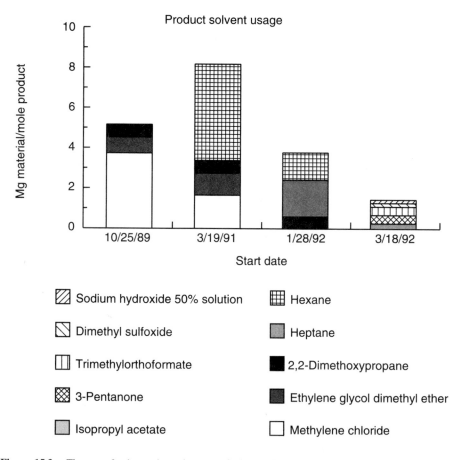

Figure 15.3   The use of solvents in a pharmaceutical manufacturing process over a period of two and one-half years. (Courtesy of T. McCarthy, Eli Lilly and Company, Lafayette, IN.)

Finally, it may be possible in many cases to eliminate the use of solvents completely, as we discuss later in this chapter for the case of the phaseout of CFCs.

15.5.5   Organics

Organics are the heart of many industrial processes, and their diversity and utility is one of the great triumphs of twentieth-century industry. The distinction that we make between "organics" and between "solvent and oils", also largely organic in nature, is that the former are intended to end up as product constituents, whereas the latter are process constituents. The loss of organics during manufacturing processes is thus a direct violation of the principle that molecules entering a manufacturing facilty should leave as components of saleable products. Residual organics will become smog-forming airborne constituents if their vapor pressures are high enough; otherwise, they will leave as adulterants in liquid- or solid-residue streams.

Most organics used in industry originate from petroleum feedstocks. We have discussed in an earlier chapter the limited supply of petroleum available to society over the medium to longer term. With this perspective in mind, minimization of use of virgin organic materials and the substitution of material produced from biomass where feasible will extend the availability of the irreplaceable petroleum resource, and should automatically be considered good design practice.

15.5.6   Acids

Acids are widely used as processing chemicals in industry and are, of course, products for some chemical manufacturers. The strong mineral acids are the most frequently employed, and the principal goal for the process designer is to minimize their use and to match them with residue streams of base, because release to municipal treatment systems or to natural waters must be at or near pH 7.

Less easy to deal with than acids used in controlled industrial environments are those generated as a consequence of resource extraction. The best approach for acid mine drainage (but one that is often very difficult given local conditions) is to prevent air and water from reaching the metallic sulfides in the first place. Alternatively, the approach should be to keep the immediate drainage flow away from natural bodies of water and to build retention basins and acid-neutralization facilities as needed.

15.5.7   Suspended Solids

An obvious consequence of suspended solids in water is a decrease in water clarity. Aesthetically displeasing, turbid waters are poor habitats for fish and other aquatic life. More subtly, the suspended particles absorb a variety of chemical substances, thus affecting chemical species lifetimes as well as providing surfaces for bacterial growth.

Sediment inputs can be minimized in various ways. Agriculturally, soil-stabilization practices are probably the most efficient. In forestry, retaining buffer zones of uncut trees near water bodies and streams is useful. Industrially, process design changes and filtration can be used. Once sediment is incorporated into flowing water, a variety of sedimentation, coagulation, and precipitation methods can be used. These methods tend to be particularly effective for particles large enough to have significant gravitational settling velocities.

15.6 DEALING WITH INDUSTRIAL GASEOUS RESIDUES

Emissions of industrial residues to the air have traditionally exceeded those to other environmental media. In 1990, for example, Table 15.1 shows that air emissions were approximately equal to the combined total of discharges to surface water, underground wells, the land surface, and landfills. Most of these emissions occurred as a consequence of one of three activities: the combustion of fossil fuels as on-site sources of energy, the use and evaporation of solvents, and the generation of volatile compounds by a chemical process, followed by venting or fugitive emissions.

The diversity of industries responsible for air emissions is substantial. Historically, the most prolific emitters have been those industries producing chemicals, metal products, plastics, and paper, as well as the transportation industry. Nearly all industries have significant air emissions, however, and great opportunities exist for industrial ecologists to decrease these emissions across virtually the entire industrial spectrum.

As with processes generating liquid effluent, there is enough commonality within industries to make some useful generalizations concerning gaseous emissions. For gaseous species (including trace metals in airborne particles) mentioned in earlier chapters, we list in Table 15.7 some of the industrial processes that commonly emit them into the atmosphere.

Table 15.7   Gaseous Species Emitted by Common Industrial Processes

Process	Gaseous Species									
	CFC, HCFC	Halons	CO_2	CH_4	N_2O	VOC	NO_x	SO_2	Trace Metals	Odorants
Cement manufacturing			+				+	+		
Chemical manufacturing					+	*		+		*
Fire prevention		•								
Foaming agent	+		+			+				
Food production										+
Leathermaking										*
Metal finishing						+			+	
Metal cleaning	•					+				
Petroleum refining				+		*	+	*		
Propellant	*		+			+				
Pulp & paper								+		•
Refrigeration	•									
Smelting, refining			*					•	•	

* = Modest impact on local, regional, or global scale; + = minor impact on local, regional, or global scale; • = major impact on local, regional, or global scale.

15.6.1   Chlorofluorocarbons and Hydrochlorofluorocarbons

This group of compounds includes the widely used chlorofluorocarbons CFC-11, CFC-12, and CFC-113, and a number of other fully or partially halogenated hydrocarbons. Their principal uses are for refrigeration, air conditioning, and metal cleaning, although in some parts of the world, CFCs continue to be used as propellants, foaming agents, and for other minor purposes.

Are there alternatives to CFC use? Partially hydrogenated CFCs (called HCFCs) have been suggested as CFC substitutes because they are susceptible to partial chemical removal prior to reaching the stratosphere, but their impacts are still thought significant enough to merit concern. As a result of ozone depletion, international agreements exist to phase out the manufacture of CFCs, HCFCs, and halons. The phaseout timetables have been revised often in recent years; those in effect as of the 1992 Copenhagen amendments to the Montreal Protocol are given in Table 15.8.

In the case of refrigeration and air conditioning, much embedded capacity is in place and "drop-in" alternatives are limited. For the short term, it is appropriate to opti-

Table 15.8   Phaseout Timetables for Halogenated Organics*

Chemical Species	Agreed Action
CFCs[†]	75% reduction of 1986 levels by 1994
	100% phaseout by 1996
Halons[‡]	100% phaseout by 1994
Other fully-halogenated CFCs	75% reduction of 1989 levels by 1994
	100% phaseout by 1996
Carbon tetrachloride	85% reduction of 1989 levels by 1994
	100% phaseout by 1996
Methyl chloroform	50% reduction of 1989 levels by 1994
	100% phaseout by 1996
HCFCs[§]	Freeze in 1996 based on 3.1% of CFC consumption in 1989 and HCFC consumption in 1989
	35% reduction of 1996 level by 2004
	65% reduction of 1996 level by 2010
	90% reduction of 1996 level by 2015
	99.5% reduction of 1996 level by 2020
	100% reduction of 1996 level by 2030
HBFCs[‖]	100% phaseout by 1996

* Production of compounds is prohibited on January 1 of the year indicated.

† Organic molecules in which all hydrogen atoms are replaced with chlorine or fluorine atoms.

‡ Organic molecules in which all hydrogen atoms are replace by halogen (F, Cl, Br) atoms, including at least one bromine atom.

§ Organic molecules in which some but not all hydrogen atoms are replaced with chlorine or fluorine atoms.

‖ Organic molecules in which all hydrogen atoms are replaced with bromine or fluorine atoms.

mize maintenance on refrigeration units to minimize the escape of refrigerants. Consideration should be given to drop-in replacement of CFCs with HCFCs, which can be done in some cases with minor mechanical modifications. This is clearly an interim solution, but has significant benefits for ozone depletion because it avoids the manufacture of additional halogenated molecules solely for "topping-up" purposes. In the longer term, alternative nonhalogenated refrigerants should be considered as they are developed. Some special applications now use liquid nitrogen or brine solutions. No matter what technique is utilized, new or modified facilities should not be designed with air conditioning overcapacity, a common practice in the past. Conservative facility design will return benefits in mimimizing both ozone depletion and energy use.

The use of CFCs and HCFCs for metal degreasing and defluxing can be minimized or eliminated by process redesign, though often with significant difficulty. The alternatives can involve aqueous or semiaqueous cleaning processes, which themselves have potential air and water emissions and frequently use more energy per unit of product. Notwithstanding the difficulties, it is clearly possible to redesign to avoid the use of halogenated hydrocarbons, and such redesign should become mandatory among process engineers.

15.6.2   Halons

This group of compounds includes the common fire extinguishing compounds $CBrFCl_2$ ("fluorocarbon 1211") and $CBrF_3$ ("fluorocarbon 1301"), as well as CH_3Br (sometimes used as a fumigant) and a variety of other bromine-substituted hydrocarbons.

Halons are used almost entirely as fire-prevention fluids, for which they are well suited. The major impact of halons on ozone depletion, however, together with increasing restrictions on their manufacture, dictate redesign of practices and equipment to minimize or eliminate their use. For example, the once-common practice of regularly testing fire-control equipment by releasing contained halons is no longer advisable. In the longer term, no buildings or equipment should be designed to use halons, and facilities that now exist should be scrupulously maintained so that leakage is minimized.

15.6.3   Carbon Dioxide

Industrial processes are major users of energy, and energy generation is the dominant source of CO_2. Other industrially related emissions of CO_2, such as its use as a propellant or foaming agent, may not represent ideal process designs from an industrial ecology standpoint but have negligible impact on the atmospheric CO_2 budget. On the other hand, the use of CO_2 pellet cleaning or supercritical CO_2 cleaning systems, which can replace chlorinated-hydrocarbon solvent-cleaning systems in many situations, would appear on balance to be environmentally preferable.

15.6.4   Methane

Methane is produced by a variety of anaerobic reduction processes, both in nature and in anthropogenic activities, and is the principal constituent of natural gas. There are many sources of methane. Most of the anthropogenic sources are related to the practice of agriculture, but others include emissions from coal mining, petroleum refining, and natural-gas distribution. Methane emissions from industrial processes other than those mentioned before appear to be small and unimportant.

15.6.5   Nitrous Oxide

Nitrous oxide is an occasional emittant from industrial processes involving nitriles or other nitrogen-containing feedstock materials, as well as from bacterial action on agricultural fertilizers. Once N_2O industrial sources are detected, they have generally proven susceptible to control (as was recently accomplished by DuPont for the manufacture of nylon). The worldwide sources of N_2O are poorly understood and appear to comprise a number of processes, none of which is dominant. Because N_2O is important as an ozone depleter and a greenhouse gas, the assessment of potential industrial sources of N_2O is important; if sources are found, process redesign is indicated.

15.6.6   Volatile Organic Carbon Compounds

Volatile organic carbon (VOC) compounds are extremely common emittants from industrial processes. The sensible approach appears often to be to minimize emissions of VOCs by designing processes to contain and reuse them wherever possible. The use of low-reactivity VOCs will minimize photochemical smog generation. In some industrial applications, it is possible to make product or process substitutions for VOC-emitting processes, and the relative impacts on air and water emissions should be carefully considered. An example is the paint industry, where the volatile contents of solvent-borne coatings have been decreased over recent years from above 80% to below 20%, and where thermal spraying of solids can often eliminate vaporizable solvent altogether. In many situations, however, industrial processes are not possible without VOCs being involved as feedstocks, solvents, or intermediates, so the designer's goals should be to minimize the quantities used, minimize the emissions from the processes, and attempt to select the species so that the environmental impacts are as benign as possible.

15.6.7   Nitric Oxide and Nitrogen Dioxide

The use of NO_x (the sum of NO and NO_2) in industrial processes is uncommon, but its generation can occur in the production of structurally related chemicals, in microbial processes, and in processes where high temperature combustion in air is involved. The uses of NO_x in industry are sufficiently diverse and of small enough scale that it is difficult to give specific advice about process redesign. Generally speaking, processes involving the emission of NO_x should be examined with the intent of minimizing or eliminating the emissions, but those actions should probably not take precedence over the elimination or minimization of the other species discussed in this chapter.

15.6.8 Sulfur Dioxide

SO_2 does not see extensive use as a feedstock material in industrial processes, but its generation and emission, together or in place of other sulfur compounds such as hydrogen sulfide and methyl mercaptan, is fairly common in industry. Among the process that generate sulfur gases are the manufacture of sulfur-containing polymers and other sulfur-containing chemicals, pulp and paper manufacture, and most processes involving the combustion of fossil fuels.

Because of the diversity of sulfur uses in industry it is difficult to give general advice concerning minimizing or eliminating emissions, other than to point out that sulfur dioxide is soluble in water and that technology for scrubbing SO_2 from stack gases is well-developed. In addition to treating exhaust flows, it may be possible to minimize or eliminate emissions by changing the characteristics of any combusted fuel or by changing process solutions or process techniques.

15.6.9 Trace Metal Emissions

Several industrial activities dominate the emissions of these species, although different activities are important for different metals. A convenient summary of current estimates is reproduced in Table 15.9. It is of interest to examine the processes that are the primary sources of the trace metal emissions to the atmosphere. Although vast amounts of material are moved during the extraction of ore from the ground, the atmospheric impacts of extraction are small. In fact, by far the dominant source of all metals from an atmospheric standpoint occurs during smelting and refining. Cement manufacture, a process involving the fragmentation of large amounts of material together with inadvertent windblown injection of portions of that material to the atmosphere, is potentially

Table 15.9   Industrial Inputs of Trace Metals to the Atmosphere (Tg/yr)

Source Category	As	Cd	Cr	Cu	Hg	Ni	Pb	Sb	Se
Metal extraction	0.04–0.08	0.001–0.003		0.16–0.80		0.80	1.7–3.4	0.02–0.18	0.02–0.18
Metals production:									
Lead	0.8–1.6	0.04–0.20		0.23–0.31	0.001–0.002	0.33	12–31	0.2–0.4	0.2–0.4
Copper-Nickel	8.5–13	1.7–3.4		14–31	0.04–0.21	7.7	11–22	0.4–1.7	0.4–1.3
Zinc-Cadmium	0.3–0.7	0.9–4.6		0.2–0.7			5.5–12	0.05–0.09	0.09–0.23
Steel and iron	0.4–2.5	0.03–0.28	2.8–28	0.14–2.8		0.04–7.1	1.1–14	0.004–0.007	0.001–0.002
Cement Production	0.2–0.9	0.009–0.53	0.9–1.8			0.09–0.9	0.02–14		

Source: Abstracted from J. O. Nriagu and J. M. Pacyna, Quantitative assessment of worldwide contamination of air, water and soils by trace metals, *Nature, 333* (1988): 134–149.

significant but not dominant for lead, and unimportant for the other metals. Coal combustion releases a number of heavy metals, including the hard-to-control mercury.

15.6.10   Odorants

Odorants are among the most instantly recognizable and least desirable of industrial emittants. Among those with the greatest olfactory impact are amines (the odor of decaying fish), sulfides (the odor of rotten eggs), and mercaptans (the odor of skunk spray). These odorants inspire vitriolic reactions from employees, citizens, and regulators because our noses have been trained by millions of years of evolution to react against them, and we do so instinctively. Thus, food-processing operations, tanneries, and the like must be alert to the need for minimizing processing times and collecting and neutralizing odorant gases. Improperly controlled composting operations can also generate significant odor problems.

   Pulp and paper mills also can be major odorant sources, because pulping involves the high-temperature digestion of wood fibers in a sulfate solution, a process that generates organic sulfides. Major efforts to minimize these emissions and develop alternative processes have been and continue to be made.

   A variety of esters, aldehydes, and other organic molecules used in or produced by industrial reactions also have the potential for release as odorants. It is often possible to substitute for an odorant a related molecule with more benign odor qualities; such substitutions generally must be studied on a case-by-case basis.

15.7   AUDITING AND MINIMIZING PROCESS RESIDUES

The first step to take in assessing process residue streams is to determine whether they contain any toxic materials (see Chap. 16). If so, it may be possible to redesign the process to substitute more benign constituents. For example, gold sulfides are replacing cyanides in some metal-plating baths and vegetable-derived compounds are replacing chromium in the tanning of leather.

   The next step in dealing with process effluent is to minimize it. Because treatment of effluent is so difficult and costly, minimization efforts have a substantial payoff. They are most effectively begun by performing detailed residue audits. Auditing residues begins with identifying whether a product or process uses any problematic constituents. Materials balances for each of those constituents need to be constructed to assess their primary pathways through the industrial system. Possible reductions in use or materials substitutions are then identified and implemented as appropriate.

   Residue streams in product manufacturing facilities often originate from cleaning operations. By reducing or eliminating cleaning steps, the resultant volume of residue is reduced. Considerable attention thus can be given to the sequencing of process steps, which can often result in one or more cleaning steps being omitted. Sequencing can also sometimes eliminate duplicate cleaning steps, as in cleaning a component prior to its storage while awaiting eventual use and again upon retrieving the component as needed for assembly.

Regenerating chemical solutions is another technique that can be used to minimize the liquid residues. Doing so can involve filtration, changes to relax purity requirements, the addition of stabilizers, redesign of process equipment, and so on. An example of the improvement that can be accomplished is shown in Fig. 15.4. A peroxide bath, initially used once and discarded, was gradually extended over a period of years until at present it is replaced only every week. The reduction in cost and decrease in liquid residues are obvious. A less obvious advantage is that a more concentrated residue stream often can be more easily and more economically recycled.

A third lifetime-extension technique for solutions is that of recuperative rinsing. This is most often used where parts are sprayed or immersed to clean them after electroplating or surface finishing. Subsequently, the rinsewater containing process chemicals can be returned to the process tank to replace fluid lost during evaporation, rather than being discarded. Recuperative rinsing can be successful where a high degree of solution monitoring and control is used to maintain quality.

After residues are minimized by preventing leaks and extending solution lifetimes, questions of reuse need to be addressed. Holding tanks are sometimes appropriate, as in the Ciba-Geigy facility in Indonesia that eliminated the discharge of dyes to local streams by storing different rinsewaters on the site and reusing them as process changeover to the appropriate dye occurred. When considering reuse, it is important that plant designers avoid the common practice of mixing residue streams. Once streams are mixed, it is often impossible to consider reuse, whereas streams that are kept separate may prove to be usable as feedstocks for other processes in the facility or may be marketable outside the corporation. For example, electronics industry solvents eventually

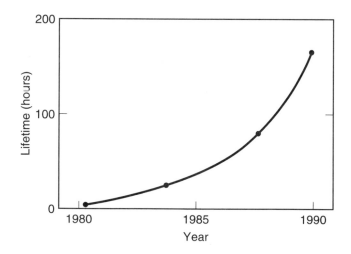

Figure 15.4   The extension in lifetime of a peroxide solution used in the manufacturing of electronic components. (Courtesy of E. Eckroth, AT&T Microelectronics, Allentown, PA.)

become sufficiently adulterated so that electronics manufacturing use is no longer possible, but they are still perfectly satisfactory as paint solvents and are sold for that purpose.

Finally, the far-sighted industrial ecologist will begin designing processes so that the residue streams become by-product streams, in other words, so that all outputs from a process are designed for optimal value from the beginning. At this stage of thinking, the manufacturer recognizes itself as not just a product producer, but a materials producer as well.

The materials balance exercise will identify a number of residue streams suitable for reduction efforts. It is useful when addressing these possibilities to have a scheme for prioritizing action. We suggest the following:

- Eliminate residue streams that contain substances under phaseout regulatory restrictions, such as CFCs, halons, and PCBs.
- Minimize the use of deionized water, the generation of which consumes substantial amounts of power and the use of which generates an additional residue stream to worry about.
- Avoid the introduction or substitution of new residue streams requiring new discharge permits, modifications to existing permits, or off-site disposal.
- Eliminate or minimize residue streams containing toxic substances, especially those found on various regulatory lists. (A nice example of such action is Monsanto's recent substitution of carbon dioxide for phosgene in its manufacture of polyurethanes.)
- Redesign factory layouts and product routings during manufacture to eliminate all unnecessary cleaning steps.
- Replace processes using organic solvents with processes using water-based solvents.
- Substitute less volatile chemicals for more volatile chemicals in industrial processes. A prime example of such a process is the rapid development of low-volatility paints.
- Change processes to eliminate the use of volatile solvents altogether. Solvent-free inks and coatings are now available for many applications, for example.
- Attempt to reduce the temperatures of manufacturing processes. Besides reducing the energy requirements, this action is likely to diminish emissions because the volatility of chemical constituents will be lower at lower temperatures.
- Avoid using heavy-metal catalysts by substituting processes that achieve the same products by reactions among environmentally benign chemicals.

15.8 TREATMENT OF INDUSTRIAL LIQUID RESIDUES

It is beyond the scope of this book to offer a primer on liquid-residue treatment, but it is appropriate to make a few comments relevant to industrial ecology on the topic. The first has to do with the use and disposal of organic compounds, especially solvents.

These materials are often products of industrial chemistry rather than being naturally produced species. As a consequence, no efficient natural degradation processes may exist. The materials must usually be destroyed (by incineration, where their energy content is recovered) or discarded in an acceptable manner. The latter alternative has often been accomplished by injection into deep wells, a practice that may be under increasing legislative control in the future.

   If process residues are liquid, they often undergo on-site treatment—settling, neutralization, and so forth. In fact, process changes made to avoid using organic solvents or CFCs will frequently result in on-site, aqueous, water-treatment problems becoming more extensive. In particular, the inputs of heavy metals present challenges. Process designers should be aware of water-quality standards (a sampling is given in Table 15.10; some of the differences emphasize the wide variations in standards from country to country). In order to achieve these surface-water standards, industrial effluents often are limited in the allowed concentrations of toxics as well as in total toxics load (the product of concentration and flow). The philosophy is that the concentration limit protects against acute effects, the load limit against chronic effects. In many cases, industrial facilities are required to return used water to public supplies in a condition purer than it was received. Such a requirement strongly encourages the complete elimination of cleaning steps, a minimization of cooling water use, and the development of processes that will not result in discharges requiring extensive treatment prior to release.

Table 15.10   A Selection of Surface-Water Chemical Standards

Constituent	Limiting Concentration ($\mu g/l$)		
	US(acute)	US(chronic)	Europe
Cyanide	22	5.2	50
Nitrate (mg/l)	10	10	50
pH	6.5–9.0	6.5–9.0	5.5–9.0
Arsenic	850	48	50
Cadmium	3.9	1.1	5
Chromium (+6)	16	11	50
Copper	18	12	50
Lead	82	3.2	50
Mercury	2.4	.012	1
Zinc	120	110	3000
Phthalate esters	940	3	
PAHs	.003	.003	.2
Methyl chloroform (mg/l)	18.4	18.4	
Trichloroethene	2.7	2.7	200

PROCESS CHANGES FOR CFC ELIMINATION

If residue streams cannot be eliminated completely, it is sometimes possible to change a process so as to substitute a more environmentally benign residue stream. The easiest way to illustrate the process of *choosing* emissions is to discuss a specific and recent example: substitution for CFCs in the defluxing operations that follow the application of solder to electronic circuit boards. In the manufacturing process, the electronic components and integrated circuits are placed in position on the board, either manually or by machine. A solder flux, that is, an acidic constituent designed to strip oxide and other contaminants from the metal leads of the components and circuits to ensure good solderability, is applied to the board, which is then passed atop a bath of molten solder. Many solder fluxes, if left on the finished circuit boards after the soldering step, eventually corrode the circuits. Hence, a defluxing step is generally included in the manufacturing process. Defluxing has customarily been performed using CFCs or HCFCs, but ozone-depletion problems are now forcing a search for alternative approaches.

Among the possible options for defluxing are the following, in order of increasing desirability:

1. Continuing with the use of CFCs.
2. Substituting HCFCs for CFCs.
3. Substituting other chlorinated solvents, such as methyl chloroform (CH_3CCl_3), for CFCs.
4. Substituting esters or other hydrocarbons as cleaning agents in place of CFCs. At present, this is not thought to be a general solution to flux-cleaning problems because those compounds are not very efficient solvents, but the approach may have merit for specific applications. Esters and other hydrocarbons are VOCs, so consideration must be given to VOC impacts.
5. Substituting semiaqueous cleaning for CFC cleaning. Terpenes, surfactants, and water, either in combination or in sequence, create cleaning agents that in some cases are superior to CFCs. The electronic product and all its exposed components must be able to "swim", that is, they must be capable of full immersion in water. Converting a product from a nonswimmer to a swimmer is often a major redesign effort. The drawbacks of semiaqueous cleaning include higher energy requirements, flammability of solution constituents, strong odors, and, because terpenes are VOCs, consideration of VOC impacts.
6. Substituting aqueous cleaning for CFC cleaning. The use of water-soluble fluxes allows plain water to remove residues from specific products. The addition of detergents and other additives can improve the defluxing but may create disposal problems. The product must be able to swim.
7. Substituting totally new nonsolvent-cleaning technologies such as CO_2 pellet bombardment.

8. Using a "low-solids" flux and eliminating the cleaning step. Low-solids fluxes contain 2–5% of solid material rather than the more traditional 15–35%. As a consequence, they leave very little residue on the surface following soldering. Drawbacks include more complex and rigorous processing control requirements, degraded visual appearance of the products, and possible problems with incompatible customer specifications.

9. Fastening components by means other than soldering. Conductive epoxies have been used in some instances to fasten components to circuit boards. Problems have been encountered with low bonding strength, decrease in conductivity upon aging, difficulty in repairing defective components, and nonsuitable mechanical properties. Intensive research on these techniques is likely to ameliorate some of the problems within the next few years.

Let us evaluate each of these options in turn. The first, continuation of the use of CFCs, is unacceptable because CFCs are the major cause of ozone depletion. (In addition, international agreements for a halt in CFC manufacturing will make them unavailable in most countries after 1995.) The second alternative, substitution with HCFCs, is undesirable. HCFCs are less potent ozone scavengers than CFCs on a molecule-for-molecule basis, but have sufficient ozone impact that restrictions on them are already beginning as part of the ozone-protection process. One does not want to choose a manufacturing process that has an excellent chance of being restricted or forbidden soon after its adoption. The third option is undesirable because most of the chlorinated solvents available are VOCs, many of which raise health concerns, have measurable ozone-depletion potentials, and require eventual (and expensive) disposal as hazardous waste. The fourth and fifth options are moderately attractive in that, whereas VOCs generate smog, smog is less critical than ozone depletion, so the substitution is an improvement from an environmental standpoint. Note that substitution of a semiaqueous or aqueous cleaning step involves the transfer of the concern from air to water, which may or may not be desirable, and generally requires more energy as well. The seventh option, new cleaning technologies without solvent, has proven highly successful for some applications, and is environmentally responsible. The eighth and ninth options involve omitting the cleaning step completely. They are clearly difficult to implement satisfactorily. Equally clearly, they are preferred from the environmental standpoint provided no major energy or reliability penalty is paid.

The rationale for the elimination of emissions of CFCs and other halogenated compounds is shown in Fig. 15.5, where the different contributors to stratospheric chlorine equivalents are shown and the anticipated reductions indicated. Even under the current protocols, it is projected that the chlorine level will stay above that at which the Antarctic ozone hole appeared until after the year 2050.

Because of the international concerns related to ozone depletion, the elimination of CFC use in industry has been vigorously pursued during the past several years. It has turned out that different solutions have been required for different industrial processes, because the degree of cleaning needed, customer requirements, and various process details prevent any single solution from being implementable across all

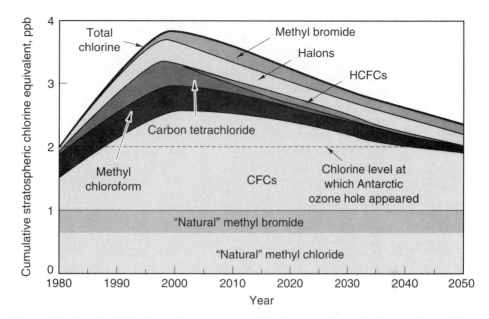

Figure 15.5   Levels of chlorine equivalents anticipated in the stratosphere under the provisions of the Montreal Protocol on Ozone Layer Depletion. The ozone-depleting effects of the bromine atoms in halons and in methyl bromide have been converted to their chlorine equivalents. (Courtesy of M. McFarland, DuPont Corporation.)

uses. This situation is probably a paradigm for many others, in that general substitutions for environmentally disadvantageous processes are unlikely and industrial activities will often need to be studied and resolved on a case-by-case basis.

SUGGESTED READING

Boyhan, W. S. Approaches to eliminating chlorofluorocarbon use in manufacturing. *Proceedings of the National Academy of Sciences of the U.S., 89* (1992): 812–814.

Nelson, K. E. Reduce waste, increase profits. *Chemtech, 20* (1990): 476–482.

Noyes, R., ed. *Pollution Prevention Technology Handbook*, Park Ridge, NJ: Noyes, 1993.

Watkins, R. D., and B. Granoff. Introduction to environmentally-conscious manufacturing. *MRS Bulletin, 17* (3) (1992): 34–38.

EXERCISES

15.1   Choose a manufacturing process, historic or modern, about which you can locate considerable detail concerning its implementation at a specific industrial facility. To the extent applicable and possible, evaluate the process using checklists 1 to 4, Appendix A, pointing out its strengths and weaknesses from an industrial ecology standpoint.

15.2 You are the industrial ecologist for a manufacturing company whose leading product is cables for personal computers. The principal components of the cable are copper wire, flexible plastic wire coating, and rigid plastic connectors. What by-product or residue streams do you anticipate? About which should you be most concerned?

15.3 Using Tables 15.3, 15.5, and 15.9, diagram the flows of nickel in the global industrial ecosystem.

15.4 Among the chemicals whose use is restricted for environmental reasons are CFC-12 (CF_2Cl_2) for air conditioners, Halon 1301 ($CBrF_3$) for fire extinguishers, organic solvents in oil-based paints, and formaldehyde-based compounds in panelboard. Prepare a 2 to 3-page report on the reasons for the restriction of one of these products, and on possible replacements.

15.5 Assume that the peroxide bath whose lifetime is shown in Fig. 15.4 is used to make 50,000 silicon wafers per year. The cost of the chemicals for the bath is U.S. $12/liter and the bath is 5 liters in volume. Ten silicon wafers per hour can be processed. How much depleted peroxide bath was generated in 1980, 1983, 1988, and 1990? At an on-site processing cost of U.S. 40¢/liter, how much was the cost in each of those years? Was the expenditure of U.S. $3,500 for a filtration system in 1988 and U.S. $9,400 for a replenishment system in 1990 justified?

CHAPTER 16 Choosing Materials

16.1 GLOBAL RESOURCE LIMITATIONS FOR MATERIALS

Ultimately, the prerogative of the designer to choose materials with physical and chemical properties suitable to the purpose at hand is limited by the supplies of those materials and their associated costs. Questions of global resource availability have seldom been concerns in the past (the Arab oil embargoes of the 1970s being obvious exceptions), but will become more and more significant at some time in the future as populations increase dramatically, as the global standard of living rises, and as resource use associated with those developments places increasing stress on materials resources.

An initial perspective on materials availability comes from the abundances of elements and compounds in the rock and soil near the surface of the planet. The elements can be divided into five classes based on elemental abundances in the upper continental crust, as shown in Table 16.1.

Table 16.1 Classes of Elemental Abundance

Abundant (> 0.1%)	Al, Ca, Fe, K, Mg, Na, Si, Ti
Common (> 100 ppm)	Ba, Mn, P, Rb, Sr, Zr
Relatively common (10–99 ppm)	Cr, Cu, Ga, Li, Ni, Pb, Sc, V, Y, Zn
Uncommon (1–9 ppm)	B, Be, Co, Cr, Th, U
Rare (< 1 ppm)	Ag, Au, Hg, Pt, Sb

This list is limited to major and selected trace elements, and does not include gases or chalcocides.

Table 16.2   Trace Metals Recovered as By-Products

Reservoir Metal	Trace Metal(s)
Al	Ga
Cu	As, Se, Te
Pb	As, Bi
Pt	Ir, Os, Pa, Rh, Ru
Zn	Cd, Ge, In, Th
Zr	Hf

Given these general abundance classes as a starting point, we can next examine the ease of extraction of the elements from their geophysical reservoirs. Many of the trace metals are available only as by-products of more abundant and more widely used metals. Those abundances therefore cannot be studied in their own right, because the rate of extraction of the principal reservoir metal will likely control the supplies and/or prices of the by-product metal. Table 16.2 lists the groupings of reservoir metals and by-products important to the availability of materials.

The relative scarcity of the elements can then be approximated by combining information on rates of use and sizes of reservoirs. These factors provide the supply index, defined as the ratio of the rate of world use of an element to the world reserves. (Reserves are those currently identified resources that can be profitably extracted with present technology.) For the major industrial metals except iron and aluminum, the supply index is shorter than 100 years and, in most cases, shorter than 50.

This sequential analysis permits us to divide the elements into five supply groups, as shown in Table 16.3, listing by-product metals in parentheses after their reservoir parent. The table clearly indicates that most of the metals widely used in industry are expected to be supply-limited in the next few decades, especially if the world's growing

Table 16.3   Classes of Supply of the Elements

Infinite supply	A, Br, Ca, Cl, Kr, Mg, N, Na, Ne, O, Rn, Si, Xe
Ample supply	Al (Ga), C, Fe, H, K, S, Ti
Adequate supply	I, Li, P, Rb, Sr
Potentially limited supply	Co[†], Cr[*], Mo(Rh), Ni[†], Pb (As,Bi), Pt(Ir, Os, Pa, Rh, Rn)[*], Zr(Hf)
Potentially highly limited supply	Ag, Au, Cu (As, Se, Te), He, Hg, Sn, Zn (Cd, Ge, In, Th)

[*] Supply is adequate, but virtually all from South Africa and Zimbabwe. This geographical distribution makes supplies potentially subject to cartel control.

[†] Maintenance of supplies will require mining seafloor nodules.

population uses these materials at per capita rates approaching those of today in the more developed countries.

A final look at the resource picture concerns materials that are recovered and used not as elements but as compounds or as mixtures of compounds. This group of materials includes most of the options for the generation of energy, as well as the forest products sector. Unlike elements, materials such as fossil fuels that are combinations of the more common elements are theoretically renewable. In practice, the time scales for renewal are too long for fossil fuels, which must be regarded, like the elements, as unrenewable resources. Although the reserves of coal and probably of natural gas are large enough that imminent concern is not warranted, those of petroleum are significantly smaller and the exhaustion of that resource must be considered a real possibility. In contrast, wood and other biomass resources are renewable with careful planning. In some parts of the world, at least, biomass renewal rates approximate use rates.

What is the proper interpretation of this information on abundance and supply? It is not that materials in potentially short supply should be avoided, especially on economic grounds. In fact, there turns out to be no robust relationship between abundance and cost, cost being so heavily influenced by stockpiling, cartelization, global economic activity, and the like. Nonetheless, it seems reasonable that materials in potentially short supply should be used only in those cases for which their properties are uniquely suited. To the extent that substitute materials can be employed, or to the extent that more efficient and acceptable means of recovery can be devised, such efforts will ameliorate materials supply limits on the growth of the technological society. Further, where a particular material is required, the minimum amount of that material should be used. In addition, there is at least the suggestion that waste-management practices that today comingle materials resources in landfills be modified to permit eventual access to minimally mixed materials, should favorable economic and technological conditions eventually make it profitable to use them. In this way, landfills would become materials storage sites for the future. In summary, therefore, we are not suggesting prohibitions so far as using materials is concerned, but rather the use of materials with care and perspective.

16.2   IMPACTS OF MATERIALS EXTRACTION

The extraction of raw materials from Earth's crust generally involves the movement and processing of large amounts of rock and soil. To recover 1 ton of copper, for example, requires the removal of some 350 tons of overburden and the processing of 100 tons of ore. As a result, extraction of materials is extremely energy-intensive and tends to be destructive of local ecological habitats. Some feel for the enormous magnitude of material that is involved is given by Table 16.4.

Among the environmentally related actions that often need to be taken at mines, wells, and other extraction sites are the following:

- Retain topsoil removed from the site so that it can later be replaced.
- Control surface runoff in sedimentation ponds.

Table 16.4 Global Materials Flows Associated with Major Minerals, 1991

Mineral	Ore (Tg)	Average Grade (%)	Residues (Tg)
Copper	910	0.91	900
Iron	820	40.0	490
Lead	120	2.5	117
Aluminum	100	23.0	77
Nickel	35	2.5	34
Others	925	8.1	850
Total	2910		2460

Source: Abstracted from J. E. Young, *Mining the Earth*, Worldwatch Technical Paper 109. Washington, DC: Worldwatch Institute, 1992.

- Line any working pits with impermeable material to reduce groundwater contamination.
- Monitor trace-metal concentrations, pH, and total suspended solids in any water discharges.
- Control acid drainage from mines (produced when sulfur in exposed tailings reacts with air and water).
- Collect water from underground operations in sumps, and pump it to the surface for treatment.
- Restore the site to its former appearance and productivity at the conclusion of extraction operations.

Given the problematic nature of materials extraction, it is apparent that extraction processes need to be sensitively performed, and that any extraction depletes the elemental reserves that one day may be crucial. In the following section, we explore an alternative available at least some of the time to designers: the use of postconsumer recycled materials.

16.3 AVAILABILITY AND SUITABILITY OF POSTCONSUMER RECYCLED MATERIALS

In contrast to extraction and processing, an efficient recycling operation may provide adequate quantities of a needed material at much lower expenditures of cost and environmental impact. Some materials have not, as yet, been efficiently recovered and recycled, but there are many that have been. Metals are recycled with reasonable efficiency and can generally be re-refined to the desired purity. Paper recycling is less prevalent, and is complicated by the fact that each stage of recycling shortens the paper fibers and restricts the material to lower-quality uses, a sequence known as *cascade recycling*. In the case of plastics, the difficulties of separating and reprocessing have made progress slower, but vigorous efforts to improve that picture are now under way.

A sometime impediment to the use of recycled materials is the specification by a designer or the designer's customer of virgin material, generally in an attempt to avoid receiving unsuitable product. Such concerns should be addressed not by specifying the source of the material, but by specifying its properties. Alternatively, one can in many cases require suppliers to provide a fixed percentage of purchased materials from post-consumer scrap sources, as Eastman Kodak Corporation has done with its plastic containers and steel drums. If these steps are taken, a more informed design will often emerge, and the number and type of suppliers of recycled materials will be increased.

16.4 MINIMIZING MATERIALS USE

No matter what materials are chosen for a product, the amount that is used can be generally minimized by careful designing involving stress analysis. Thinner walls and supporting members often can be found suitable by such techniques, expecially if common physical design rules are applied. These include the following:

- Where sheets of metal or plastic are used, achieve strength by providing support with bosses (protruding studs included for reinforcing holes or mounting subassemblies) and ribs, not by using thick sheets.
- Use a greater number of smaller supporting ribs rather than a few large ribs.
- Design bosses so that they play useful roles in supporting the main structural elements of the product.
- Gussets (supporting members that provide added strength to the edge of a part) can aid in designing thin-walled housings.

After the minimization of first-try materials has been studied, consider as well the substitution of nontraditional materials. A simple example of the success of this approach is the recent announcement by AlliedSignal Corporation of an automotive catalytic converter using only palladium as the active agent. Because palladium is significantly more abundant and cheaper than platinum and rhodium, formerly used with palladium in such catalysts, both resource sustainability and market advantage are achieved with the new approach.

A more common situation is that a product will contain many materials and have the potential for substitution of a number of them. This is the case with automobiles, where over the past decade the use of carbon steel, iron, and zinc die castings has dropped significantly and high-strength steel, aluminum, copper (mostly electrical), and plastics use has risen substantially (Table 16.5). The overall weight of a typical vehicle has decreased over that period by about 11%.

Table 16.5   Material in a Typical U.S. Automobile (kg)

Material	1978	1988	% change
Carbon steel	870	654	−25
High-strength steel	60	105	74
Stainless steel	12	14	19
Other steels	25	20	−19
Iron	232	207	−11
Plastics	82	101	23
Fluids	90	81	−10
Rubber	67	61	−8
Aluminum	51	68	32
Glass	39	38	−2
Copper	17	22	32
Zinc castings	14	9	−33
Other	62	57	−9
Total	1621	1437	−11

Source: Data apply to sedans, vans, and station wagons and are from H. A. Stark, ed., *Ward's Automotive Handbook.* Detroit: Ward's Communications, 1988.

16.5 TOXICITY OF CHEMICALS

In addition to supply limitations, materials choices may be limited by toxicity concerns. Other things being comparable for a specific application, a designer's objective should be to select materials that have the least significant toxic properties. Government environmental agencies generally define those materials that merit concern from a toxicity standpoint, and those lists are a good starting point for the physical designer. In Table 16.6 are seventeen chemicals or chemical groups targeted for reduction in the U.S. EPA's Industrial Toxics Project, a voluntary effort by industrial corporations to reduce emissions of targeted chemicals in the United States relative to 1988 levels by 33% in 1992 and 50% in 1995. The list includes both product chemicals and process chemicals. Cadmium, chromium, lead, mercury, nickel, and their compounds are sometimes used industrially such that the metals end up as part of the products that are produced, often as platings or coatings. Most of the remaining materials in Table 16.6 are process chemicals that can be used as either solvents or cleaners. Chlorinated solvents and monoaromatic species comprise most of the listed items; cyanide solutions (generally employed in metal plating) also appear. The physical design team thus needs to consider two facets of materials choices involving toxic materials: the potential for materials substitution in products and the potential for process changes.

More detailed lists of suspect chemicals could be given. An example is reproduced in Appendix B: the nearly 200 species identified in the 1990 amendments to the U.S. Clean Air Act as "hazardous air pollutants". The same chemical groups appear

Table 16.6   Chemicals Identified in EPA's Industrial Toxics Project

Benzene	Cadmium and compounds
Carbon tetrachloride	Chloroform
Chromium and compounds	Cyanides
Dichloromethane	Lead and compounds
Mercury and compounds	Methyl ethyl ketone
Methyl isobutyl ketone	Nickel and compounds
Tetrachloroethylene	Toluene
Trichloroethane	Trichloroethylene
Xylenes	

on Table 16.6: heavy metals and their compounds, cyanides, and halogenated solvents. A number of pesticides are included, as are several organic nonhalogenated compounds that have been shown to have toxic properties. Also present are solids such as asbestos (magnesium silicate minerals) whose toxicity is related to their physical properties, not their chemical ones. Good toxicology is obviously an important prerequisite to an adequate appreciation of the implications of these species lists.

Radionuclides, whether natural or artificial, have the potential to cause substantial health problems and are closely controlled as a consequence. Typical annual radiation doses are given in Table 16.7; they demonstrate that natural and medical sources comprise virtually the entire concern for anyone not occupationally exposed to radionuclides. Nonetheless, small amounts of radionuclides are used industrially in specific nonmedical applications such as radioluminous products (watches, clocks), smoke detectors, electronic devices, antistatic devices, and scientific instruments. The governmental restrictions on the extraction, use, and disposal of radioisotopes are probably sufficient to ensure that the materials are seldom used unless their properties are essential to the particular product involved.

Because the restrictions on manufacture, use, and disposal of products containing significant amounts of radionuclides are so onerous, the use of radionuclides should be avoided by designers unless specific characteristics make such use necessary. If used, the amounts should be stringently minimized; a good example is the improved targeting

Table 16.7   Estimates of Average Radiation Dose to the U.S. Population, 1990

Source	Annual dose (mrem)
Natural (cosmic rays, soil constituents)	93.0
Medical (radiography, pharmaceuticals)	34.0
Nuclear power	0.2
Miscellaneous (see text)	1.1

Source: Committee on Environmental Improvement. *Cleaning Our Environment: A Chemical Perspective*, 2nd ed. Washington, DC: American Chemical Society, 1978.

of pharmaceutical uses of radionuclides so that the required medical benefits can be achieved with ever smaller quantities of radioactive material.

Even if a material on one or more hazardous lists is important to a product or process and can be used with safety in a manufacturing facility, it is necessary to consider the implications of disposal of the inevitable residues. Disposal regulations vary widely around the world and undergo constant change, but a suitable reference list of materials with disposal restrictions has been provided by the International Electrotechnical Commission; it is reproduced as Appendix C. Many of the materials on the toxic or hazardous lists occur here also, but included in addition are classes of compounds having particularly undesirable physical characteristics such as instability, high flammability, high acidity, and high basicity. Finally, one should avoid the use of any ozone-depleting substances subject to phaseout under international protocols (see Chaps. 3 and Appendix D).

The use of a toxic material or a material chemically closely related to a toxic material as a feedstock or intermediate in an industrial process generally requires that it be shipped to the manufacturing facility and stored there until consumed. Storage of a toxic material or its precursors was the major cause of the worst chemical accident in history: the deaths of some 2000 people due to the dispersal from a storage tank of methyl isocyanate in Bhopal, India, on December 3, 1984. An alternative design approach is the generation of a required chemical from nontoxic precursors as it is needed. For example, on-demand generation of arsine, a toxic gas widely used in the manufacture of electronic materials, has been demonstrated by Jorge Valdes and co-workers at AT&T Bell Laboratories. The technique is based on an arsenic metal cathode in an electrolytic cell and is shown schematically in Figure 16.1 and in a production model in Figure 16.2.

A final caution for design engineers selecting materials is to attempt to anticipate future restrictions on materials whose use is not now constrained. In this connection, concerns are growing regarding the use of chlorine and chlorinated organic compounds. Among the issues raised are the tendency of chlorinated organics to be health hazards, manifesting this property through cancer production and modifications to endocrine function in both humans and animals. Individual compounds show a wide range of behavior on specific toxicity tests, however, and producers and users of chlorine-containing materials defend the undeniable utility of the materials and suggest a compound-by-compound approach. It seems unlikely that a comprehensive ban on chlorine-containing compounds will ever result. Nonetheless, the astute practitioner of industrial ecology may wish to consider the following:

- Investigate alternatives to chemicals that contain chlorine in the final product.
- Attempt to develop alternative synthesis routes for processes in which chlorinated compounds are currently used as intermediates.
- Minimize the use of chlorinated compounds in processes where their utility is especially suitable. (The use of chlorine as a bleaching agent has been sharply reduced over the past several years by the pulp and paper industry.)

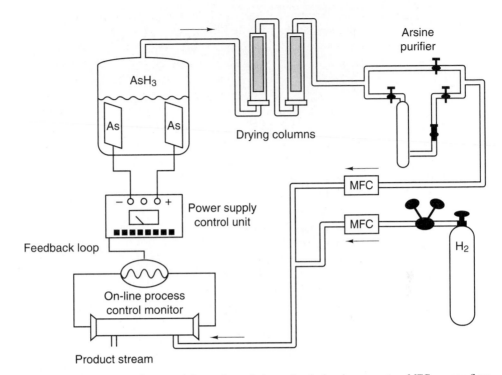

Figure 16.1   A schematic diagram of the on-demand electrochemical arsine generator. MFC = mass flow controller. (Courtesy of J. L. Valdes, AT&T Bell Laboratories.)

16.6 MATERIALS SELECTION RECOMMENDATIONS

When considerations of supply and toxicity are merged to give a common perspective, one finds a clear preference for some materials over others. Because their supplies are ample (and/or the potential for recycling is good) and because they have no significant toxicity problems, we recommend that designers investigate using the following materials: Al, Br, C, Fe, H, Mn, N, O, S, Si, and Ti, with the provisions that Br should not be used in any volatile form and that S not be used where it will be subject to loss during a combustion process. Conversely, because they promise to be in short supply and/or have significant toxicity problems, we recommend that designers attempt to limit or avoid the use of the following: Ag, As, Au, Cd, Cl, Cr, Hg, Ni, Pb, and petroleum. Copper, nickel, and zinc may come under supply limitations at some point, but it seems premature to advise against the use of these versatile, inexpensive materials as long as specific hazards related to their use are avoided.

The materials selection process can be summarized in four short goals for physical designers:

Figure 16.2   The prototype model of the on-demand electrochemical arsine generator. (Courtesy of J. L. Valdes, AT&T Bell Laboratories.)

1. Choose abundant, nontoxic, nonregulated materials if possible. If toxic materials are required for a manufacturing process, try to generate them on site rather than by having them made elsewhere and shipped.
2. If possible, choose natural materials rather than synthetic materials.
3. Design for minimum use of materials in products, in processes, and in service.
4. Try to get most of the needed materials through recycling streams rather than through raw materials extraction.

BISMUTH SHOTGUN SHELLS

In Chap. 10 we presented the industrial lead budget for the world, pointing out that some 3% of annual lead use in 1988 was for ammunition. Soon thereafter, lead-loaded shotgun shells were banned for waterfowl hunting because substantial numbers of birds were dying from lead poisoning after ingesting pellets dispersed into the environment. The initial replacement for these shells was shells loaded with steel pellets. These shells were not a success, primarily because the density of steel is so much less than that of lead. This difference had several consequences: different forward allowances were required while aiming at moving birds (old techniques are hard to relearn), the steel shot dispersed more rapidly so that effective shooting ranges decreased, and modest initial dispersal of the pellets within the gun barrel caused damage to some older weapons.

A design for environment solution to this situation was clearly called for, and one emerged from John Brown of Ontario, Canada: the substitution of bismuth for lead. Bismuth is nearly as dense as lead, so the shells perform similarly to those they replace, but bismuth is nontoxic, so the principal negative characteristic of lead is avoided. The shells are now being manufactured by the Bismuth Cartridge Company of Dallas.

It is worth pointing out that the substitution of bismuth for lead is not a long-term solution, because bismuth and lead occur in the same ore bodies and bismuth cannot be recovered from its natural deposits without recovering lead as well. Thus, the bismuth shotgun shell substitution works only because ammunition is a minor use of lead. Were most uses of lead to be phased out, the supply of bismuth would be constrained. Further, ammunition is obviously an application in which the pellets are lost when used. No recovery of the pellet material is possible, so that even nontoxic pellet supplies only can be sustained if other bismuth uses permit recovery and recycling.

SUGGESTED READING

Allen, D. T., and N. Behmanesh, Wastes as raw materials, in *The Greening of Industrial Ecosystems*, B. R. Allenby and D. J. Richards, eds., pp. 69–89. Washington, DC: National Academy Press, 1994.

Ayres, R. U., Industrial metabolism, in *Technology and Environment*, J. H. Ausubel and H. E. Sladovich, eds., pp. 23–49. Washington, DC: National Academy Press, 1989.

Goeller, H. E., and A. Zucker. Infinite resources: The ultimate strategy. *Science, 223* (1984): 456–462.

Herman, R., S. A. Ardekani, and J. H. Ausubel. Dematerialization, in *Technology and Environment*, J. H. Ausubel and H. E. Sladovich, eds., pp. 50–69. Washington, DC: National Academy Press, 1989.

Hileman, B. Concerns broaden over chlorine and chlorinated hydrocarbons. *Chemical and Engineering News, 71* (16) (1993): 11–20.

World Resources Institute. *World Resources 1990/1991*. New York: Oxford University Press, 1990.

EXERCISES

16.1 In 1991, nearly 2.5 Pg of residues were produced worldwide as a consequence of ore processing. If the typical density of this ore is 4.5 g/cm^3, how long would be a line of trucks carrying this ore if each truck holds 2.5 m^3 of ore and is 5 m long?

16.2 In Chap. 14 it was shown that the operation of nuclear power plants has almost no impact on the environment, and Table 16.7 shows that the estimated dose to humans from that activity is negligible. What are the arguments against nuclear power? Does the virtual ban on nuclear power in a number of countries represent a rational response to world energy needs given the benefits and problems? Given nuclear power's "image problem", do you anticipate that radionuclides may eventually be banned from any product no matter how useful they are to its function?

16.3 Ignoring cost as a factor, which one of each of the following pairs of materials should be preferred by product and process designers? Why? (a) Titanium or tin, (b) toluene or heptane, (c) tin or bismuth, (d) ethanol or ethylene glycol, (e) formic acid or sulfuric acid, (f) CFC-113 or HCFC-123.

17 Product Packaging, Transport, and Installation

17.1 INTRODUCTION

Where detailed assessments have been made in Europe and in the United States, some 30% of all municipal solid waste has been found to be packaging material. Indeed, it has been estimated that about one-third of all plastics production is for short-term disposable use in packaging. For many products—convenience food items, for example—packaging is the primary residue of consumer use. The use of toxic materials such as heavy metals in packaging inks may be a first-order environmental impact for such products. Proper packaging of products of all sorts, from large-volume chemicals to small consumer personal-care items, thus plays an important role in maintaining environmental sustainability. It is worth noting that many environmental standards programs, such as Germany's "Blue Angel" program, require the use of packaging that is fully recyclable, has maximum recycled content, does not contain any toxics, contains no unnecessary pigments, and, if paper, is not bleached.

An important perspective is that some 40% of all U.S. goods and services are not purchased by individual consumers, but by other business or governmental agencies. Packaging in these cases is on a "corporation-to-corporation" basis, and the potential for negotiating packaging reductions and improvements is substantial.

Although packaging is an important topic in industrial ecology, it is often not within the normal province of the physical designers who supervise the processes or products that generate the packaging need. In our age of specialization, product packaging and shipment is often under the control of an engineer specializing in packaging or transportation technology. The product and process designers must thus work closely with these specialists to see that environmentally responsible products are shipped in environmentally responsible packages.

17.2 GENERAL CONSIDERATIONS

Product packaging should always aim to use the minimal number of stages. For some products, no packaging at all may be required. In other cases, only primary packaging, that is, that which is in physical contact with the product, is needed. Less certain is the need in many cases for secondary packaging, a supplementary interior stage, or tertiary packaging, the outer shipping container and associated material. However, different applications impose different packaging requirements. Some food packaging, for example, is quite complex: a potato chip bag may be a "sandwich" of seven or eight different components, each with a separate function. To the degree that any packaging stage can be eliminated or simplified, the residue stream will be reduced and shipping and storage expenses for the producer and consumer will be minimized.

The use of packaging does not necessarily involve the generation of residues that must be discarded or reused, because some packaging can serve as an integral part of the product it protects. For example, glass jam jars that serve as beverage containers after the jam is used have existed for many years, and part of the packaging of AT&T's current computer keyboards can double as a dust cover during the product's lifetime. Packaging usually must be recycled or reused, however. That is most easily done if the packaging is made from a single material, such as a cardboard carton assembled without staples. Next best are packaging designs that use more than one material but make them easily separable, such as the bottle cap of a different material than the bottle itself or the styrofoam insert in a cardboard box. Less desirable are commingled materials that are difficult to separate cleanly, such as the polyethylene overwrap fastened by adhesive atop a screwdriver mounted on a cardboard backing. Worst of all are dissimilar packaging materials bonded together so that separation is essentially impossible, such as the aluminized bags often used for electronic circuit boards. The U.S. Council of Northeast Governors has suggested an order of precedence for approaches to packaging. In decreasing order of preference, it is as follows:

- No packaging
- Minimal packaging
- Consumable, returnable, or refillable/reusable packaging
- Recyclable packaging

This list is only a guide, because innovative packaging solutions for specific products may outweigh the precedence order, but the order is a good initial perspective for the packaging engineer.

Manufacturers should expect to have to work with their customers in deciding how to package their products. To those customers, packaging entering their facility or home is material that they would rather not have. To the extent that packaging engineers can develop packaging that is easier for them to recycle or reuse, the product becomes more attractive. Alternatively, many customers and some countries are now encouraging suppliers to take back the packaging on items sold by them. Such take-back arrangements can take several forms. One is the negotiated agreement, in which containers for consumed supplies—chemical drums, for example—are returned to the supplier when a new shipment is delivered. The second is the legislated requirement such as in Germany, where all sellers are required to accept from customers the packaging in which their products were delivered. This exchange can occur at purchase, as with external packing, or at a later time, as with the tube holding toothpaste. In either the negotiated or legislated case, when a corporation gets back its own packaging, it has a very strong incentive to minimize the amount of that packaging and to make it easy to recycle or reuse.

17.3   SOLID RESIDUE CONSIDERATIONS

The first consideration in designing solid packaging for products is to minimize it as much as possible consistent with other packaging requirements. Among minimization options that have been implemented are the substitution of soft pouches and paper cartons for rigid plastic bottles and cans in consumer products such as detergent and coffee. For fabric softeners and other liquids that can be sold in concentrated form, customer dilution can permit significant reductions in packaging size; if marketed properly, many consumers are willing to take this small extra step. Analogous approaches in new packaging and concentration of commodity chemicals can and are being used for industrial and commercial products, such as the growing practice in Europe of shipping bulk chemicals in plastic-lined paper bags rather than in drums.

Compared with the volume of packaging that has traditionally been used, shrinking the amount may not be necessarily difficult. An example is shown in Fig. 17.1, which illustrates old and new packaging for a personal computer keyboard. The 30%

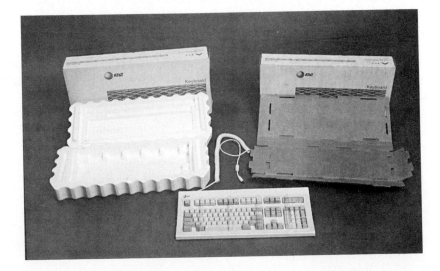

Figure 17.1   Packaging for the AT&T Model 6386 Keyboard. Left: 1988 packaging, with cardboard exterior and expanded polystyrene foam insert. Right: 1990 packaging, all cardboard. This packaging is 30% smaller than the packaging it replaces. (Courtesy of AT&T.)

volume difference was achieved after a series of tests to verify that the new packaging was sufficiently protective. This change alone justified the effort, as the smaller package was much cheaper to move, ship, and store. In addition to the volume reduction, the materials diversity of the packaging was substantially decreased by using a cardboard interior cushioning in place of styrofoam. A factorywide project with similar goals by NCR's facility in Oiso, Japan, recently produced reductions in package costs alone of 3–30%.

A list of causes for overpackaging has been presented by the Institute of Packaging Professionals in Herndon, Virginia:

- An overly cautious approach to the protection of the packaged contents.
- Increasing the package size to deter shoplifting.
- Overly conservative environmental test specifications.
- Requirements of packaging machinery.
- Decorative or representational packaging.
- Increasing packaging size to provide space for regulatory information, customer information, or bar coding.

Most of these causes are capable of being surmounted by thoughtful packaging design, should the designer attempt to do so. For example, the use of electronic theft protection systems can mitigate the need for overpackaging.

Once the amount of packaging is minimized, the next thing to consider is whether the packaging can be reused. Reusable packaging is not limited to foam "peanuts"; it includes as well innovations like Ametek Corporation's "couch pouch", a composite of polypropylene sheet foam and polypropylene fabric that can be repeatedly reused for shipping furniture.

A general rule for designing packaging that is environmentally responsible is to mimic nature. To the extent possible, using natural materials avoids problems with toxicity and degradability. In the case of many food products, edible packaging is an option that should be considered.

Packaging materials must eventually be discarded, so they should be made of materials that have good recyclability. To the maximum extent possible, the materials used should be themselves made of recycled rather than virgin stock. This is generally easy for forest products items such as cartons, instruction cards, customer information booklets, and the like. It is less easy for packages that come into intimate contact with the product, such as plastic bottles holding either food items or manufacturing chemicals. Some of these containers can be reused after cleaning; some can be recycled only in degraded form. In any case, plastic packaging materials should be thermoplastics, not thermosets, and their composition should be clearly marked (see Appendix E for standardized marking codes).

It has long been an act of faith to assume that forest products have less environmental impact than do plastic products. This assumption, based entirely on the belief that forest products degrade more rapidly than plastics, has now been shown to have a questionable foundation because very little biodegradation occurs in most landfills. Even if biodegradability favored forest products, the topic is not central if designers plan for cyclization rather than disposal of residues. Plastics are usually lighter than the forest products materials they replace, consume much less energy in manufacture, and often have better physical properties. However, plastics are not renewable resources on reasonable time scales unless they are produced from biomass precursors. Hence, forest products will continue to play important roles, particularly when long-term resource sustainability is considered.

In addition to minimizing the packaging of a single product item, designers can often minimize the number of packages that are needed. A common example is the sale of window cleaner and liquid soap refills in a size inappropriate for use but suitable for replenishing containers in service. The use of refills also eliminates the need to manufacture, use, and discard spray heads for each bottle of product.

A final rule in packaging is to put package recycling information on (or in) the container and to help provide an infrastructure for the return of packaging for reuse or recycling.

17.4 LIQUID- AND GASEOUS-EMISSION CONSIDERATIONS

Although liquid and gaseous emissions are not major problems in connection with product packaging, it is important that those issues not be forgotten in cases where they may be important. A simple example is the steel drum used for shipment of liquid chemicals. Traditionally, the disposal of these drums involved the disposal as well of significant amounts of residual chemicals. New drum designs now allow virtually all of the chemicals within to be drained before disposal. (Even better, where possible, are reusable drums.)

Another current example of emissions from packaging is the heavy metals traditionally used to provide color in printing inks. This practice contributes to heavy-metal pollution during the manufacture of the packaging and upon packaging disposal, generally by leaching of the metals into groundwater or surface water. A final example is the use of some types of blown foam packing, which has traditionally involved the use and subsequent loss of chlorofluorocarbons. This CFC use has now virtually ceased in many countries, but some developing countries have retained this environmentally damaging type of packing material.

17.5 TRANSPORTATION

In many ways, it is inherently somewhat contradictory to advise minimum use of materials in packaging and then to advise safe and trouble-free shipping, because products must be protected in some way from such in-transit conditions as shock, vibration, condensation, corrosive gases and liquids, and temperature extremes. One way to approach the problem is to realize that only large products are packaged and shipped individually;

most of the rest are shipped in quantity in larger cartons, shipping containers, plastic sheeting, and the like.

The effect of the realization that product packaging and transportation packaging are distinct is that designers have two opportunities to approach packaging from an industrial ecology viewpoint. Thus, packaging and transportation engineers need to work together to optimize the combination of multiple-product packaging for shipment and individual product packaging for the final consumer. For example, it may be the case that good shock protection in the product package can minimize the need for shock protection of the shipping package, or vice versa. The decision on appropriate packaging may depend on the mode of product shipment and on the environmental stresses that the product encounters. As in many other instances, working together with the transportation firm provides another avenue for optimizing a design (a packaging design, in this case) from the industrial ecology perspective.

Transporting products, whether by company-owned vehicles or by transportation contractors, offers additional opportunities to minimize environmental impacts. If a product is itself hazardous, contains hazardous constituents, or is potentially subject to spilling or venting, transportation routes should be chosen to minimize possible contact with humans or with sensitive environmental areas. Drivers should be well-trained in avoiding problems or dealing with them if they do occur. Finally, no matter what the product, delivery offers the opportunity to collect packing materials such as plastics, cardboard, and pallets and reuse or recycle them.

The impacts of transporting components into a facility and then transporting products from it to customers are too frequently regarded as not being directly relevant to a product's environmental impacts. In many cases in today's economy, components and subassemblies will be produced in many countries around the world, then shipped to one or several locations for final assembly into a product that is, in turn, shipped around the world. Japan's otherwise efficient JIT (just-in-time) manufacturing and parts stocking systems have significantly increased the number of delivery vehicles in Tokyo, as well as their daily duty cycles. The result is a substantial and undesirable increase in traffic congestion and direct contributions to air pollution.

TRANSPORTING CHEMICALS BY RAIL

A large fraction of bulk chemicals move from manufacturer to customer by train. Because chemicals can be spilled during loading and unloading operations and because rail accidents occasionally occur, an important facet of the life-cycle assessment for those chemicals involves the degree to which the impact from those activities can be avoided or minimized. In the United States, manufacturers have entered into partnerships with the railroads to promote environmentally responsible transportation of their products. In the case of the Southern Pacific Railroad, joint training programs with chemical manufacturers and their customers help to ensure that cars are loaded and unloaded properly. For the transport phase of the operation, the railroad keeps current logs of disaster-response organizations along all its routes, together with maps showing emergency equipment locations and their contents, together with access routes to its right-of-way.

The Union Pacific Railroad has a similar program, supported by hazardous-materials agents who perform random security checks on chemical-carrying cars moving along their rail system. They have worked with communities to prepare local emergency-response plans, and have begun a detailed derailment prevention program.

As a result of these activities, accidents attendant to chemicals moving by rail are much less frequent than was formerly the case, and response to incidents that do occur is much quicker and more efficient.

17.6 INSTALLATION

In the case of large or complex equipment, installation is often accomplished by the manufacturer. Depending on the product, the installation step can carry with it the potential for environmental degradation. Examples include underground storage tanks for fluids, pipelines for liquids and gases, and the laying of intercontinental communications cables. The most straightforward advice for these situations is to minimize environmental disruptions and to avoid sensitive areas as sites for large projects, especially those that produce significant emissions during use. The ideal industrial ecology solution, however, is to design products or construct societal networks that avoid the necessity for such installations altogether. An example, now in rapid deployment and yet to be fully realized, is cellular, mobile telephony. By using digital radio signals for many calls, the designers may be moving toward a world in which communications need not move by buried or elevated wire and cable at all. Similar gains can be achieved by gradual movement toward more concentrated but eminently livable urban areas, a concept that we discuss briefly in the final chapter of this book.

17.7 DISCUSSION AND SUMMARY

Packaging and shipping are the stage at which the manufacturer has the opportunity to indicate to the customer the environmentally responsible characteristics of its products. The performance of many manufacturers in this regard has not been salutary. Their approach too often has been to trumpet on their product packages unjustified or at least unsupported claims for environmental benefits. In contrast, products that are responsibly designed are seldom advertised or sold with enough information to inform and instruct the consumer. We recommend that the outer package and any more detailed information, such as operating manuals, discuss why the manufacturer considers a design to be environmentally responsible and how these activities compare with what might be done.

In packaging as in other aspects of industrial ecology, designer decisions are aided if a structured method of comparing alternatives is available. Merck and Company assigns numerical ratings to such factors as use of recycled material, capability of the packaging to be recycled, energy used to produce the package, amount of material required, toxicity of the packaging materials, and potential energy recovery from incinerating the packaging. Competing packaging designs are then ranked for environmental impact, and special justification must be presented if the highest-ranking design is not chosen. A more highly structured but narrower approach is that of the Swiss firm Migros-Genossenschafts-Bund, whose computer analysis program allows comparisons of energy use and air and water impacts of alternate packaging concepts. The result of either of these programs is a straightforward, understandable way to encourage environmentally responsible packaging.

SUGGESTED READING

Russell, P., et al. *Handbook for Environmentally Responsible Packaging in the Electronics Industry.* Herndon, VA: Institute of Packaging Professionals, 1992.

Stilwell, E. J., R. C. Canty, P. W. Kopf, and A. M. Montrone, *Packaging for the Environment: A Partnership for Progress.* New York: American Management Association, 1991.

EXERCISES

17.1 Critique the packaging of six food products in a grocery store, selecting them for maximum diversity. Discuss the good and bad points of each approach to packaging, using checklist 6, Appendix A, to the extent applicable and possible. If competing products adopt different approaches, point out the good and bad aspects of each.

17.2 Discuss options for packaging of the following consumer products: motor oil, grapefruit juice, toothpaste, magazines, shirts, and decongestant tablets.

17.3 In many cases, packaging systems reengineered for environmental reasons have proven to be less costly and easier to transport than the ones replaced. Thus, firms have been paying too much money for packaging that was also not environmentally responsible. Discuss why this may have been the case and suggest incentives that could modify the situation.

17.4 Visit one or more hardware or "home center" stores and compare different approaches to packaging several common items such as nails, light bulbs, and screwdrivers. Consider the need for brand identification, theft protection, and other nonenvironmental factors. What are the good and bad points of each product's packaging? Can you think of packaging techniques better than any of those you saw?

CHAPTER
18

Environmental Interactions During Product Use

18.1 INTRODUCTION

Design with the environment in mind involves much more than designing so that products may be manufactured with minimal environmental impact. Another important consideration, sometimes an overriding consideration, is the amount of environmental impact produced by products when they are used. Unlike the extraction or manufacturing environments, which are under the direct control of corporations, the use and maintenance of a product after it passes to the customer is currently constrained only by the product design. This circumstance places special responsibilities on the designer to envision aspects of design that minimize impacts during the entire useful life of the product.

18.2 SOLID-RESIDUE GENERATION DURING PRODUCT USE

There are a number of types of solid-residue generation during product use. For example, products that in themselves may be long-lived and recyclable may nonetheless generate solid residue from consumables. Common examples include typewriters (ribbons) and cameras (plastic containers from photographic film). To the degree that solid residues from consumables can be eliminated by innovative design, the beneficial impact on the environment is obvious. In any case, consumables should have little or no inherently toxic materials within them.

Less desirable than eliminating solid residues from consumables, but still meritorious, is the design of consumables to encourage efficient recycling. There are two requirements for such a recycling program. One is a design that permits ready recycling once the consumable item has been returned. The second is a procedure and supporting infrastructure to encourage the user to return the item for recycling. An example of the latter is the approach used by Hewlett-Packard Corporation (HP) to recycle the cartridges from its laser printers. Users are encouraged by clear instructions to return the cartridges to HP for regeneration and reuse. An additional incentive is a charitable donation by HP for each cartridge returned. Parcel delivery systems provide much of the necessary infrastructure.

18.3 LIQUID-RESIDUE GENERATION DURING PRODUCT USE

Some products generate regular or aperiodic liquid residues as a result of use. Examples include soil- and detergent-laden water from washing machines, coolants from large industrial motors, and lubricants from internal-combustion engines. The ideal designs in this connection are those that require the customer to use no consumable fluids at all. Next best is a design that minimizes the quantity of fluids and uses only fluids with a modest environmental impact. Finally, strong efforts should be made to recycle fluids. An example of the latter program is that originated by Saf-T-Clean Corporation, a supplier of solvents to independent automotive garages. Saf-T-Clean supplies solvents in special containers, reclaims them when dirty, completes the necessary shipping manifests, and cleans and reuses the solvents as part of an integrated systems approach to solvent cleaning of metal parts. It is worth noting that this system is largely the result of organizational initiative, not technological initiative. It is also worth noting that many such programs are quite profitable.

A final consideration is that liquid products are often held in underground or aboveground storage tanks prior to use. Leaks from storage tank plumbing or loss of integrity of the tank itself can result in unplanned dissipative emissions of liquids to water or soil. For such products, containers and/or products should be designed to minimize the potential for inadvertent loss from storage facilities. For example, where possible and safe, tanks and pipes should be above ground to allow for easy visual detection and repair of malfunctions or leaks.

18.4 GASEOUS-RESIDUE GENERATION DURING PRODUCT USE

Products whose use involves such processes as the venting of compressed gas or the combustion of fossil fuels require the industrial ecologist to explore design modifications in an attempt to minimize or eliminate these emissions. The automobile is perhaps the most common example of such a product, and one whose cumulative emissions are very substantial, but anything that emits an odor during use is, by definition, generating gaseous residues; examples include vapors from carpet adhesives, polymer stabilizers from plastics, and vaporized fluids from dry cleaners. Replacements for the volatile chemical constituents will often be available if the designer looks for them.

Table 18.1 Equivalent CO_2 Emissions (g/km) from a Volvo 740 GL Using Various Fuels

Fuel	Extraction	Refining	Distribution	Operation	Total
Diesel	17	10	7	205	239
Gasoline	18	15	8	225	266
Methanol	13	51	12	186	263
Methane	10	7	41	187	245

An example of exploring options for reducing the impacts of products in use is provided by Volvo Car Corporation, which has studied emissions from alternative fuels in its Model 740 GL. The analysis included the complete life cycle of the vehicle, although most emissions occur during operation rather than during extraction, refining, and distribution. The result (Table 18.1) is that on a CO_2 basis, the engines using diesel fuel and methane are superior to those using methanol or gasoline.

18.5 ENERGY CONSUMPTION DURING PRODUCT USE

A number of products use energy when in operation. The energy can be electrical, as with refrigerators or hair dryers, or furnished by fossil fuel, as with power lawn mowers or chain saws. Major advances have produced lower energy consumption during product use, and legislation in an increasing number of countries will enhance such efforts. Energy-efficient designs sometimes involve new approaches; the result providing not only lower-cost operation but also improved product positioning (from a sales standpoint), particularly in areas of the world that are energy-poor.

Energy use can be a function of the way in which a product is designed to use replenishable supplies. To continue our example with Volvo automobiles, Table 18.2 summarizes the energy use of different types of engines. It is interesting to compare this table with Table 18.1, which shows gaseous emissions from the same engines. Diesel engines are seen to be far more energy-efficient than other types, as well as being relatively low CO_2 emitters. From an environmental standpoint, there appears to be much to

Table 18.2 Energy Consumed (MJ/km) by a Volvo 740 GL Using Various Fuels

Fuel	Extraction	Refining	Distribution	Operation	Total
Diesel	0.15	0.14	0.08	2.52	2.89
Gasoline	0.17	0.22	0.10	2.87	3.36
Methanol	0.12	0.85	0.15	2.42	3.54
Methane	0.12	0.12	0.29	2.87	3.40

recommend the diesel engine, although a trade-off between energy efficiency and toxicity also exists here: The U.S. EPA has determined that a number of gaseous and particulate compounds in diesel emissions are probable human carcinogens.

18.6 INTENTIONALLY DISSIPATIVE PRODUCTS

Many products are designed to be dissipative in use, that is, to be eventually lost in some form to the environment with little or no hope of recovery. Examples include surface coatings such as paints or chromate treatments, lubricants, pesticides, personal care products, and cleaning compounds. Attempts are being made to minimize both the packaging volume and the product volume in some of these situations; the recent introduction of superconcentrated detergents is an example. Alternatively, some liquid products that are dissipated when used can be designed to degrade in environmentally benign ways. Over the past several years, this approach has successfully been adopted for a number of pesticides and herbicides. A recent demonstration of design for biodegradability is the development of a biodegradable synthetic engine oil, designed specifically for inefficient two-cycle engines that in operation emit approximately 25% of the gasoline–oil mixture unburned to the environment.

Another common example of a potentially dissipative product is fertilizer for crops, where any excess spread on fields is dissipated to local and regional groundwaters and surface waters. The amount of fertilizer used by farmers shows great variation, as seen in Fig. 18.1. For the same yield, fertilizer application ranges over some two orders of magnitude, or, to look at the data in another way, the same level of fertilization can produce tenfold variations in yield. Much of this variation is due, of course, to the qualities of the different soils and differences in climate, but it is generally agreed by experts that farmers in the developed countries tend to use more fertilizer than can be justified by agricultural data and farmers in the developing world tend to use too little (often because of the expense). To the extent that excess fertilizer is used, it has a triple negative impact: the use results in excessive extraction of raw materials, financial penalties accrue to the farmers, and dissipative release of the excess fertilizer can have negative impacts on proximate water supplies.

A seldom recognized but clearly dissipative use of heavy metals is exemplified by the zinc applied to steel during the galvanizing process. Because the purpose of the zinc is to protect the steel by sacrificially corroding, it accomplishes its purpose only by being dissipated to the environment.

The concept of intentionally dissipative products is a poor one, but better alternatives are not always readily available. Many gear trains require lubrication, for example. To the degree that they can be sealed so that lubrication need not be replenished often or ever, the dissipative nature of the gear train lubricant is ameliorated. Cadmium provides an example of dissipative design related to purity, because zinc oxide is used as a constitutent of automobile tires, which wear gradually during their use, and because cadmium is an impurity in the refining of zinc. To the degree that the zinc is not well purified,

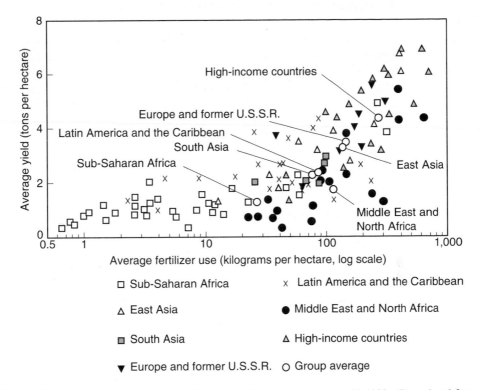

Figure 18.1 Fertilizer input and cereal yields for countries throughout the world, 1989. (Reproduced from *World Development Report 1992*. Copyright 1992 by The International Bank for Reconstruction and Development/The World Bank. Reprinted by permission of Oxford University Press, Inc.)

dissipative emissions of zinc oxide from tire tread wear produces dissipative emissions of cadmium as well.

An example of efforts to minimize the environmental impacts of a dissipative product is that of Procter & Gamble's research on disposable diapers. Such diapers make up about 2% of the U.S. solid-residue stream. Cloth diapers, the alternative, consume so much energy and water in repeated washings that the environmentally desirable alternative for diapers appears to depend on local costs for energy, water, and solid-residue disposal (the actual analyses are quite sensitive to the assumptions used). Biodegradable diapers do not constitute a solution, because it has been shown that little biodegradation occurs in a modern, well-covered landfill. Procter & Gamble's approach is to develop a diaper that can be broken down in the municipal composting facilities now becoming increasingly available. As of the publication of this book, more than 80% of the material in P&G disposable diapers is compostable, and efforts are being made to achieve 100%.

18.7 UNINTENTIONALLY DISSIPATIVE PRODUCTS

Unintentional dissipative emissions occur when products that are exposed to the environment suffer destructive degradation. Calcium carbonate in disintegrating concrete provides an example. Another is that of cadmium present as an impurity in the zinc used for galvanizing steel. As the zinc sacrificially corrodes, as it was designed to do, both the zinc and its cadmium impurity are dissipated to the environment. Unintentionally dissipative products are often difficult to recognize, and each product engineer needs to ask the question, "What can possibly happen to the materials in my product while it is in use?" Loss during shipment or storage also falls in the unadvertent dissipating category, but such losses can be minimized or prevented by good planning and diligent oversight.

18.8 DESIGN FOR MAINTAINABILITY

A final consideration in product use is related to the requirements placed on product maintainability by the designer. Several key principles are involved:

- Components and subassemblies should be repairable and/or replaceable, preferably by the customer.
- Upgrading the system should be possible with exchange of modular parts, and should not involve the purchase of redundant components.
- Residues generated as a result of routine maintenance or repair should not contain toxic substances, and the amounts should be minimized.
- Manufacturers should ensure that infrastructures exist for the proper handling of maintenance residuals.

Maintainability of in-service products may involve the use of cleaners or lubricants. To the degree possible, products should be designed so that maintenance procedures involving dissipative materials are performed infrequently; a good example of this approach is the lengthening intervals required by automotive manufacturers between oil changes. Where the replacement of worn components or subassemblies can be anticipated, those components should be made so that they are easy to remove and replace. If possible, manufactured systems should be designed to indicate when maintenance is necessary rather than to rely on (necessarily conservative) maintenance schedules to ensure satisfactory performance. If a system requires maintenance involving materials with substantial environmental impact, such as CFCs in air conditioning units, the design and product support should encourage maintenance only by trained technicians so that loss can be minimized and recovery and recycling optimized.

Design for maintainability often involves a greater degree of commitment to customers than would otherwise be the case. This commitment implies the provision to customers of worry-free product function rather than an abdication of product responsibility by the manufacturer. Modular design and the replacement of defective or worn modules encourage maintenance contracts or other cooperative agreements, as does a commitment to proper treatment of liquid or gaseous residues generated as a part of maintenance procedures. These close relationships can encourage not only proper maintainability but also proper treatment of packaging and proper recycling of obsolescent products. It is apparent that a systems approach to industrial ecology and customer relationships is the manner in which intelligent corporations should proceed.

ENERGY CONSUMPTION AND SOLID-RESIDUE GENERATION
DURING CUSTOMER USE OF WOMEN'S KNIT POLYESTER BLOUSES

Although many products consume no energy and produce no residues when in use, some cause the bulk of their environmental impact after manufacture and before disposal. A common example is clothing, which undergoes many washing cycles during its lifetime, consuming energy in heating water for washing and air for drying as well as requiring resources for the production and disposal of the detergents. An assessment of these environmental interactions for the case of a women's knit polyester blouse has been prepared by Franklin Associates for the American Fiber Manufacturers Association.

Figure 18.2(a) shows the total energy requirements per million wearings (this turned out to be a convenient comparitive unit) for the blouses. The energy used during manufacture, Fig. 18.2(b), is divided as would be anticipated, resin production and fabric production requiring roughly equivalent amounts and other activities being much less important. Manufacturing energy is less than a fourth of that required for consumer use. The latter is allocated in Fig. 18.2(c), and is seen to be almost entirely due to laundering operations. In fact, if energy alone were a consideration, blouses should be replaced after every fourth wearing. Solid residues also occur predominantly during the customer-use phase; these are the municipal sludge attributable to the washing operation and the ash related to off-site energy generation.

What changes in customer-use patterns could improve the environmental responsibility of the blouse design? The study found that use of a cold wash cycle, thus eliminating the need for heating water, reduced overall laundering energy consumption by 60%. A less extreme alternative is lowering the water temperature but retaining a warm wash; a 10% temperature reduction reduced laundering energy use by 14%. Line drying was also a very beneficial activity, reducing overall energy consumption by 31%.

What do these results have to say to the product designer? An obvious answer is that she or he should try to develop blouses that can be cleaned effectively in cold water, perhaps the easiest change that can influence consumer behavior. A second potential change is to modify the product so that it line-dries quickly or dries mechanically in a shorter period of time while retaining an attractive appearance. Probably less effective, but still worth doing, is encouraging in the product labeling the use of lower-temperature water if a warm-water cycle is used, and of air drying.

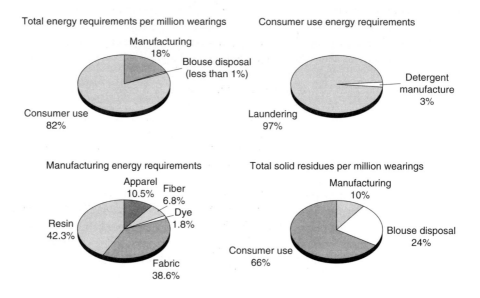

Figure 18.2 Data from an inventory assessment analysis of women's polyester blouses. (upper left) Allocation of total energy requirements per million wearings. (lower left) Allocation of manufacturing energy requirements. (upper right) Allocation of consumer-use energy requirements. (lower right) Allocation of total solid residues per million wearings. (Reproduced with permission from the American Fiber Manufacturers Association, *Resource and Environmental Profile Analysis of a Manufactured Apparel Product: Woman's Knit Polyester Blouse.* Copyright 1993 by American Fiber Manufacturers Association.)

SUGGESTED READING

Morehouse, Jr., E. T. Design for maintainability. *Design for Environment White Paper No. 7.* Washington, DC: American Electronics Association, 1992.

Saf-T-Clean Corporation. Pamphlets, available from Ronald H. Mulholland, Corporate Accounts Manager, Chicago, IL.

Stahel, W. R. The utilization focused service economy—Resource efficiency and product life extension in utilization, in *The Greening of Industrial Ecosystems*, B. R. Allenby and D. J. Richards, eds. Washington, DC: National Academy Press, 1994.

Stigliani, W. M., and S. Anderberg. Industrial metabolism at the regional level, in *Industrial Metabolism—Restructuring for Sustainable Development*, R. U. Ayres and U. E. Simonis, eds. Tokyo: United Nations University Press, 1993.

Volvo Car Corporation. *The Volvo Car Corporation's Fuel Database*, Environmental Report 26. Gothenburg, Sweden, 1991.

EXERCISES

18.1 For one week, audit the emissions to air, water, and soil directly relating from your activities. Could changes in the designs of the products that produced those emissions have minimized them? How?

18.2 As a user of many emissive products, you are a valuable data source. Based on the data you and your classmates generated in Ex. 18.1, determine which products have emitted the most material over the 1 week monitoring period. What changes in the collective consumption or use patterns of the class would reduce these emissions?

18.3 Conduct an audit of the in-use impacts of a photocopy machine convenient to you, using checklists 7 to 13, Appendix A, as possible and appropriate. Can you suggest design changes to minimize these impacts?

18.4 Conduct an audit of the in-use impacts of a washing machine convenient to you, using checklists 7 to 13, Appendix A, as possible and appropriate. Can you suggest design changes to minimize these impacts?

18.5 Conduct an audit of the in-use impacts of a passenger or freight train for which you can acquire data. Include any dissipative emissions that are present. Can you suggest design changes to minimize these impacts?

Design for Recycling

19.1 INTRODUCTION

The concept of industrial ecology is one in which the cyclization of materials at their highest possible purity and utility level is of prime importance. This cyclization can only occur if materials from products that have reached the end of their useful life reenter the industrial flow stream and become incorporated into new products. As Kumar Patel, currently with the University of California at Los Angeles, puts it, "The goal is *cradle to reincarnation*, since if one is practicing industrial ecology correctly there is no grave". The efficiency with which cyclization occurs is highly dependent on the design of products and processes; it thus follows that designing for recycling (DFR) is one of the most important aspects of industrial ecology.

The consequences of not having considered DFR in earlier periods of industrial design are graphically illustrated by a 1991 Carnegie Mellon University research project on personal computer disposal. It was estimated that by year 2005 some 150 million obsolete PCs, none with readily recoverable materials, would be landfilled. The required landfill volume is more than 8 million cubic meters, and the landfill cost some U.S. $400 million. If we consider also washing machines, refrigerators, automotive plastics, and all the other products now in use and not designed for recycling, the embedded stock of potentially unrecoverable materials is enormous. Hence, DFR may not only be advisable, it may be crucial to the ability of societies to survive on a reasonable basis.

In Chap. 8, we made the point that the lower down one goes toward the beginning of the chain of materials flows, the more energy must be invested to recover a unit of material. One is therefore advised not merely to make it possible to recycle a particular material, but to degrade that material as little as possible and avoid losing its embedded utility. Having stated this goal, it is nonetheless true that it is often difficult to avoid some level of degradation in a recycled material. In all cases, reuse of a material, even at a degraded level, is far better than discarding it. Hence, the reuse of polystyrene cafeteria trays as foamboard insulation or of polyethylene terephthalate soda bottles as carpet fiber is to be commended even while researchers attempt to develop levels of reuse that retain more of the embodied utility of the original material.

The point on embedded utility is made diagrammatically in Fig. 19.1, where four approaches to product maintenance and modification are shown. This concept may differ slightly for different types of products, but the basic principles are product-independent. From an industrial ecology standpoint, the preferred approach in the figure is to practice preventive and therapeutic maintenance for as long as possible, including upgrading to capture efficiency and performance gains resulting from technological innovation. Sooner or later, however, the impracticality of further maintenance or the availability of products with superior capabilities will call for major renovation or replacement. At that point, the characteristics embodied in the product by the physical designer will determine how high up the materials chain the recycling of product materials can be accomplished. The ideal design permits renovation and enhancement to be accomplished by changing a small number of subassemblies and recycling those that are replaced. Next best is a design that requires replacement of the product but permits many or most of the subassemblies to be recovered and recycled into new products. If subassemblies cannot be reused, attempts should be made to design components for recovery and use through several product cycles. Usually the least desirable of the alternatives pictured in the figure is removal of the product followed by recovery of the separate materials in it (or perhaps some of the embedded energy, if the product is best incinerated) and the injection of the materials or energy back into the industrial flow stream. Disposal of the product without the possibility for any of these recycling options is not an acceptable alternative from the industrial ecology viewpoint.

19.2 GENERAL CONSIDERATIONS

A directly practical reason for all industries to practice design for recycling is the trend for governments and other consumers to require or give preference to products incorporating the DFR philosophy. In 1991, for example, U.S. Government Executive Order 12780 was promulgated. It requires all agencies of the government (when combined, the agencies are the country's largest consumer) to buy products made from recycled materials and to encourage suppliers to participate in residue recovery programs. In the same year, the State of New York issued a Request for Proposal for personal computers for its offices in which it stated that recyclability would be a factor in choosing the successful bidder. The German "Blue Angel" environmental seal routinely includes such requirements in its assessments, and the U.S. Air Force is considering similar stipulations as part of its procurement process. Such actions provide a graphic and easily communicated rationale for DFR.

Another good reason to recycle materials is that many of them are of substantial commercial value. In the case of automobiles, for example, the aluminum has a scrap

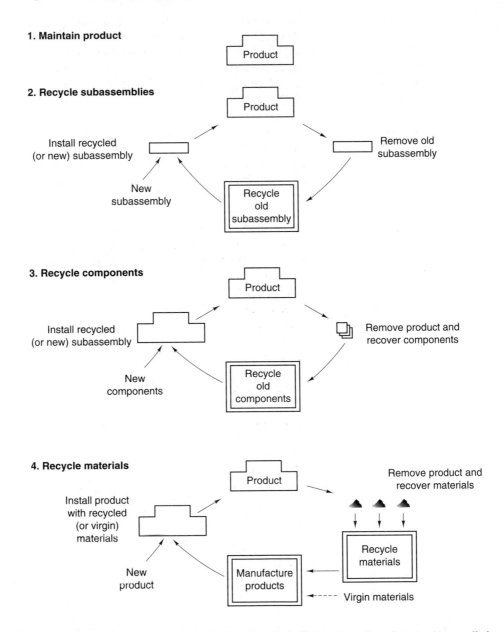

Figure 19.1 The heirarchy of preference in recycling of industrial products. Recycling should generally be accomplished as high up the chain as possible. Dashed lines indicate alternatives discouraged in the practice of industrial ecology. (See text for details.)

value of about U.S. $7000 per ton, the rubber about U.S. $260 per ton, and the steel about U.S. $100 per ton. There is essentially no value in the plastics and glass, which suggests in part that the cyclization of those materials needs to be better addressed.

Perhaps the most important consideration in DFR is to minimize the number of different materials and the number of individual components used in the design. To get a sense of the importance of this recommendation, picture yourself responsible for recycling hundreds of television sets or photocopy machines or airplanes every week. If you need to locate, sort, clean, and provide efficient recycling for two or three metals and two or three plastics, you are far more likely to be successful than if you must deal with five metals, four alloys, twelve plastics, and miscellaneous items such as glass or fabric. The functional and aesthetic demands of design sometimes make it difficult to limit materials diversity too greatly, but materials minimization should be a central focus for every physical designer.

A second general goal is to avoid the use of toxic materials. This topic has been discussed earlier with respect to the extraction or manufacture of toxic materials and their dissemination during industrial processes, but plays an important role also in the product recycling arena, where the presence of toxics is a deterrant to detailed disassembly, eventual reuse, or, if necessary, safe incineration and energy recovery. (These activities are aided if heavy metals and halogenated organics are avoided as product constituents.) When toxic materials must be utilized in a design they should be easily identifiable and the components that contain them readily separable, as are cadmium–nickel batteries and mercury relays.

Another general recommendation for the designer is not to join dissimilar materials in ways that make separation difficult. A simple example of a product not designed for recycling is the glass container for liquids whose top twists off while leaving a metal ring affixed; small cutting pliers are required for the conscientious housekeeper to properly sort the materials. More complex variations on this theme are metal coatings applied to plastic films, plastic overmolded over metal or over a dissimilar plastic, and the "up-scale" automobile dashboard, which is often a potpourri of metal, wood, and plastic. Any time a designer uses dissimilar materials together, she or he should picture whether and how they can eventually be easily separated, an important concept because labor costs tend to be a significant barrier to recycling.

When planning for product end-of-life, two complementary types of recycling should be considered: *closed-loop* and *open-loop*. As seen in Fig. 19.2, closed-loop recycling involves reuse of the materials to make the same product over again (sometimes called *horizontal recycling*), whereas open-loop recycling reuses materials to produce a different product (sometimes called *cascade recycling*). (Typical examples are aluminum cans to aluminum cans in the first instance, office paper to brown paper bags in the second.) The mode of recycling will depend on the materials and products involved, but closed-loop should generally be preferred unless (as with steel) the material can repeatedly undergo either open-loop or closed-loop recycling.

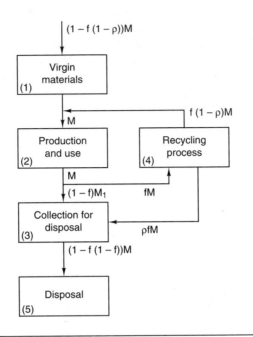

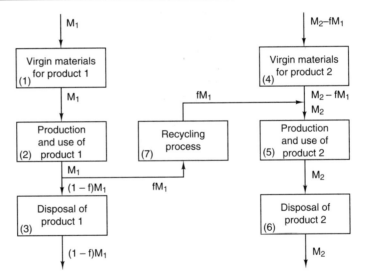

Figure 19.2 Closed-loop (top) and open-loop (bottom) recycling of materials. In the diagrams, M refers to mass flows, f to the fraction of the flow delivered to the recycling process, and ρ to the fraction of that flow rejected as unsuitable for recycling. (B. W. Vigon, D. A. Tolle, B. W. Cornaby, H. C. Latham, C. L. Harrision, T. L. Boguski, R. G. Hunt, and J. D. Sellers, *Life-Cycle Assessment: Inventory Guidelines and Principles*, EPA/600/R-92/036, Cincinnati: U.S. Environmental Protection Agency, 1992.)

19.3 REMANUFACTURING

Most products designed for long life do not wear out all at once: a mechanical part may fail, an essential fluid may leak out or become contaminated, or a critical component may become obsolete. As we noted in Fig. 19.1, recycling should preferentially take place as high up the embedded information chain as possible. An efficient way to accomplish this goal is by remanufacturing.

Remanufacturing involves the reuse of obsolescent or obsolete products by retaining serviceable parts, refurbishing usable parts, and introducing replacement components

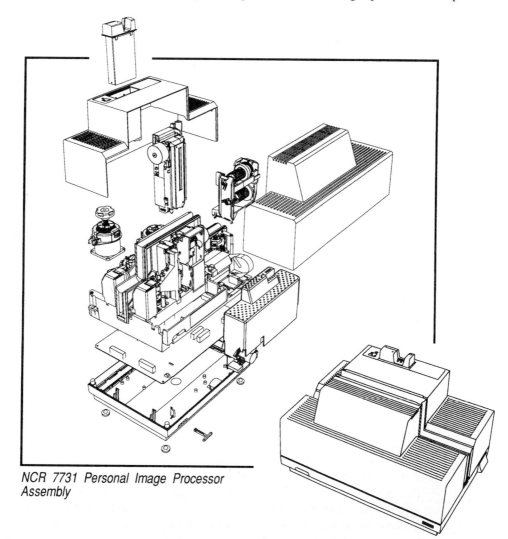

NCR 7731 Personal Image Processor Assembly

Figure 19.3   The NCR 7731 Personal Image Processor. This optical imaging product features modular design so that it can be adapted or refurbished to fit a customer's evolving needs without the necessity for total product disposal or replacement. (Courtesy of A. Hamilton, NCR Canada Ltd.)

(either identical or upgraded). Such a process is almost always cost-effective and almost always environmentally responsible. It requires close relationships between customer and supplier, frequently on a lease contract basis; these relationships are often competitive advantages in any case. Remanufacture also requires thoughtful design, because the process is often made possible or impossible by the degree to which products can be readily disassembled and readily modified.

A general concept of design for recycling is to make designs modular. If a designer can anticipate that a certain portion of a design is likely to evolve or to need repair or replacement and other portions of the product probably will not, the portion likely to change can be designed in modular fashion so that it, and it alone, will need to be replaced and recycled. The use of plug-in circuit boards in modern television sets is a good example of this philosophy. Another is NCR's Personal Image Processor, shown in Fig. 19.3, which embodies a high level of modularity throughout.

19.4 RECYCLING METALS

Pure metals are supremely recyclable, and many of them have been historically recycled to a very high degree. The recycling process involves the reentry of metal scrap into the refining process, often after a purification step largely involving the removal of oxides and other corrosion products.

Metals recycling is complicated by the use of mixed metals from different basic extraction processes. In the chapter on choosing materials, Table 16.2 showed the metals that commonly occur together and for which common extraction and purification processes have been developed. If possible, therefore, the use of metals should emphasize a single metal or metal grouping. This guideline is especially important if a small amount of a metal is used with a large amount of another metal, such as when steel is plated with cadmium. When the material is recycled, the plated metal is generally difficult and uneconomical to recover and tends to be discarded. A second example is when steel automotive scrap mixed with copper wire is recycled. Copper impurities in steel have substantial negative effect on its mechanical properties; aluminum wire is preferred if steel and an electrical wire are to be recycled together.

The desirability of avoiding the mixing of materials streams is neatly illustrated in Fig. 19.4, which shows that the selling prices of virgin materials vary approximately linearly with their degree of concentration in the matrix from which they are extracted. Also shown on the figure are points for a number of metals that are currently being extracted from residue streams rather than from virgin sources. In most cases, the residue streams are richer (i.e., less diluted) than virgin material matrices, so that efficient recycling operations can be expected to be financially rewarding. (Ironically, it is environmental regulations in many cases that make such desirable recycling uneconomical.) In the cases of arsenic and cadmium, recycling is now being done because of toxic hazard and not because of economic viability; that viability could be enhanced by efforts to decrease the dilution of those materials in residue streams.

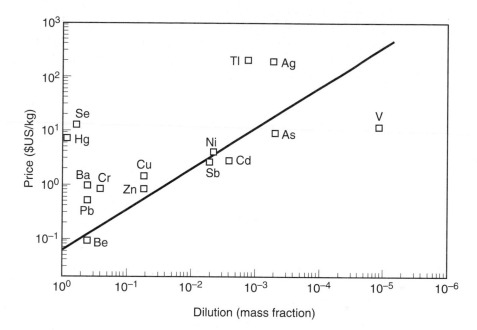

Figure 19.4 The relationship between dilution and price. The solid line shows the relationship between the price of virgin materials and their dilution in the matrices from which they are extracted. The individual points are for metal residues undergoing recycling. Additional discussion is contained in the text. (Reproduced with permission from D. T. Allen and N. Behmanesh, Preprint: Wastes as raw materials, in *The Greening of Industrial Ecosystems*, B. R. Allenby and D. J. Richards, eds., Washington, DC: National Academy Press, 1994.)

19.5 RECYCLING OF PLASTICS

Given careful attention to design and materials selection, many of the plastics in industrial use can be recycled. This is particularly true of thermoplastics, which can be ground, melted, and reformulated with relative efficiency. Among the thermoplastics for which recycling facilities now exist are polyethylene terephthalate (PET), polyvinyl chloride (PVC), polystyrene (PS), and the polyolefins [such as high-density polyethylene (HDPE), low-density polyethylene (LDPE), and polypropylene (PP)]. The utility of recycling these materials is a function of their purity, which implies that the use of paint, flame retardants, and other additives should be minimized or avoided if at all possible. Having plastics of many different colors in a product limits recyclability options as well. A further consideration is that designs should minimize the potential for plastics to become coated with oil or grease in use, since in-use adulteration also limits the efficacy of recycling.

Recycling is much more difficult with thermoset plastics, a group that includes phenolics, polyesters, epoxies, and silicones. Thermosets form crosslinked chemical bonds as they are created; recycling consists of reduction to lower molecular weight species by pyrolysis or hydrolysis. These processes are endothermic, however, and much of the embedded utility in the thermosets is thereby sacrificed. Incineration for energy recovery is preferable to landfilling, but represents a complete degradation of the material and is generally a recycling process of last resort.

When planning for recyclability of a product containing plastics, the thermal stability of the base resin and the other constituents must be considered, because reprocessing will involve thermal cycling of the material. It is usually the case that additives such as adhesives, paints, and coatings that are not removed prior to reprocessing tend to degrade in molding and extrusion machines. The result is outgassing, which inhibits or prevents the reprocessing cycle. Another potential difficulty occurs when mixed polymers are recycled. Degraded mechanical properties of the resulting material often result, but each individual case must be examined; a very small polyethylene part on a polycarbonate housing is satisfactory, for example, because a small degree of impurity does not compromise polycarbonate recycling.

No matter how efficiently a plastic can be recycled, if the relevant information concerning it is known, the multiplicity of plastics in use often makes it difficult to tell one from another, especially if recycling occurs a decade or more after product manufacture. To alleviate this problem, international standards have been developed for the marking of plastic parts. Although several versions have been promulgated, the most widely accepted is that of the International Standards Organization (ISO). Its recommendations are summarized in Appendix E. A firm rule for physical designers is that no plastic part of significant size should be used without its identity being marked upon it, and that marking should follow the ISO standard.

19.6   RECYCLING FOREST PRODUCTS

Of all the raw materials that humankind utilizes, the one that best approaches the cyclization goal of industrial ecology may be forest products: paper, cardboard, wood, and so forth. The raw material itself can be regenerated within a few decades and the processed material is often utilized successfully for very long periods of time, as in housing timbers and frames. Following use, a significant fraction of forest products is recycled. The most highly developed recycling system is the several-stage process for paper. At each recycling stage, the fibers in the paper become shorter and the acceptable use is more restricted, a normal cycle being from white bond to colored bond to newspaper to grocery bags to toilet paper.

A significant limitation on forest product recycling occurs when the forest product materials are combined with dissimilar materials, such as by adding plastic coatings to hot drink cups or using heavy-metal preservatives in timber. Most recycling systems find commingled dissimilar materials difficult to handle, with the result that adulterated forest products are often landfilled rather than returned to the materials stream.

19.7 FASTENING PARTS TOGETHER

The way in which parts are fastened together has a great effect on whether or not a product will be recycled after its useful life, independent of whether its materials were wisely selected. The challenge to the designer is to create a product that is rugged and reliable during service, but easily dismantleable when it becomes obsolete. Generally speaking, if a product is designed to be easily manufacturable (i.e., a minimum of fasteners, commonality of fasteners, modular components and few of them, and so forth), much of the design for recycling has probably already been accomplished. Given that general advice, designers should be aware of the environmental advantages and disadvantages of different types of fastening techniques.

Screws are perhaps the simplest of the recommended conventional ways to fasten components together. As few as possible should be used, and the number of different sizes and types should be minimized. (If a product is being assembled from modules produced by different suppliers, special attention to specifications is required to encourage uniformity.) An especially difficult design approach from the recycling standpoint is the use of threaded inserts embedded into plastic; these generally require heating to dislodge the inserts, which even then often cannot be separated by the magnetic approaches commonly used in recycling facilities.

In place of more traditional fastening approaches, a variety of quick-release connectors are available. Clips and similar fasteners are generally made of metal, but an increasingly common approach is to use sheets of hook-and-loop fasteners, especially for large panels such as automobile outer doors. Hook-and-loop fasteners are available in both normal and industrial strengths, are secure in use, and are readily detached after use. A more precise technique, requiring very accurate machining, is to use parts that snap tightly together, perhaps with break-out inserts, thus avoiding the use of any fasteners at all. Labels and corporate logos preferably should be embossed. If formed of a different material applied or inserted into a product, they should be readily removable.

Fastening techniques to avoid are those that make it difficult and time-consuming to disassemble a product. Rivets fall in this class, as do chemical bonds between similar or (worse) dissimilar plastics. Welds between metals are also very difficult to deal with in a recycling facility.

19.8 PLANNING FOR RECYCLABILITY

19.8.1 Design for Disassembly

Designing durable goods for disassembly may seem like an oxymoronic phrase, but such design is being done. An example of design for disassembly is the teapot recently brought to market by Polymer Solutions, Inc., a joint venture of GE and Fitch Richardson Smith, which makes the injection molded parts in the United States and uses British heating elements and switches (Fig. 19.5). Because the parts snap together, engineers found that the tolerance requirements were much more stringent than older assembly

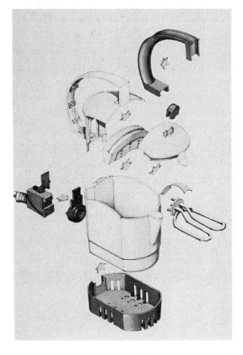

Figure 19.5   The teapot designed for dis-assembly by Polymer Solutions, Inc., for Great British Kettles Ltd. (Courtesy of Fitch, Inc., Columbus, Ohio.)

methods if leaking were to be prevented. In the short term, the enhanced tooling costs were a negative factor, but in the longer term, they are providing the impetus to develop the ability to fabricate products with much more precise and thus more desirable properties.

The demands of design for disassembly can often inspire great ingenuity. Another is a recent effort by BMW, whose Z-1 sports car has an all-plastic skin designed to be completely disassembled from its metal chassis in 20 minutes. The skin components are of recyclable thermoplastic supplied by GE Plastics Corporation. An unexpected side benefit of this design is that it has proven much easier to repair, because damaged components can be readily removed and replaced.

There are two methods of disassembly. One is *reversible disassembly*, in which screws are removed, snap-fit parts are unsnapped, and so forth. The second is *irre-*

versible disassembly, in which parts are broken or cut apart. If parts are designed with one of the two options in mind and the goals of rapid and efficient separation of components in mind, either option may be suitable. Otherwise successful design choices often can be disasters at disassembly time: Parts may require special tools to put together that are unlikely to be available at dissassembly or inserts or coatings may contaminate otherwise usable materials after brute-force disassembly. As with many other aspects of DFE, the simple and common is generally to be preferred.

The design-for-disassembly scenario can be appreciated by plotting the cost of different disposal options against the number of steps required for product disassembly (Fig. 19.6). If the product is to be landfilled, the highest cost generally occurs if no disassembly at all is performed, because the volume and difficulty of handling the product is at the maximum. That cost decreases as some disassembly is performed, but levels off before many steps have occurred. A contrasting behavior results if disassembly and reuse of modules or materials are to be performed. The cost of doing so obviously increases with the number of disassembly steps, and does so rapidly as the remaining modules become smaller and disassembly becomes more difficult. The designer can minimize end-of-life cost if the product can be more or less completely disassembled in only a few steps. Conversely, landfilling becomes the financially preferable option if many disassembly steps are required.

One aspect of designs that adds greatly to disassembly time is if it is difficult to identify the materials from which a product has been made, the functions of its modules, and other characteristics. Such considerations are of little concern if a manufacturer receives its own products for recycling, but may be a major recycling roadblock in facilities dealing with the products of numerous industrial organizations. To alleviate this difficulty, Sony's technology center in Stuttgart, Germany, has proposed that all products incorporate in their design a "Green Port", that is, an electronic module that contains retrievable product materials data in tamperproof form. The module would be made to an industrywide standard and addressable through a diagnostic connector. It seems likely that some variation of this idea will eventually be implemented, at least for reasonably expensive and long-lived products.

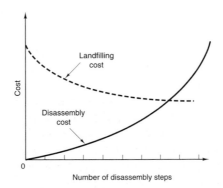

Figure 19.6   The conceptual relationship between the number of steps required for dissassembly of a product and the cost of end-of-life disposition options.

19.8.2 "Just-in-Case" Designs

Among the more unusual topics that enter into design for environment are those concerning products designed in the hope and expectation that their use may seldom or never be required. Spare parts are an obvious example, but entire products can fit the definition as well. Examples of the latter include emergency equipment such as air packs or medical supplies, sprinkler systems for fire minimization, and backup safety devices for elevators. The range of technology in such equipment extends to the most sophisticated, as in intercontinental ballistic missiles with computer-guided control systems.

Just as society needs "green products" in its everyday activities, so, too, it needs "green spares", "green rarely used items", and (an intriguing oxymoron) "green weapons". In other words, just because an item is designed with a reasonable expectation that it will sit around for a decade or two without being used and then be discarded, but must work the first time if needed, the designer of that item is not excused from the responsibilities of design for environment. Such features as materials selection, modularity, and design for disassembly need to be given special attention for "just-in-case" products. This topic has received almost no attention to date, but the enormous inventory of such materials and products involved suggests that the time for focused activity is well overdue.

19.8.3 Priorities for Recyclability

As final guidance to the designer, we reproduce a priority list of recycling options developed by Rosenberg and Terry in the American Electronics Association guide. In essence, this list is a word version of our Fig. 19.4, and a reiteration of our advice to maintain as much embedded information as possible in materials reaching the end of their useful lives. However, it makes sense to recycle only if the energy, environmental, and labor costs are such that recycling is preferable to not recycling, or to recycling while relinquishing embedded information. An example is that reusing or recycling beverage containers should be done only if it is easy and inexpensive to collect, transport, and reutilize the materials. Another example is that with old assemblies or equipment not designed for reuse or remanufacture, it is seldom possible to do more than recover the materials, and even that may be difficult and costly. Recycling decisions must be made from a logical perspective, therefore, and often that perspective requires a low-grade disposal of old, non-DFE items. Therefore, the following priority list is a guide, not an imperative.

Usually most preferable:	Reduce materials content
	Reuse components/refurbish assemblies
	Remanufacture
	Recycle materials
	Incinerate for energy (if safe)
Usually least preferable:	Dispose of as waste

19.8.4   Does It Make Sense to Recycle?

It is not automatically a sound policy to recycle products, and it is not automatically sensible to recycle only in a certain way. As with other aspects of industrial ecology, trade-offs are always present, and a comprehensive analysis is often needed to determine the most sensible approach to a specific problem. In particular, performing recycling should not result in a greater environmental impact than not performing recycling.

The impacts of the transportation involved in the recycling operation are often a major consideration when deciding on a recycling strategy. Even otherwise environmentally preferable activities, such as post-consumer product-takeback systems, can founder if the environmental costs of transportation are not considered. Germany's packaging-takeback system, for example, initially discouraged local processing by banning local plastics incineration for energy production. The result was large and environmentally undesirable shipments of light plastic materials around the country. Even greater transportation impacts per unit of packaging often occur when recycling of glass bottles is carried out far from the locations where the residue is generated by consumption. A case-by-case analysis is generally required to determine the best solution to any individual situation; this analysis must consider the full range of impacts, the value of the recovered materials, and the alternatives to low-grade recycling or none at all. Although deciding not to recycle a specific material or type of product seems intuitively wrong, a careful analysis will sometimes show it to be the best solution from an overall standpoint, at least in the short term.

WHAT TO DO WITH OLD TIRES?

Old tires are a good object lesson in disposal and reuse. The numbers of them are daunting—the United States alone throws away 250 million tires every year. For decades these tires were dumped in landfills and other less suitable places, but limited landfill capacity and a feeling that there must be better alternatives are gradually changing that approach.

Retreading tires is useful in lengthening service life, but merely delays the inevitable. Upon eventual discarding, a fraction of today's old tires are sent to modern facilities that shred and separate them into three flow streams: small tire chunks, steel shards, and crumbs. The steel is readily recyclable. The crumbs are burned for energy (each tire contains more than eight liters of recoverable petroleum). The chunks see a variety of uses—for running tracks, rubber boots, and rubberized asphalt, to name a few. As a possible alternative to the shredding operation, BOC Ltd. has developed a method to recycle scrap tires by first freezing them with liquid nitrogen and then grinding them efficiently while in the frozen state.

Although adequate technology is becoming available for tire recycling, appropriate economics may not yet be in place. Tire disposal costs have never been internalized; unlike aluminum beverage cans, tire users do not currently pay a recyling deposit upon initial purchase. The result is that any recycling that occurs happens only if it is profitable, and in many cases it is not. Part of the reason for this situation is that legislation often prohibits the use of tires as feedstock for incinerators (though the petroleum from which tires are made is an approved fuel) or makes it difficult to transport old tires into a city or across a state line to a recycling facility. It is obvious that the recycling industry, economists, and politicians all have roles to play if the issue is to be properly addressed.

The tire design engineer also has a role to play: designing tires for the environment. Today's recyclers are dealing with tires designed with no consideration for their eventual disposal. Perhaps a tire's composition can be changed to make it burn more efficiently while releasing few or no toxics. Perhaps a tire can be made so as to be more quickly and easily separated into its components. Perhaps a tire can be reformulated so as to be more readily transformed into a new product. In the case of old tires, the recycling engineers have made substantial progress and the economists and politicians are beginning to think about the situation. The design engineer has yet to start.

SUGGESTED READING

Henstock, M. E. *Design for Recyclability*. London: Institute of Metals, 1988.

Kirkman, A., and C. H. Kline. Recycling plastics today. *Chemtech, 21* (1991): 606–614.

Lave, L. B., C. Hendrickson, and F. C. McMichael, Recycling decisions and green design. *Environmental Science and Technology, 28* (1994): 19A–24A.

Lund, R. T. Remanufacturing. *Technology Review*, Feb./Mar. (1984): 19–28.

McPhee, J. Duty of care. *The New Yorker*, June 28, 1993, pp. 72–80.

Rosenberg, W., and B. Terry. *Design for Materials Recyclability*, AEA DFE White Paper No. 4. Washington, DC: American Electronics Association, 1992.

Stahel, W. R. The utilization-focused service economy—Resource efficiency and product life extension in utilization, in *The Greening of Industrial Ecosystems*, B. R. Allenby and D. J. Richards, eds., pp. 178–190. Washington, DC: National Academy Press, 1994.

EXERCISES

19.1 In a closed-loop recycling system, mass flow M is 5000 kg/hr, f is 0.7, and ρ is 0.1. Diagram the system and indicate all flow rates on the diagram.

19.2 In an open-loop recycling system, mass flows M_1 and M_2 are 8000 kg/hr and 6000 kg/hr and f is 0.6. Diagram the system and indicate all flow rates on the diagram.

19.3 In the open-loop system of Problem 19.2, assume that the recyling process rejects 15% of the material provided to it. Diagram this altered system and indicate all the flow rates on the diagram.

19.4 You are the designer of a table to be used for sorting fruit in a field near a cannery. The table is to have a steel surface and wooden legs. The surface is to be covered with a soft foam top to reduce fruit damage. It is expected that the foam top and the legs will need to be replaced periodically, and the cannery owner, who expects to purchase several hundred tables, wants component replacement to be quick and efficient. With the help of your local hardware store (if needed), design the table for optimum disassembly.

19.5 Choose a fairly complicated manufactured item, such as a personal computer, a bicycle, or a microwave oven, and perform a DFR analysis on two competing products using checklist 14, Appendix A, to the extent applicable and possible. What facets of the designs are favorable from a recycling standpoint? What are unfavorable? Which would you hope was chosen if you owned a recycling facility and expected to receive a few dozen each week for several years?

19.6 What are the appropriate ISO markings for plastic parts made of the following: (a) polyethylene terephthalate; (b) epoxy and polycarbonate; (c) polycarbonate with 15% glass bead filler; (d) polyethylene and polyvinyl chloride, filled with 10% carbon fiber and 15% mineral powder?

CHAPTER	**The Improvement Analysis**
20	**for Products, Processes,**
	and Facilities

20.1 THE IMPROVEMENT ANALYSIS STAGE OF LIFE-CYCLE ASSESSMENT

Of the three formal stages of life cycle assessment, we have thus far discussed inventory analysis and impact analysis, the first stage being reasonably well conceptualized, the second in the early stages of becoming so. The third stage, improvement analysis, is where "the rubber meets the road": That is, where the results of the analyses are translated into the specific actions that benefit the industry-environment relationship. The Society of Environmental Toxicology and Chemistry (SETAC), which originated the staged life cycle assessment concept, defines improvement analysis as follows:

> A systematic evaluation of the needs and opportunities to reduce the environmental burden associated with energy and raw materials use and environmental releases throughout the whole life cycle of the product, process, or activity. This analysis may include both quantitative and qualitative measures of improvements, such as changes in product, process, and activity design, raw material use, industrial processing, consumer use, and waste management.

In practice, it has proven difficult for corporations to carry out detailed life cycle inventories, more difficult to relate those inventories to a defendable impact analysis, and still more difficult to translate the results of the first two LCA stages into appropriate actions. There are several reasons for these problems:

- Comprehensive life-cycle inventories are expensive and time consuming, partly because the acquisition of quantitative information may require on-site analytical measurements or detailed reviews of files and records.

- Many LCA methodologies in current use are applicable to a limited subset of commercial products. This limitation is often not readily recognized. Techniques appropriate for assessing hot drink cups and diapers are not readily transferable to complex items such as computers, nor should they be expected to be.

- Impact analyses are inevitably contentious, in part because they involve value judgments in comparing and balancing different impacts. Accordingly, numerical assignments of impact are often not accepted as adequate guidance.

- It is not always appreciated that the various types of life-cycle assessments have been developed with different purposes in mind, and are not apposite for the full range of possible assessments. For example, a study intended to support an advertising or legal claim for a product must be more rigorous than one performed for internal use to stimulate environmentally preferable choices by the product design team.

- It is difficult to rank one new product design against another, an old product against a new product, or an old factory against a new factory, or to identify possible options and alternatives, based only on sporadically-performed life cycle inventory analyses and impact assessments with different approaches. However, making these comparitive rankings are exactly what corporate managements wish to do.

Dealing with these problems and at the same time producing improvement analyses that are useful to the decision-makers with whom one wishes to interact are difficult tasks at best. Our experience is that the LCA process works best when it is purposely done in modest depth and in a qualitative manner by an industrial ecology expert. The goal is to do the LCA rapidly, say, two days for a typical product, one week for a typical facility. In the remainder of this chapter, we describe techniques that can accomplish this goal and produce improvement analyses that have the potential to be promptly implemented.

20.2 EFFICIENT ASSESSMENT TOOLS

It has often been said that "What gets measured gets managed". Although it is easy for managers to support the principle of DFE, committing the organization to incorporating DFE as a matter of course requires a standard measuring system. A suitable system should have the following characteristics:

- It should lend itself to direct comparisons among rated products.
- It should be usable and consistent across different assessment teams.
- It should encompass all stages of product, process, or facility life cycles and all relevant environmental concerns.
- It should be simple enough to permit relatively quick and inexpensive assessments to be made.

20.2.1   The Product Assessment

The five life-cycle stages in a typical manufactured product are shown in Fig. 20.1. Stage 1, resource extraction, is performed by suppliers, drawing on (generally) virgin resources and producing materials and components. Stage 2 is the manufacturing operation directly under the control of the manufacturer. Stage 3, packaging and shipping, will usually be under the control of the manufacturer, although complex products containing many components and subassemblies may involve a global web of suppliers and contractors. Stage 4, the customer-use stage, is influenced by how products are designed and by the degree of continuing manufacturer interaction. (A leased product under maintenance contract maximizes the interaction, for example.) In Stage 5, a product no longer satisfactory because of obsolescence, component degradation, or changed business or personal decisions, is refurbished or discarded. This last stage, which traditionally has been performed by other than the manufacturer, will increasingly become a manufacturer's function as lease and product takeback activities expand. Stage 2 and perhaps Stage 3 have been customarily regarded as those where industry's environmental responsibility lies, but the evolving vision within corporations and without is that an

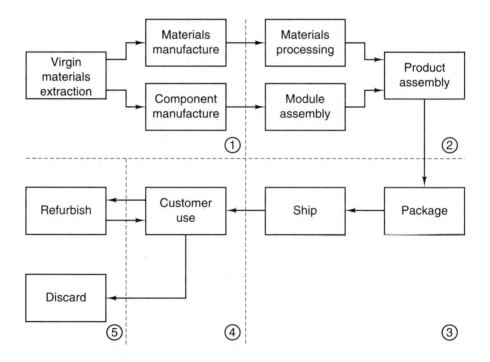

Figure 20.1   Activities in the five life-cycle stages (circled numbers) of a product manufactured for customer use. In an environmentally responsible product, the environmental impacts at each stage are minimized, not just those in Stage 2.

environmentally responsible product minimizes its external environmental impacts at all five of its life stages.

The abridged assessment system that we recommend has as its central feature a 5×5 assessment matrix, the environmentally responsible product assessment matrix, one dimension of which is life-cycle stage and the other of which is environmental concern (Fig. 20.2). The topics treated by the matrix are those discussed in the previous chapters in this section of the book. In use, the DFE assessor studies the product design, manufacture, packaging, in-use environment, and likely disposal scenario and assigns to each element of the matrix an integer rating from 0 (highest impact, a very negative evaluation) to 4 (lowest impact, an exemplary evaluation). In essence, what the assessor is doing is providing a figure of merit to represent the estimated result of the more formal LCA inventory analysis and impact analysis stages. She or he is guided in this task by experience, a design and manufacturing survey, appropriate checklists, and other information. Unlike some forms of abridged LCA (Chap. 11) which attempt to be quantitative but selective, the process described here is purposely semi-qualitative and utilitarian.

In arriving at an individual matrix element assessment, or in offering advice to designers seeking to improve the rating of a particular matrix element, the assessor can refer for guidance to the underlying set of checklists and protocols. The relationship between the matrix and the supporting information is shown in Fig. 20.3. An initial example of the contents of elements of the product improvement checklists is given in Appendix G.

The environmentally responsible product matrix					
	Environmental concern				
Life stage	Materials choice	Energy use	Solid residues	Liquid residues	Gaseous residues
Resource extraction	1,1	1,2	1,3	1,4	1,5
Product manufacture	2,1	2,2	2,3	2,4	2,5
Prod. packaging and transport	3,1	3,2	3,3	3,4	3,5
Product use	4,1	4,2	4,3	4,4	4,5
Refurbishment, recycling, disposal	5,1	5,2	5,3	5,4	5,5

Figure 20.2   The environmentally responsible product assessment matrix. The numbers are the designations of the matrix elements.

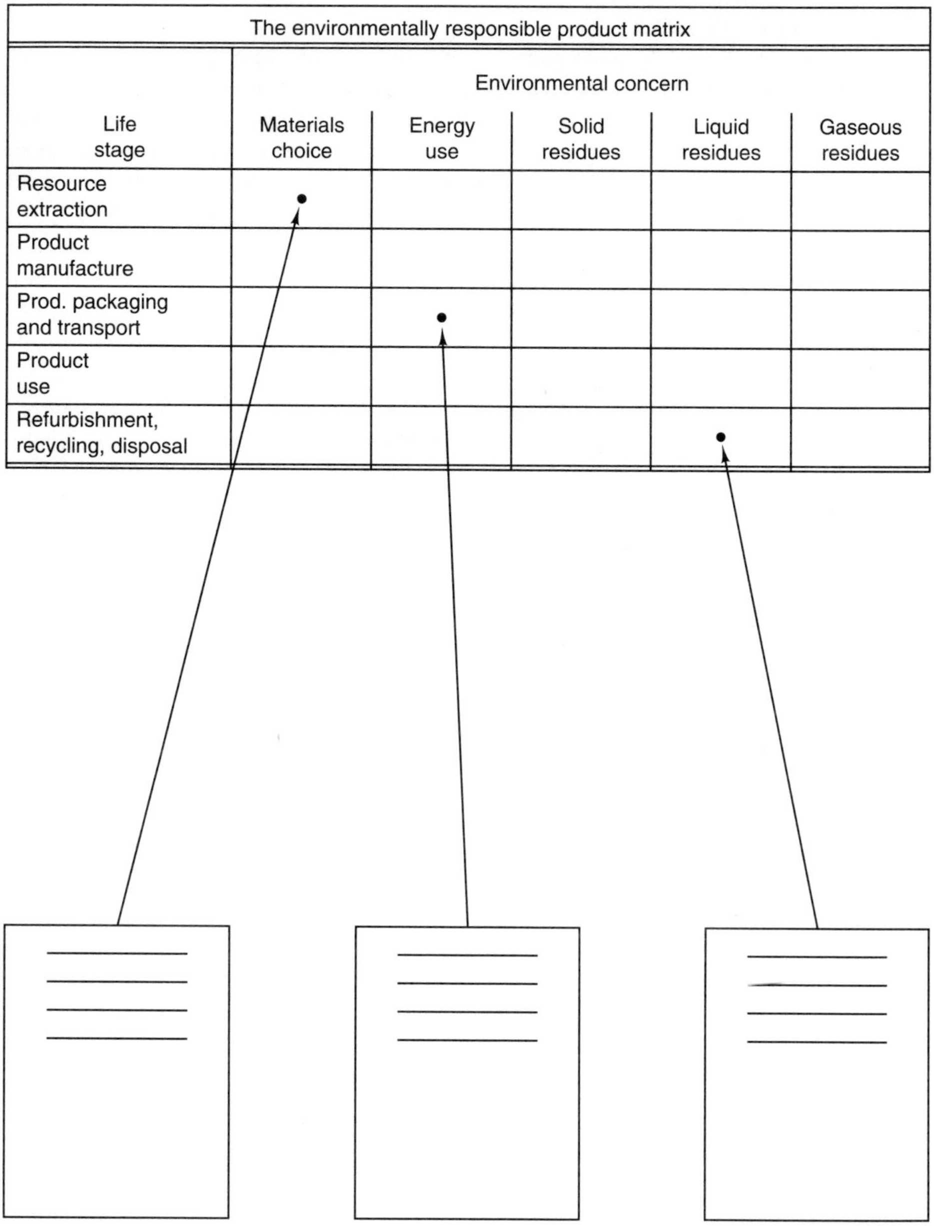

The environmentally responsible product matrix					
	Environmental concern				
Life stage	Materials choice	Energy use	Solid residues	Liquid residues	Gaseous residues
Resource extraction	●				
Product manufacture					
Prod. packaging and transport		●			
Product use					
Refurbishment, recycling, disposal				●	

Figure 20.3   The relationship between the environmentally responsible product assessment matrix and the underlying checklists and protocols that drive its evaluation.

Once an evaluation has been made for each matrix element, the overall environmentally responsible product rating (R_{ERPT}), is computed as the sum of the matrix element values:

$$R_{ERPT} = \sum_i \sum_j M_{i,j} \qquad (20.1)$$

Because there are 25 matrix elements, a maximum product rating is 100.

The matrix displays provide a useful overall assessment of a design, but a more succinct display of individual DFE design attributes is provided by the "target plot" shown in Fig. 20.4. To construct this plot, the value of each element of the matrix is plotted at a specific angle. (For a 25-element matrix, the angle spacing is 360/25 = 14.4°.) A good product or process shows up as a series of dots bunched in the center, as would occur on a rifle target in which each shot (using non-lead bullets, of course!) was aimed accurately. The plot makes it easy to single out points far removed from the bull's-eye and to mark their topics out for special attention. The product design teams and the facility operations organizations can then consult the checklist for that matrix element for information on improving individual environmentally-responsible product assessment matrix element ratings.

The assignment of a discrete value from zero to four for each matrix element implicitly assumes that the DFE implications of each element are equally important. An option for slightly increasing the complexity of the assessment (but perhaps increasing its utility as well), is to utilize detailed environmental impact information to apply weighting factors to the matrix elements. For example, a certain product might be thought to generate most of its impacts during product operation and few during resource extraction, so the product operation row could be weighted more heavily than before and the resource extraction row weighted correspondingly lighter. Similarly, a judgment that global warming constituted more of a risk than did liquid residues might dictate an enhanced weighting of the energy use column and a corresponding decreased weighting of the liquid residue column.

20.2.2 The Process Assessment

The growing willingness of manufacturers to accept responsibility for the environmental stewardship of their products has not been matched by a focused effort to make manufacturing processes more environmentally responsible across their life cycle, except for some "pollution prevention" activities. This situation exists in part because of the scarcity of usable metrics to guide the activity. The environmental assessment of processes across their life span is particularly important, since processes often remain in place for decades once they are developed and designs for new products will be developed with those processes in mind. Detrimental environmental impacts of processes are thus locked in for much longer times than is the case with the design and manufacture of individual products. Furthermore, since the implementation of a process generally involves the manufacture and installation of process equipment, both the impacts of the process materials (raw materials, solvents, and so on) and those related to the manufacture of the process equipment itself must be considered.

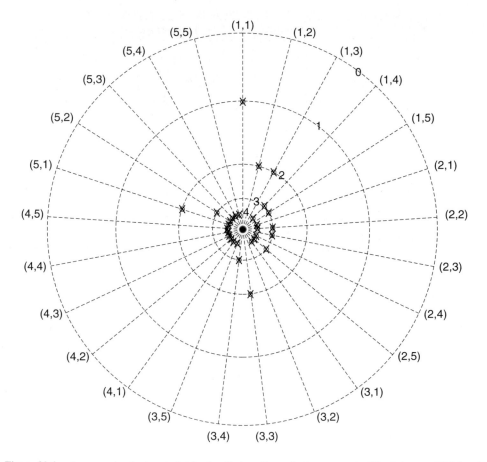

Figure 20.4   An example of a target plot for the display of a product assessment. (The data are artificial and were devised only for illustrating the display technique.)

The assessment system that we presented above for products can be readily adapted to the case of processes. As before, it features a 5 × 5 matrix, one dimension of which is life cycle stage and the other of which is environmental concern (Fig. 20.5). The environmental concerns are the same as those for products, but the life cycle stages are different. Their characteristics are described below.

20.2.2.1   Resource Extraction.   The first stage in the life cycle of any process is the extraction of resources from their natural reservoirs. The extractive activities to be considered are those used to produce the consumable resources used throughout the life of the process being assessed. In these considerations, recycled materials are nearly always preferable to virgin materials because they (1) avoid the environmental disruption that virgin material extraction involves, (2) generally require less energy in recycling than would be required for virgin material extraction, and (3) avoid landfilling or

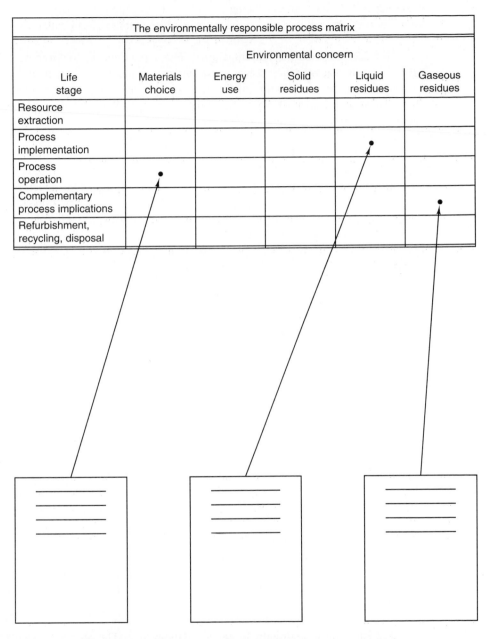

Figure 20.5   The environmentally responsible process assessment matrix.

other disposal of the material being recycled. In addition, the recycling of materials often produces less solid, liquid, or gaseous residues than do virgin materials extractions.

20.2.2.2 Process Implementation.

This topic treats the environmental impacts that result from the activities necessary to implement the process, principally including the manufacture and installation of the process equipment and other resources that are required. As a consequence, this row of the matrix has a strong commonality with the matrix for evaluating a product rather than a process.

20.2.2.3 Process Operation.

A process should be designed to be environmentally responsible in operation. Such a process would ideally limit the use of toxic materials, minimize the amount of energy required, avoid or minimize the generation of solid, liquid, or gaseous residues, and ensure that any residues that are produced can be used elsewhere in the economy. Effort should be directed toward designing processes whose secondary products are saleable to others or usable in other processes within the same facility. In particular, the generation of residues whose toxicity renders their recycling or disposal difficult should be avoided. Since successful processes can become widespread throughout a manufacturing sector, they should be designed to perform well under a variety of conditions.

An unrealizable goal but a useful perspective is that every molecule that enters a manufacturing process should leave that process as part of a saleable product. One's intuitive perception of this goal as unrealistic is inaccurate: certain of today's manufacturing processes, such as molecular beam epitaxy, come close, and more will do so in the future.

20.2.2.4 Complementary Process Implications.

It is often the case that several manufacturing processes form a symbiotic relationship, each assuming and depending upon the existence of others. Thus, a comprehensive process evaluation needs to consider not only the environmental attributes of the process itself, but also those of the complementary processes that precede and follow. For example, a welding process generally requires a preceding metal cleaning step, which traditionally required the use of ozone-depleting chlorofluorocarbons. Similarly, a soldering process generally requires a post-cleaning to remove the corrosive solder flux. This step also traditionally required the use of chlorofluorocarbons. Changes in any element of this system—flux, solder, or solvent—usually requires changes to the others as well if the process is to continue to perform satisfactorily. The responsible process designer will consider to what extent his process imposes environmentally difficult requirements for complementary processes, both in their implementation and their operation.

20.2.2.5 Refurbishment, Recycling, Disposal.

The process designer must be aware that all process equipment will eventually become obsolete, and must therefore be designed to optimize disassembly and reuse, either of modules (the preferable option) or materials. In this sense, process equipment is subject to the same considerations and recommended activities that apply to any product—use of quick disconnect hardware, identification marking of plastics, and so on. Many of these design decisions are made by the corporation actually manufacturing the process equipment, but the process designer can control or frustrate many environmentally responsible equipment recycling actions by his or her choice of features on the original process design.

20.2.2.6 The Rating Matrix.

In arriving at an individual matrix element assessment, or in offering advice to designers seeking to improve the rating of a particular matrix element, the assessor can refer for guidance to underlying checklists and protocols. An example of the contents of elements of such checklists and protocols is given in Appendix H, though different process types may require specific additions to these lists. Once an evaluation has been made for each matrix element, the overall environmentally-responsible-process rating (R_{ERPS}), is computed as the sum of the matrix element values:

$$R_{ERPS} = \sum_i \sum_j M_{i,j} \qquad (20.2)$$

There are 25 matrix elements, each with a maximum value of 4, so a maximum process rating is 100.

20.2.3   The Facility Assessment

A corporation designing its products, processes, and services with the DFE philosophy in mind should, of course, apply a similar approach to its facilities. The goal in this regard is to design or adapt facilities so that their environmental responsibility will constitute a competitive advantage for the corporation on both short and long time scales. Although DFE approaches to products achieve some of this goal, they leave other topics unaddressed. These facility-related topics, together with assessments of the products and processes of a facility, constitute the environmentally-responsible facility (ERF) evaluation.

ERF assessment need not and should not be applied only to manufacturing facilities, but rather to any facility engaged in products or services—oil refineries, auto body shops, fast food restaurants, office buildings, and so forth. The assessment will obviously be more complex in some cases than in others, but managers of all types of facilities should strives to ERF status.

20.2.3.1 Site Selection, Development, and Infrastructure.

A significant factor in the degree of environmental responsibility of a facility is the site selected and the way in which that site is developed. If the facility is an extractions or materials processing operation (oil refining, ore smelting, and so on), the location will generally be constrained by the need to be proximate to the resource. A manufacturing facility usually requires access to good transportation and a suitable work force, but may be otherwise unconstrained. Service facilities, in many cases, must be located near customers. Office buildings may be located virtually anywhere.

Manufacturing plants have traditionally been in or near urban areas. Such locations have the advantages of drawing on a geographically concentrated work force and of using the existing transportation and utility infrastructures. It may be possible to add new operations to existing facilities, avoiding many of the regulatory difficulties of establishing a wholly new plant site. One of these difficulties for an urban site, generally an insuperable one at present in some countries, is a legal system that forces purchasers of property formerly used for commercial or industrial purposes to assume any environmental liabilities attributable to the actions of the previous owner or owners. The result has been that urban industrial areas, in many ways the ideal sites for industrial facilities from an industrial ecology standpoint, are virtually impossible to use. The governmental and legal systems need to devise a means around this difficulty, which clearly produces less than optimum results. At the same time, they need to provide protection against the preferential siting of particularly hazardous or environmentally-undesirable facilities in minority locales, the so-called "environmental equity" issue. The enlightened design of environmentally-responsible facilities should avoid many of the difficulties that could arise from these dual concerns.

For facilities of any kind built on land previously undeveloped as industrial or commercial sites, ecological impacts on regional biodiversity can be anticipated, as well as added air emissions from new transportation and utility infrastructures. These effects can be minimized with attention to working with existing infrastructures and developing the site with the maximum area left in natural form. Nonetheless, given the current overstock of commercial buildings and facilities in many countries, such "green field" choices are hard to justify from an industrial ecology perspective.

As discussed in earlier chapters, it is sometimes possible for a facility to use a residue stream from one process as a feed stream to another, to use excess heat from one process to heat another, and so on. Such actions constitute steps toward a *facility ecosystem*. Chemical manufacturing plants, in particular, have made good progress along these lines. Opportunities also exist to establish portions of *industrial ecosystems* when facilities of different corporations agree to share residual products or residual energy. Such an approach is encouraged by geographical proximity. For example, an AT&T manufacturing plant in Columbus, Ohio, is about 1 km from a solid-waste landfill that had emitted methane gas as a by-product of the biodegradation of landfilled material. AT&T made arrangements to purchase the gas from the landfill facility and pipe it to its plant, where it furnishes up to 25% of the necessary energy for manufacturing. At the same time, a source of methane, a greenhouse gas, is eliminated.

More complex arrangements are possible, especially if planning is done before facilities are built. These cooperative operations involve establishing close relationships with suppliers, customers, and neighboring industries, and working with those partners to close the materials cycles of industrial ecology. In the same way that close relationships promote just-in-time delivery of supplies and components, so those relationships can be strong aids in the implementation of a corporation's environmentally responsible approach to manufacturing. An outstanding example of the partnership approach exists in Kalundborg, Denmark, where 20 years of effort have culminated in the interactive network shown in Fig. 20.6. Four main participants are involved: the Asnaes Power

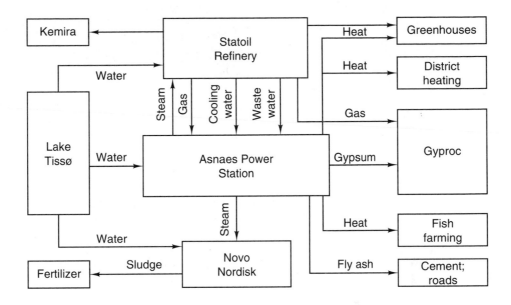

Figure 20.6   The industrial ecosystem at Kalundborg, Denmark.

Company, a Novo Nordisk pharmaceutical plant, a Gyproc facility for producing wallboard, and a Statoil refinery. Steam, gas, cooling water, and gypsum are exchanged among the participants, and some heat also is used for fish farming and home and greenhouse heating. The residual products sulfur, fly ash, and sludge, not usable in the immediate vicinity, are sold for use elsewhere. None of the arrangements was required by law; rather, all were negotiated independently for reasons of better materials prices or avoidance of materials disposal costs. It is probably accurate to refer to this cooperative project as an early model of an *industrial ecosystem*. The Kalundborg experience provides a model for industrial ecology anywhere, especially where industrial activities occur in close proximity to one another.

20.2.3.2 Principal Business Activity-Products and Principal Business Activity-Processes.
An essential element of an environmentally responsible facility is the production of environmentally preferable products and the use of environmentally preferable processes. These may be evaluated in two ways: by performing a facility mass-balance analysis or by using the product and process matrix systems previously discussed.

Manufacturers (using the term broadly to include such industrial activities as agriculture, forestry, and commodity chemical production) often wish to know the flows not of a particular process or product stream but of the facility as a whole. At least in concept, facility budgets follow the same approaches as used for less-integrated budgets. An example is shown in Fig. 20.7, which schematically illustrates a facility in which a

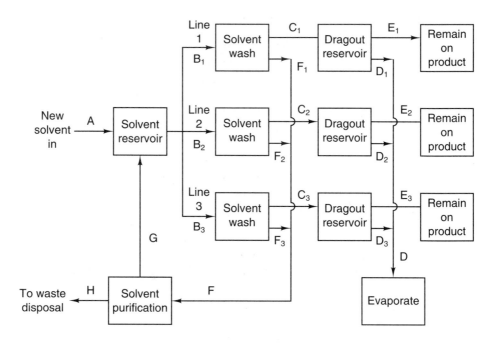

Figure 20.7   Schematic diagram of the flow streams involved in a budget analysis of a manufacturing facility with several solvent washing processes.

solvent wash is used on three different production lines running at different rates, the solvent being drawn from a common reservoir and returned to a common purification device. By analogy with the process budgets in Chap. 9, the facility mass balance equation for the solvent can be written as

$$A = \sum_{\lambda}(D_\lambda + E_\lambda) + H \tag{20.2}$$

where A is the rate of solvent supply, D is the rate at which it remains on products, E is the rate at which it evaporates, H is the rate at which disposal occurs. λ is the production-line index. The goal of the mass-balance analysis is, of course, to attempt to devise ways in which to minimize D, E, and H in order that A may be as low as possible.

One could equally well set up materials budgets around product lines, an activity that might be done if one wished to discover, for example, which of several product lines was generating the largest volume of a residual material common to several products. For agriculture or forestry, such mass balance equations can be developed for pesticide or herbicide inputs, or even for carbon flows.

An alternative approach is to perform matrix analyses of major products and processes related to the facility, and use the results of the analyses to develop the needed rankings.

20.2.3.3 Environmental Interactions Related to Facility Operations.   The impact of any facility on the environment is heavily weighted by transportation. As with many other aspects of industrial ecology, tradeoffs are involved. For example, just-in-time delivery of components and modules has been hailed as a cost-effective and efficient boon for manufacturing. Nonetheless, it has been estimated that the largest contribution to emissions in Tokyo smog comes from trucks making just-in-time deliveries. The corporations delivering and those receiving these components and modules bear some degree of responsibility for those emissions.

It is sometimes possible to reduce transport demands by improved scheduling and coordination, perhaps in concert with nearby industrial partners or by siting facilities near to principal suppliers. Options also exist for encouraging ride sharing, telecommuting, and other activities that reduce overall emissions from employee vehicles.

Material entering or leaving a facility also offers opportunities for useful action. To the extent that the material is related to products, it is captured by the product LCA assessments. Facilities receive and disperse much nonproduct material, however: food for employee cafeterias, office supplies, restroom supplies, and maintenance items such as lubricants, fertilizer, and road salt, to name just a few. An ERF has a structured program to evaluate each incoming and outgoing materials stream and to tailor it and its packaging in environmentally-responsible directions.

Facility energy use requires careful scrutiny as well, as opportunities for improvement are always present. An example is industrial lighting systems, which are responsible for between 5–10% of air pollution emissions. As with many environmentally-related business expenditures, lighting costs are often lumped in with "overhead" and not precisely known, yet the use of modern technology has the potential to decrease electrical expenditures for lighting by 50% or more. To promote these changes, the U.S. EPA has initiated the "Green Lights" program, which encourages the use of high-efficiency fluorescent ballasts and lamps, automatic turnoff of lights when not in use by means of occupancy sensors, and mirrorlike reflectors in existing fluorescent systems. Corporations agreeing to voluntarily participate in this program commit themselves to surveying their lighting and to upgrade systems where pollution reductions, lighting quality improvement, and profit goals are all met. Several states and several hundred corporations have agreed to participate.

20.2.3.4 Facility Refurbishment, Transfer, and Closure.   Just as environmentally-responsible products are increasingly being designed for "product life extension", so ERFs should be. Buildings and other structures contain substantial amounts of material with significant levels of embedded energy, and the (especially local) environment disruption involved in the construction of new buildings and their related infrastructure is substantial. In the U.S., for example, the Bureau of Mines notes that construction materials constitute the largest societal material use by far: in 1990, some 2.53 billion metric tons of material were consumed, of which about 70% was construction materials. Clearly an ERF must be designed to be easily refurbished for new uses, to be transfered to new owners and operators with a minimum of alteration, and, if it must be closed, to permit recovery for reuse and recycling of materials, fixtures, and other components.

20.2.3.5   The Rating Matrix.   Just as with products and processes, one can use the matrix assessment technique to evaluate the environmental responsibility of a corporate facility. The system we recommend is identical in concept to that for products or processes. Its central feature is a 5×5 environmentally responsible facility assessment matrix, one dimension of which is environmental concern and the other of which is facility activities (Fig. 20.8). As with products and processes, the assessor studies the different facility activities and their impacts and assigns to each element of the matrix a rating from 0 (highest impact, a very negative evaluation) to 4 (lowest impact, an exemplary evaluation).

In arriving at an individual matrix element assessment, or in offering advice to designers seeking to improve the rating of a particular matrix element, the assessor can extract additional detail from a facility checklist system, which provides checklists and evaluation techniques by which the environmentally responsible facility assessment matrix values are determined. The detailed contents of the checklists will be specific to the type of facility being assessed, rather than generic as is the environmentally responsible facility assessment matrix. For the perspective of the reader, however, a basic checklist system for a manufacturing facility is given in Appendix I. The overall environmentally responsible facility rating (R_{ERF}) is the sum of the matrix-element values:

$$R_{ERF} = \sum_i \sum_j M_{i,j} \qquad (20.3)$$

Because there are 25 matrix elements, a maximum facility rating is 100.

The environmentally responsible facility assessment matrix is principally a management tool. After the overall rating for a facility is determined, the use of a "target plot" will permit points far removed from the bull's-eye to be singled out for special attention.

20.3   LIMITATIONS AND BENEFITS OF IMPROVEMENT ANALYSIS

Unlike classical inventory analysis and perhaps impact analysis, overall life cycle assessment as presented in this chapter is less quantifiable and less thorough. It is also inestimably more practical. A survey of the modest depth that we advocate, performed by an objective professional, will succeed in identifying 70 or 80 percent of the useful DFE actions that could be taken in connection with corporate activities, while consuming sufficiently small levels of time and money that it has a good chance of being accomplished. In our view, it is far better to conduct a number of LCAs of modest depth than to conduct one or two in great depth. A key ingredient in a successful LCA of this type is the expert who performs it. This person, whether from inside the corporation or outside it, must be experienced, knowledgeable about the types of products, processes, and facilities being reviewed, and reputable enough so that the recommendations are thoroughly integrated into corporate planning processes.

Improvement analysis is the ultimate goal of all industrial ecology activities. As with most ecological situations, however, the actions that are taken as a result will be

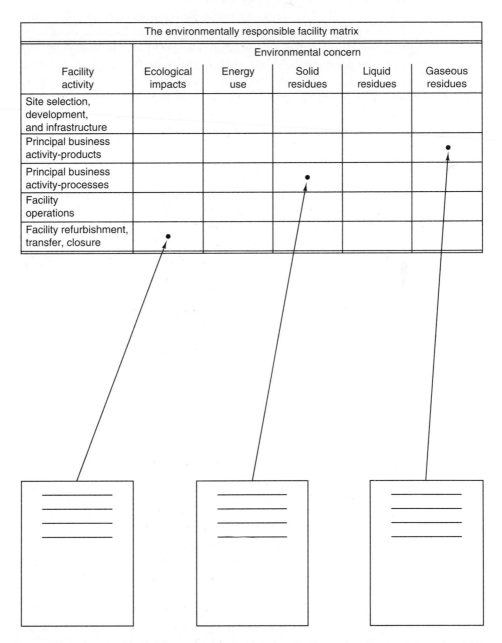

Figure 20.8   The environmentally responsible facility assessment matrix. The considerations and desirable actions pertaining to each matrix element are described by checklists and recommendations specific to the type of facility under evaluation.

compromises based on a variety of tradeoffs. One should not enter into a life-cycle analysis of a product, process, or facility with the idea that all possible actions that are identified for consideration can be accomplished. Non-environmental objectives and constraints are always present and their benefits must be balanced against those recommended by the LCA analysis. However, one will become informed about the existence of a number of design or operations choices that might be modified and can then implement all that are practical. The result will in each case be much more environmentally sustainable than if nothing were done. As a consequence, it will assist the corporation in building its long term environmental performance.

SUGGESTED READING

Hernandez, A. *Just-in-Time Manufacturing: A Practical Approach.* Englewood Cliffs, NJ: Prentice Hall, 1989.

Knight, P. A rebirth of the pioneering spirit. *Financial Times* (London) Nov. 14, 1990.

Nelson, K.E. Practical techniques for saving energy and reducing waste, in *Industrial Ecology and Global Change*, R. Socolow, C. Andrews, F. Berkhout, and V. Thomas, eds. Cambridge U.K.: Cambridge University Press, 1994.

EXERCISES

20.1 Assume that the product inventory results of Exercise 9.3 are assessment matrix results and draw target plots for each. Discuss the relative value of the plots and of the matrix summation techniques you devised in Exercise 9.2.

20.2 Choose a convenient group of several facilities and explore possibilities for infrastructure interaction. (a) In a report on the results, comment on whether additional carefully selected facilities or transportation of certain materials in or out would enhance the industrial ecosystem. (b) Evaluate the robustness of the system to a disruption in any participant's operations, and discuss how any system-wide disruption impact could be minimized.

20.3 Locate a facility that is willing to let you conduct a facilities assessment. Possible facilities include university cafeterias, hardware stores, appliance repair shops, and small manufacturing concerns. Prepare an assessment matrix and target plot and share the results with the facilities manager and the class.

PART V: FORWARD-LOOKING TOPICS

CHAPTER
21

Organizational Opportunities and Constraints

21.1 THE BENEFITS OF ECO-EFFICIENCY

Many of the world's largest corporations are coming to the realization that business activities and environmental concerns are not antithetical, but rather are essential components of excellent organizational approaches. That is, the topic has moved from one of altruism to one of competitiveness. Stephan Schmidheiny, Chairman of the Business Council for Sustainable Development, has coined the term "eco-efficiency" to describe corporations that produce ever more useful goods and services while continuously reducing resource consumption and pollution. He lists six reasons why it makes sense for corporations to do so:

- Customers are demanding cleaner products.
- Environmental regulations are stringent and will become even more so.
- Employees, particularly the best ones, prefer to work for environmentally responsible companies.
- Banks are more willing to lend money to companies that prevent pollution.
- Insurance companies are more amenable to covering environmentally responsible companies.
- New economic instruments—taxes, charges, and tradable permits—are rewarding environmentally responsible companies.

Given these undeniably good reasons for action, what are the prospects that calls for response will be heeded? The answer lies with the interplay among the traditional structures of corporations, their response to external stimuli, and the changing demands of customers. It is worth exploring some of those issues to get a feeling for how rapidly and universally corporations are likely to respond to the environmental challenges they face.

21.2 INFORMATION FLOW INTO CORPORATIONS

Information flows toward corporations from many directions and many different constituencies. Bruce Paton of Hewlett-Packard Corporation has made the point that information from some sources is assimilated and acted upon readily and information from other sources is largely ignored. His vision is reproduced in Fig. 21.1. From the standpoint of industrial ecology, the important message of the figure is that the wishes of customers and the pressures of competition are efficiently incorporated into industrial decision-making systems. Government regulations, although perhaps not welcome, are generally recognized as a common constraint about which corporations are informed and to which they adhere. The interests and information from environmental scientists, non-governmental organizations, and individuals are only heard dimly. It is unrealistic to

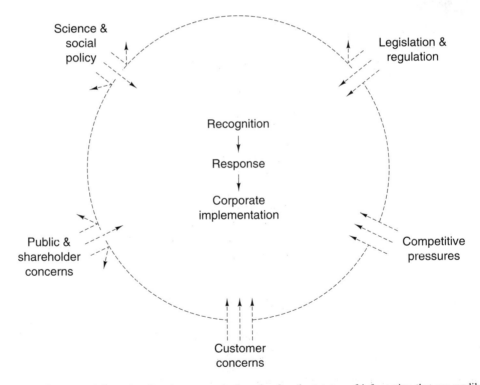

Figure 21.1   The information flow into a corporation, showing those types of information that are readily recognized and responded to, and those types that go unrecognized. (Adapted with permission from B. Paton, Design for environment: A management perspective, in *Industrial Ecology and Global Change*, R. Socolow, C. Andrews, F. Berkhout, and V. Thomas, eds., Cambridge, UK: Cambridge University Press, 1994.)

expect much change in these responses. The appropriate question to ask, rather, is how this information flow structure can be used to promote industrial ecology.

Perhaps the most efficient promotion technique is through listening to customer preferences. Customer voices are readily heard by industrial corporations, and customers of all kinds, individual, corporate, and governmental, are rapidly becoming more environmentally sophisticated. Hence, the desires of customers for products manufactured under some or all of the guidelines espoused by industrial ecology will provide strong impetus to corporations to produce products of that type. Examples of this stimulus include a recent request by New York State for bids on personal computers, which stated that environmentally responsible design and manufacturing would be a consideration in the awarding of contracts, and a U.S. Government announcement that it will begin to incorporate the environmental attributes of products into all of its purchasing activities.

Just as customers can place demands on manufacturers, so manufacturers can place demands on their suppliers, a process known as "supply-line management". To the extent that these interrelationships are promoted, the practice of industrial ecology has the potential to spread much more rapidly than would otherwise occur.

21.3 CORPORATE-RESPONSE STRATEGIES

Although corporations differ greatly in their responses to external stimuli, it is well established that those corporations that aggressively work to maintain profits by focusing on their customers are most likely to make the most forward-looking and aggressive business decisions. They are also somewhat more willing than average to adopt ideas and technologies developed outside their own corporations. These characteristics fit those firms that are now beginning to implement industrial ecology within their corporate structure, as opposed to those who are content to comply with regulations but to go no further. In this context, consider the matrix in Figure 21.2. Issues falling in the upper left corner, such as paying the salaries of employees or buying needed raw materials, are always done if the corporation is to stay in business. Issues falling in the lower right corner, such as the production of goods for which the potential financial return is negligible, are seldom done. Issues falling in the lower left corner, such as compliance with government regulations, are generally accomplished, but with reluctance. Issues falling in the upper right corner are those that challenge executives; their resolution is unlikely to have much impact on short-term corporate success, but could have major impact on long-term success. Industrial ecology falls in this quadrant, though it hovers near the quadrant division point. To the extent that corporate operations can deemphasize environmentally related activities in the lower left quadrant and replace them with environmentally related activities in the upper right quadrant, and to the extent that public policies can move corporate concerns along the same diagonal (as in the U.S. substitution of the "Clean Car Initiative" for Corporate Automotive Fuel Efficiency (CAFE) requirements), the corporation will be aiding its long-term success in a world in which governments and individuals are increasingly interested in long-term environmentally responsible corporate behavior.

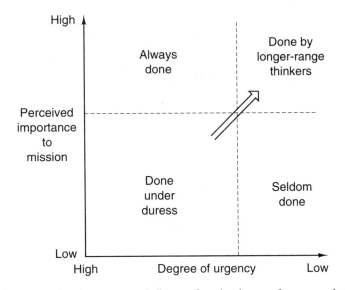

Figure 21.2   Corporate actions in response to challenges of varying degrees of urgency and perceived importance to the corporate mission. (Adapted from a diagram devised by S. Rayner, Battelle Pacific Northwest Laboratories.)

Corporations are becoming increasingly multinational as they seek expanding or newly opened markets for their products. Markets are often opened only if the entering multinational corporation provides jobs and does some portion of manufacturing within the country. As a consequence, it has become not uncommon for corporations to own or share ownership in manufacturing facilities in a score of countries and to have sales offices in many more.

Although the day of the rapacious multinational corporation has not passed, it is waning as a consequence of an increasing sense of corporate environmental responsibility, increased monitoring of corporate activities, increasing uniformity of environmental regulations from country to country, and potential liability if corporate standards differ in different locations. An increasing number of corporations are applying similar stringent environmental standards to their operations throughout the world, no matter what the local regulations (and within the constraints of local environmental management infrastructures). To the extent that multinational corporations transfer advanced technology throughout their global operations and require local suppliers to adopt environmentally responsible practices, the multinationals have the potential to encourage industrial ecology at the local and regional level everywhere.

21.4   PHASING-IN INDUSTRIAL ECOLOGY

21.4.1   Major Goals and Principles

The information presented heretofore in this book has implicitly assumed that a product or a process is being designed with no constraints imposed by embedded design protocols or restrictions, by an existing manufacturing process facility, or by the existing organization structure of the corporation. In practice, hardly anything starts from scratch, and designers are faced with implementing industrial ecology by phasing it into existing operations. In such circumstances, instead of questions such as "What is the best possible solution to my design challenge if industrial ecology is taken into account?", the questions are along the lines of "What can be done to be responsive to the concerns of industrial ecology given my existing product designs and existing manufacturing capabilities?" The most efficient way is to consider the existing situation in light of the goals and principles stated earlier in this book, the most significant and all-encompassing of which are the following:

- Every molecule that enters a specific manufacturing process should leave that process as part of a saleable product.

- Every erg of energy used in manufacture should produce a desired material transformation.

- Industries should make minimum use of materials and energy in products, in processes, and in services.

- Industries should choose abundant, nontoxic materials when designing products.

- Industries should get most of the needed materials through recycling streams (theirs or those of others) rather than through raw materials extraction, even in the case of common materials.

- Every process and product should be designed to preserve the embedded utility of the materials used. An efficient way to accomplish this goal is by designing modular equipment and by remanufacturing.

- Every product should be designed so that it can be used to create other useful products at the end of its life.

- Every industrial landholding or facility should be developed, constructed, or modified with attention to maintaining or improving local habitats and species diversity, and to minimizing impacts on local or regional resources.

- Close interactions should be developed with materials suppliers, customers, and representatives of other industries, with the aim of developing cooperative ways of minimizing packaging and of recycling and reusing materials.

Working with these goals and principles, and with the techniques given herein, corporations can gradually bring their operations into modes that are increasingly consistent with industrial ecology.

21.4.2   Setting Priorities

The process of implementing a program to phase-in industrial ecology is aided by the use of Pareto diagrams, as shown in Fig. 21.3. Pareto diagrams come in various forms, but the general approach is to choose a dependent variable of substantial concern and

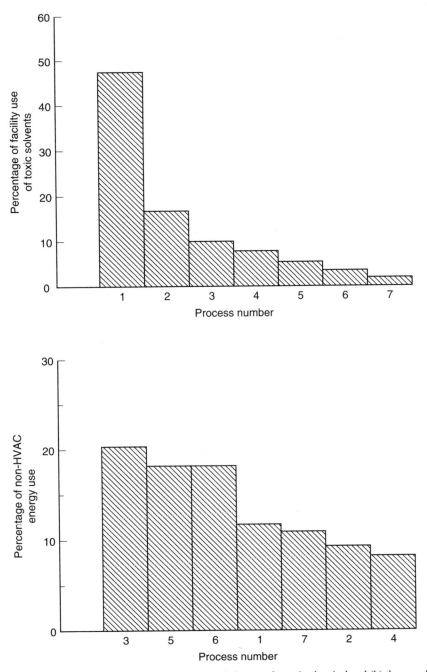

Figure 21.3   Hypothetical Pareto diagrams for (a) the use of a toxic chemical and (b) the use of energy in a manufacturing facility containing seven manufacturing processes.

compare the level of involvement of different corporate entities or components with that variable. In Fig. 21.3(a), for example, the use of a toxic substance in manufacturing is compared for several manufacturing processes within a single manufacturing facility. (The data are artificial, generated here for purposes of demonstrating the principle.) One immediately sees that Process 1 uses far more of the toxic substance than do the other processes, so its use in Process 1 is the highest priority for amelioration and, to the extent permitted by capital requirements, phase-in should begin by addressing the problems having the highest environmental impact. In contrast to solvent use, the use of energy within the facility, shown in Fig. 21.3(b), is less diverse, so efforts to reduce energy use can probably begin with the most convenient or least expensive process to work with rather than worrying too much about the order in which the processes are studied. The Pareto diagrams thus assist the industrial manager in prioritizing actions to phase industrial ecology into manufacturing operations.

21.4.3   Developing Uniform Corporate Standards

The growth of multinational corporations has resulted in the existence of many firms that manufacture products in facilities all over the world. To an increasing degree, these firms are rejecting the approach of minimal compliance with local environmental standards and instead are developing uniform corporate standards for environmental performance. Generally, these standards are equivalent to those of the most stringent country in which the firm operates. A frequent result is that the corporation's industrial operations are conducted in such a way that they are substantially ahead of local regulations. In addition to anticipating legislation that may eventually occur, this approach permits a firm to ship products or intermediate assemblies freely throughout the world without worrying whether a product made in one country will meet the standards for manufacture, packaging, disassembly, and so on of another.

Uniform standards, however, cannot be applied in a uniform way throughout the world, because local situations and options differ. For example, the Chevron Corporation explores for and produces petroleum in Nigeria, which has no facilities for the disposal of hazardous waste. Accordingly, the corporation uses water-based lubricating fluids in its drilling rather than the more efficient, but more hazardous oil-based compounds that are traditional. Thus, the ideal approach for a corporation is to adopt a uniform environmental perspective for its operations worldwide, but to encourage the intelligent application of those standards by taking local conditions into account.

21.4.4   Property Acquisition and Use

Property acquisition and use is one of the principal areas in which industry has the potential to influence two primary environmental concerns: loss of habitat and decrease in biodiversity. Each time a corporation buys or sells property, builds a new facility, or remodels an existing one, the potential exists for contributions to preserving local, regional, and global habitats and biodiversity. (The impacts of property decisions by the global forestry and agriculture sectors are obvious.) It should be standard practice, for example, that corporations establish new manufacturing operations in geographical areas where manufacturing has already occurred rather than on undisturbed land. (However, it is an unfortunate aspect of liability law in some countries such as the United States that this environmentally beneficial practice is strongly discouraged.)

A number of corporations whose operations involve large landholdings, such as forest products firms and petroleum exploration companies, have performed reasonably well in this regard. In the case of firms where the use and alteration of land is less obvious, effort is often needed. That effort can be modest, as in returning large percentages of "corporate campus" lawns to their native state, or it can be innovative and substantial, as with Merck's swap of Costa Rican land-preservation funds for the privilege of screening soil, plants, and microorganisms from that region for new molecules that may eventually lead to new medicines.

Unlike other aspects of industrial ecology, which are generally under the operational control of product and process designers and operating engineers, property acquisition and use is generally the responsibility of corporate officers. Ideally, the latter should see thoughtful property acquisition and use as their opportunity to contribute toward the implementation of industrial ecology by their corporation.

21.4.5 Energy Use

There can be no question that opportunities for energy savings abound in industry and that most of those opportunities produce financial gains, not costs. These financial gains are made all the more welcome by the simultaneous reduction in environmental impact, because profligate energy generation is one of the most hazardous long-term activities from an environmental standpoint. Among the energy-conserving activities that are advisable are the following:

- Replace light fixtures with compact fluorescent lighting, reflectors, and occupancy sensors, and utilize natural light to the extent practicable.
- Reduce the building cooling loads by improved heat pumps, window treatments, and chillers.
- Install more efficient motors and electrical drive systems.
- Install high-efficiency product and building furnaces.
- Adopt energy management procedures such as residual heat use in cogeneration facilities.

21.4.6 Solid Residues

Solid-residue generation occurs ubiquitously throughout manufacturing, and opportunities nearly always exist for reduction of the solid-residue stream. In addition to the relatively obvious generation of solid residue as a by-product of manufacturing, an additional consideration is the packaging residue from product components entering the facility from outside suppliers. Mutual efforts by suppliers and those supplied can often result in substantial reductions in packaging while still maintaining component quality and reliability.

Another solid-residue area sometimes deserving attention is the problem of "dying" or "dead" components inventory. This is inventory that is seldom or never needed because the products to which it applies are obsolete, inherently reliable, or repairable elsewhere, so that the components that are retained will eventually require disposal. Again, relations with suppliers are important to optimizing just-in-time delivery of components as needed.

21.4.7   Liquid Residues

The first step in dealing with liquid-residue streams is to examine all processes involving solvents. The experience of AlliedSignal's Kansas City Division is that straightforward conservation, including operator training, can generally reduce solvent usage by 5 to 20%. Substitution of readily available alternative technologies can save an additional 40 to 60%. The remaining 20 to 30% is more difficult and expensive to deal with, and should be the subject of long-range plans, not the initial phase-in program.

After minimization to the degree possible, liquid residue from a facility should be tailored for minimal environmental impact. Minimization is accomplished in phase-in by two major thrusts: the recycling of by-products and residue streams and negotiating with suppliers of liquid process chemicals to encourage them to accept your residual liquids for recycle and reuse. Neither course is easy; both are productive.

Minimal impacts are produced when liquid residues contain little or no material that has significant environmental consequences. In particular, the use of toxic heavy metals (aluminum, arsenic, cadmium, chromium, copper, mercury, nickel, lead, silver, tin) almost always leads to problems during disposal, either in the manufacturing facility or when the product is eventually recycled. One process worth mention is the plating of one metal on another. Plating inevitably produces toxic residue streams and creates materials that are difficult to recycle in environmentally benign ways. To the extent possible, heavy-metal use should be avoided.

Steps such as these are becoming increasingly common throughout industry. One bit of evidence is the number of voluntary commitments by corporations to the U.S. EPA's 33/50 program, which we described in Chap. 16. In this program, begun in 1990, corporations agreed to reduce their releases and offsite transfers of 17 toxic chemicals 33% from the 1988 level by the end of 1992 and 50% from the 1988 level by the end of 1995. Corporations joining after 1992 set their own target date for 33% reduction and agree to meet the 1995 date for 50% reduction. As Fig. 21.4 shows, the number of participating corporations is increasing rapidly, and the program as a whole by 1993 was halfway to its 1995 goal of a 700 million pound reduction in the targeted toxic emissions.

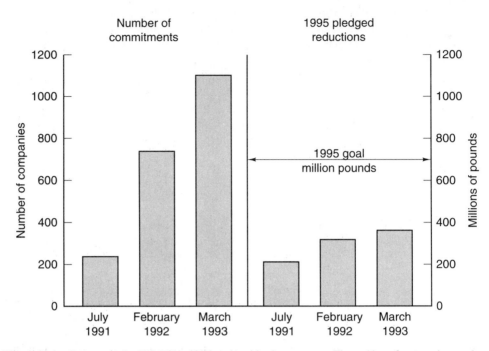

Figure 21.4   Progress in the U.S. EPA's 33/50 toxics reduction program. The number of corporations making voluntary commitments to the program is shown on the left, the pledged reductions in effluents anticipated by the end of 1995 on the right. (Courtesy of J. S. Hoffman, U.S. EPA.)

21.4.8   Gaseous Residues

When dealing with gaseous residues, it is well to recall the maxim of one of the earlier chapters, that to the extent possible every molecule entering a manufacturing facility should leave that facility as part of a salable product. Hence, we can introduce another maxim here: "If you can smell it, fix it". Any odor is evidence that material is escaping into the air rather than ending up as product.

Gaseous residues often originate from the use of volatile solvents or from various washing operations. A first step is to see if these potential sources can be eliminated completely. If not, the emissions from them should be contained to the maximum extent possible and remaining emissions captured and recycled back into the process. A last resort, but better than none at all, is to capture emissions and to dispose properly of the resulting scrubbing liquid.

21.4.9   The Product-Use Environment

Even if products have not originally been designed with optimum product-use characteristics, redesign is sometimes possible and should be implemented whenever the opportunity occurs. One way to do this is to minimize product packaging, to use as few materials as possible in the packaging, and (if feasible) to offer incentives for customer return of packaging.

A second phase-in option for the product-use environment is the redesign of a product to minimize generation of residues during product use. For example, a design for computer printers requiring disposal cartridges could be revised to permit recyclable cartridges, and a program of recycle and reuse established. Alternatively, the lifetimes of consumables could be extended by design revisions. Such design changes are often possible without a major change in the manufacturing procedure, and contribute to a lessened environmental impact for the product.

21.4.10   The After-Product-Use Environment

Just as designs can be modified to make them more environmentally friendly during use, they can be modified to make them more recyclable after they reach the end of their useful life. A modification that is simple to accomplish if injection molds for plastics are being revised is to add standard ISO identification symbols to the exterior of plastic parts, thus permitting easy identification and aiding recycling. Other options include materials substitution or the revision of designs to make disassembly and materials separation easier and more straightforward, following the guidelines given in earlier chapters in this book.

A final activity with positive consequences for industrial ecology is the initiation of take-back programs for products. Doing so closes the loop on the recycling activity. If no ready route exists for the use of products that have been recovered, innovative uses may be able to be worked out with industrial partners. In the same way, facilities should be alert to the possibility of using the take-backs or residue streams of others. It is a maxim of industrial ecology that "If you are not reusing, you are not recycling".

21.5   THE CHANGING STRUCTURE OF DEMAND

The relationship of individual decisions to the level of environmental quality throughout the last half century is symbolized by Fig. 21.5(a). In this diagram, the ordinate (the Y-axis) measures the degree of concentration of decision makers. Individual owners of automobiles, for example, fall near the lower end of that axis, owners of electricity generating facilities near the top. The abscissa (the X-axis) measures the concentration of emissions sources, sources such as ore smelters falling to the right, corrosion and runoff from galvanized fencing falling to the left. Once an environmental impact has been identified, it is obviously easier to achieve voluntary or mandatory corrective activity if only a few sources and a few decision makers are involved than if ownership is ubiquitous and the sources are widely distributed. (This statement seems to be contradicted by our preference for increasingly distributed energy generation, but in that case the transformation, driven by economics, is to cleaner technologies, an overall beneficial trade-off.)

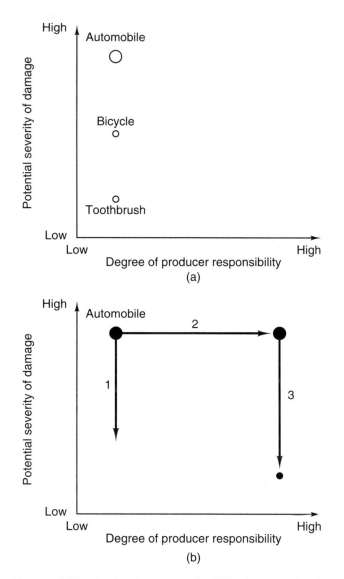

Figure 21.5   The potential for changing the structure of satisfying human needs and wants. (a) The current situation. (b) The potential future situation. The examples in the diagram are discussed in the text.

Figure 21.5 suggests that industrial ecology is enhanced if trajectories of development could be designed to encourage movement along the arrow from lower left to upper right, thus concentrating sources of environmental impacts in the hands of fewer decision makers and in fewer locations. Doing so involves what social scientists term "altering the structure of demand". A picture of the alterations that would be desirable is shown in Fig. 21.5(b), where the potential severity of damage to the environment from a product or a process is plotted against one of several measures of the degree of producer responsibility.

Consider first the toothbrush, located in the lower left of Fig. 21.5(a). The potential severity of its environmental damage is low, so even though toothbrush decision makers are a very diffuse group, little harm is done by that ownership pattern. The situation is nearly as benign with a bicycle, because most of its impacts result from raw materials extraction and improper disposal after its useful life. To the extent that proper recycling of the bicycle is followed, its environmental impacts will be minimal.

A different situation obtains for the private automobile, located on Fig. 21.5(a) as the large dot in the upper left corner. The potential severity of its damage is high, largely as a result of in-use emissions, and the manufacturer has little or no responsibility for that damage. In the last decade of the twentieth century, movement on the plot is expected to occur as shown by arrow 1 in Fig. 21.5(b) as a consequence of materials substitution, more efficient combustion, and other product modifications. Haphazard owner maintenance and uncertain owner recycling of automotive components will limit this downward motion. However, substantial change is anticipated in the first two decades of the twenty-first century as a result of corporate and political actions. The result will be to move the trajectory along arrow 2, where automotive ownership is placed in fewer hands. Once that change in the structure of demand is accomplished, many factors within and outside the restricted ownership circle will produce motion along arrow 3 toward lower impacts. Although movement along arrows 2 and 3 seems at first to be unlikely, it turns out to be natural once human preferences are considered. Customers do not buy a thousand kilograms of metal, a hundred kilograms of plastic, and an assortment of mixed materials because they have an inate desire to own them; they buy a bundle of functionalities manifested in the form of an automobile.

The anticipated transformation in the structure of demand will have at least two interesting effects. The first is that companies that used to think of themselves as manufacturers—providers of things—will become service companies. To stick with our automobile example, it seems likely that vehicles will increasingly be leased rather than owned. Customers will pick up a leased vehicle, use it as they need to, and return it to automobile lease centers; next time, in another location, they pick up another car. The leasing agency, meanwhile, is responsible for the flow of materials into the vehicle, its manufacture, all maintenance and life-extention activities, and eventual dismantling of the automobile for appropriate recycling of subassemblies, components, and materials.

An example of evolving structures of demand in automobile ownership is provided by the VivallaBil car co-op in Örebro, Sweden. This organization is made up of 25 households that among them own a half-dozen automobiles. Reservations must be made for use of an automobile, a daily charge is assessed, and the vehicle must be returned with a full tank of gasoline. Several additional automobiles are rented by the co-op in the summer, when demand is higher. Conversely, during the daytime, when co-op members drive less, some of the vehicles are rented to the municipality, which uses them for community business and thus avoids buying its own vehicles. The result of the co-op is that members plan their activities more and drive less. Those involved have clearly decreased their transportation expenses at the cost of some inconvenience to themselves. If incentives for such activities should be added by municipalities, as has been suggested, car co-ops may burgeon within metropolitan regions in the next decade.

Leasing equipment from the manufacturer instead of purchasing it is an option that has been available to business for some time, photocopying machines and electronic test equipment being examples. Consumer leases are becoming increasingly common, automobiles being one example, telephones another. Leased equipment is returned to the lessor, who retains ownership and responsibility for it. To the extent possible, that equipment is refurbished and returned to the pool of equipment available for leasing. In contrast, most equipment that is purchased is consigned to disposal when obsolete. The sequence is shown in Fig. 21.6(a).

It is possible, of course, for a manufacturing company to voluntarily take back equipment that it has sold. This step is being taken now by some European corporations as they seek to be prepared for extensive mandatory product take-back expected in the future. Product take-back ordinances are currently under discussion in Austria,

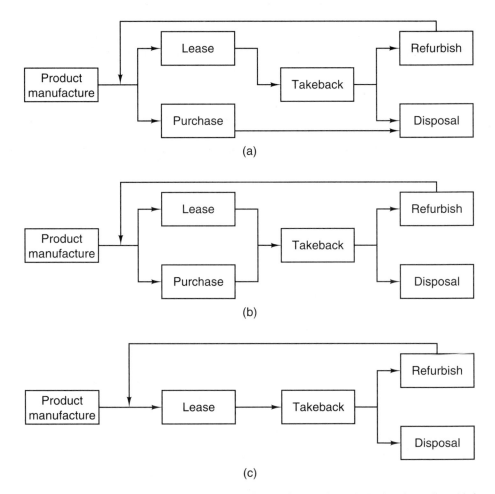

Figure 21.6   Lease-purchase arrangements and their relationship to product take-back and recycling: (a) the current situation; (b) the evolving situation; (c) the potential future situation.

Germany, and The Netherlands, and Switzerland has instituted a voluntary agreement with many of the same features. The result of any of these actions is a flow diagram of the form of Fig. 21.6(b).

Ultimately, corporations are likely to decide that, for large and/or complex products, purchase no longer should be an option. At this stage, the flow diagram of Fig. 21.6(c) will become operative. This situation will represent a complete transformation from the present structure of demand, and will very likely become the standard mode of operation in the twenty-first century for all but the simplest products and consumables.

The transition from sales to service will be an enormous culture shock for many manufacturing companies, involving as it will the need to supplement, or even replace, technologically driven corporate cultures with service-oriented, marketing cultures. One can imagine that the conglomerates of the twenty-first century will be structured from the remains of corporations unable to manage the transition.

21.6 MECHANISMS FOR INDUSTRIAL ECOLOGY MANAGEMENT

21.6.1 Introducing DFE to the Firm

It is helpful to think of industrial ecology as the sum of two focused activities (Fig. 21.7). The first, pollution prevention, is largely directed at taking current processes and products and minimizing their environmental impact. The typical time scale for the actions and their effects is a year or two, and typical actions include leak prevention,

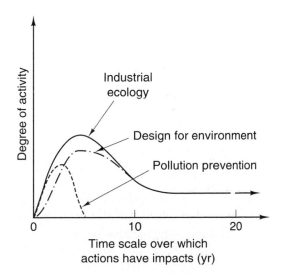

Figure 21.7   The action-impact distribution for pollution prevention, design for environment, and industrial ecology.

energy conservation, and packaging improvements. The second activity, design for environment, has a longer time horizon and often deals with products and processes prior to their introduction. Its typical actions are the development of modularity, minimization of materials diversity, and process substitutions. These two thrusts, taken together, constitute the industrial ecology activities of a corporation.

At this initial stage in the evolution of industrial ecology and its practical application, it is fair to say that only a few firms, almost all of them large transnational manufacturing corporations, have begun to experiment with implementation of DFE. No firm is yet at the point of full implementation. Accordingly, it is important that students of industrial ecology—whether they intend to work in firms, in government, or in nongovernmental organizations—understand the issues involved in implementation of DFE.

The most important point to recognize is that integrating environment and technology throughout a firm's operations represents significant and fundamental change for the organization. It amounts to environmental issues becoming strategic for the firm in the same sense that competitive and economic conditions already are. A related point is that the principal constraints and barriers to implementation of DFE are likely to be cultural and organizational, not technical. It is the need to implement fundamental change within a complex organization, not the development of technologies that are necessary for that change, that will challenge the industrial ecologist.

These points suggest several approaches:

1. Anticipate that implementation of DFE will be neither quick nor easy. In any firm of reasonable size, it will require years of effort and a substantial commitment of time, energy, and top-level executive support.

2. Identify means by which the magnitude of change can be minimized. For example, specific DFE design tools should be similar to existing design tools so that they are familiar, and they should be easy for those who are not environmental experts to use. Indeed, this philosophy led us to create the checklists found in the appendices of this book. Where possible, embed the necessary changes in systems that will automatically lead to environmentally preferable results when used as usual. For example, the expertise necessary to support "green" management accounting systems should, to the extent possible, be resident in the accounting system and not require additional generation of data or separate analyses by the management users of the system.

3. Present industrial ecology and DFE initiatives in the language and incentive structure of the target audience. When speaking with business managers, express the need to implement DFE in terms of competitive advantage and market access. (For example, "Failure to eliminate brominated fire retardants from our plastic housing may limit our ability to market our product in the European Union.")

4. Develop a strategic plan for the integration of technology and environment throughout the firm. An ad hoc initial approach may be necessary to begin the process, but only through the process of creating a more formal plan can necessary work items, required resources, critical paths, timelines, and organizational changes be identified. The integration of the DFE introduction effort into existing

financial, technological, and business plans is necessary if environmental considerations are to be regarded as strategic considerations rather than overhead.

Initial experience indicates that firms, or their individual business units, take one of two approaches to beginning to implement DFE. The first is to focus on a single dimension, or small set of dimensions, of DFE. Corporate users of this approach include Xerox, Cannon, or Lexmark, all of whom manufacture printers and copiers and all of whom are concentrating a significant effort on the return of used components and refurbishment of existing equipment. Similarly, many firms in the chemical industry are focusing their initial efforts on pollution prevention in manufacturing activities, a desirable but very limited goal. The second approach is to begin implementation of DFE principles as a comprehensive system for a single product or small set of products. In electronics, IBM and AT&T are examples of firms taking this latter approach. The automobile companies, working through consortia addressing different technological issues—the development of efficient, inexpensive and powerful batteries, product takeback and materials recycling, the development in the United States of a new generation of automobile—also are beginning to adopt a comprehensive DFE approach to deal with the very difficult trade-offs their products entail.

In practice, of course, the two approaches are compatible, not mutually exclusive, and in virtually all cases some combination of the two will be used. Which one is best for a given firm depends on its size, the complexity and characteristics of its products and product lines, its relations with suppliers and customers, and its culture and existing organization.

21.6.2   Generic and Specific DFE Activities

One fundamental and complementary set of activities by which industrial ecology can be introduced in manufacturing and processing firms focuses on evaluating the structural role of the firm in the life cycles of the products, processes, and materials with which it deals, then using that knowledge to create environmentally preferable practices based on the firm's position in the web of commerce. We call these activities *generic* design for environment. The second approach is by implementing *specific* design for environment practices as part of the firm's product and process design procedures. Among the structural mechanisms brought into play are the following:

Standardized components lists are an important part of many design systems and customer specification documents. Their purpose is to simplify and standardize the design process somewhat; they also, however, offer a powerful lever to improve the environmental aspects of designs without considering each design itself. For example, specification of open relays in electronics design often requires the use of volatile organic solvents in manufacturing, because open relays cannot be exposed to aqueous cleaning systems without damage. An alternative, sealed relays, are somewhat more complicated to manufacture but are compatible with aqueous cleaning processes. Substituting sealed relays on the standard components list would restrain a designer from using a component that, inadvertently, would create a requirement for environmentally undesirable cleaning procedures.

Many components are purchased from outside suppliers. Thus, the *standard purchasing contract* many firms already use can be a ready-made tool for improving the environmental impact of components and the industrial practices of suppliers. For example, when AT&T was encouraging its suppliers of packaging material to avoid making foam by using CFCs, it modified all its contracts to require that non-CFC alternatives be used wherever possible.

Customer specifications and standards are sometimes a significant cause of unnecessary environmental impacts. For example, the single most important roadblock to the American electronics industry's efforts to reduce use of ozone-depleting substances, especially chlorofluorocarbons, was for some time U.S. military specifications and standards that required the use of CFCs. Such standards are seldom reviewed from an environmental perspective and were frequently established before environmental issues were an important factor. Accordingly, they often embody overly conservative, outdated, environmentally inappropriate technologies. Nonetheless, it is an extraordinarily complex and time-consuming task to change customer standards, especially where reliability and performance of substitute techniques and technologies are critical (such as in avionics, defense or communications systems). Alternatives will require substantial "proving-in" to ensure adequate performance. Change is vital, however, because standards and specifications do not exist by themselves; they are often incorporated indirectly or by reference in contracts, bids, and subcontracts with vast networks of suppliers. Frequently, standards known to be thoroughly tested (as are military specifications) will be adopted by third parties not associated with the standard-setting body.

Corporate environmental management structures must be modified to acknowledge that environmental issues are strategic to the firm in the same sense as competitive or economic considerations. In many corporations, such moves require a significant rethinking of how technology-environment interactions are managed.

It is common for environmental issues to be regarded as "overhead"; that is, essentially external to the firm's real business. Environmental management in most firms is organized accordingly; in some cases, in fact, almost all environmental activity is contracted out to independent consultants. Thus, environmentally related costs, which could be reduced or eliminated by good design of products or processes, can be buried in accounting systems as overhead and be essentially invisible to the managers of the business. What is not seen is not managed. Linkages between the environmental experts and the research community, the purchasing department, the design community, the strategic planning organizations, and other critical internal groups are often virtually nonexistent. More subtly, existing environmental personnel in many companies do not have the skill sets—business and technical expertise, strategic vision—required to help the firm internalize environmental issues and begin to practice industrial ecology.

Once the bulk of the generic DFE actions have been taken, one can proceed to *specific DFE*, which focuses on the tools, rule sets, data, and checklists that are necessary to reduce a vast amount of information into a form that is easily integrated into individual product and process design activities. In many cases, the eventual goal is to have specific DFE applications computerized and embedded into existing mechanical design and manufacturing systems such as computer-aided design/computer-aided manufacture

(CAD/CAM) programs. The industrial ecology checklists provided in Appendix A are examples of specific DFE tools, and the electronic solder alternatives case study presented in Chapter 12 and Appendix F is an example of the application of a specific DFE tool to a specific design issue.

21.7 BUSINESS INITIATIVES

In 1988, the Chemical Manufacturers Association in North America began a cooperative program known as "Responsible Care". The program was designed to improve the chemical industry's health, safety, environmental quality performance, and communications with the public concerning products and plant operations. Thus, the aim embodies much of the same philosophy as industrial ecology.

Responsible Care, which has been adopted by more than 200 corporations, many of them multinationals, is implemented by 10 principles and 6 management practice codes. In Tables 21.1 and 21.2, we reproduce portions of those guides that are relevant to industrial ecology. It is readily apparent that the Responsible Care initiative contains within it much that is pertinent to the theme of this book. In that context, it is particularly significant that chemical industry organizations in some 35 countries have now adopted their own versions of Responsible Care.

Table 21.1   Selected Principles of the CMA Responsible Care Program

Develop and produce chemicals that can be manufactured, transported, used, and disposed of safely.

Make health, safety, and environmental considerations a priority in all planning for all existing and new products and processes.

Extend knowledge by conducting or supporting research on the health, safety, and environmental effects of products, processes, and waste materials.

Table 21.2   Selected Requirements of the CMA Product Stewardship Code

Establish and maintain a system that makes health, safety, and environmental impacts key considerations in designing, developing, and improving products and processes.

Consider the health, safety, and environmental programs of suppliers and require them to provide appropriate health, safety, and environmental information and guidance.

Encourage distributors and direct product receivers to set up and implement proper health, safety, and environmental practices.

A second initiative, undertaken in 1990, was the formation of the U.S. American Electronics Association Design for Environment Task Force. This group has fostered research on and support for DFE initiatives in the electronics sector in the United States, and generated a series of White Papers by industry experts on DFE issues (see Table 21.3). These informational materials have been widely disseminated, both in the United States and in other countries around the world.

A third initative of interest was begun in 1991 by the International Chamber of Commerce. This organization, based in Paris, developed a "Business Charter for Sustainable Development" and presented it at an international industrial conference on environmental management. The charter has 16 principles, as given in Table 21.4. Several are particularly relevant to the topics in this book, including the emphases on energy efficiency, sustainable use of resources, minimizing residue generation, and modifying products and services. As of August 1992, more than 1000 corporations around the world, including 150 of the "Fortune 500", had signed the charter and agreed to incorporate its principles in their operations.

Table 21.3   AEA DFE White Papers *

What is Design for Environment? (B. R. Allenby, AT&T)

DFE and Pollution Prevention. (B. R. Allenby, AT&T)

Design for Disassembly and Recyclability (R. G. Grossman, IBM)

Design for Environmentally Sound Processing (J. C. Sekutowski, AT&T)

Design for Materials Recyclability (W. Rosenberg, COMPAQ, and B. Terry, Pitney-Bowes)

Cultural and Organizational Issues Related to DFE (B. R. Allenby, AT&T)

Design for Maintainability (E. T. Morehouse, Jr., U.S. Air Force)

Design for Environmentally Responsible Packaging (K. Rasmussen, General Electric)

Design for Refurbishment (J. C. Azar, Xerox)

Sustainable Development, Industrial Ecology, and Design for Environment (B. R. Allenby, AT&T)

*Copies of White Papers can be obtained from the American Electronics Association, Washington, DC (202-682-9110).

Table 21.4   Principles of the Business Charter for Sustainable Development

Recognize environmental management as among the highest corporate priorities.

Integrate environmental policies and practices fully as a key element of management.

Continue to improve business's environmental performance.

Educate and motivate employees to carry out their activities in an environmentally sound way.

Table 21.4 (continued)

Assess environmental impacts before starting a new project or decommissioning an old facility.

Develop and provide products and services that do not harm the environment.

Advise customers on the safe use, transportation, storage, and disposal of products provided.

Develop and operate facilities and undertake activities with energy efficiency, sustainable use of renewable resources, and waste generation in mind.

Conduct or support research on the impacts and ways to minimize the impacts of raw materials, products or processes, emissions, and wastes.

Modify the manufacture, marketing, or use of products and services so as to prevent serious or irreversible environmental damage.

Encourage the adoption of these principles by contractors acting on behalf of a signatory company or organization.

Develop and maintain emergency preparedness plans in conjunction with emergency services and relevant state and local authorities.

Contribute to the transfer of environmentally sound technology and management methods.

Contribute to the development of public policy and government-business programs to enhance environmental awareness and protection.

Foster openness and dialogue with employees and the public regarding potential hazards and impacts of operations, including those of global or transboundary significance.

Measure environmental performance through regular environmental audits and relay appropriate information to the board of directors, shareholders, employees, authorities, and the public.

21.8 SUMMARY

It is important that corporations treat the process of design for environment as an opportunity, not a constraint. Like most opportunities, it will benefit from commitment and from hard thinking. It is one thing to attempt to minimize or eliminate the use of an environmentally harmful chemical from a manufacturing process. It is another thing altogether to do so while maintaining or enhancing product quality, time to market, cost, and all the other factors that must be considered in a competitive economy, but an important aspect of DFE is that its use enhances competitiveness rather than inhibits it. A side benefit of implementing DFE is that efforts to improve a process for one reason often end up improving many aspects not connected with that reason, simply because everything about the process is getting a close look and a broad review. It is no coincidence that the corporations taking the lead in industrial ecology are those that are often in the lead in other measures of corporate efficiency as well.

Once the commitment is made to implement industrial ecology in a corporation, the perspective must become ingrained within corporate thinking. Just as designers and managers instinctively work hard to remove every extraneous nickel of cost from a product entering a highly competitive market, so they need to instinctively work hard to optimize all aspects of the product dealing with industrial ecology. In the long run, the corporations that do so will be among those that grow and prosper; those that do not will wither.

SUGGESTED READING

Allenby, B. R., and D. J. Richards, eds. *The Greening of Industrial Ecosystems.* Washington, DC: National Academy Press, 1994.

Arnfalk, P., and A. Thidell. *Environmental Management in the Swedish Manufacturing Industry: Fact or Fiction?* Lund: Swedish National Board for Industrial and Technological Development (NUTEK) and Lund University, 1992.

Schmidheiny, S. *Changing Course: A Global Business Perspective on Development and the Environment.* Cambridge, MA: The MIT Press, 1992.

Smart, B., ed. *Beyond Compliance: A New Industry View of the Environment.* Washington, DC: World Resources Institute, 1992.

Todd, R. Zero-loss environmental accounting systems, in *The Greening of Industrial Ecosystems,* B. R. Allenby and D. J. Richards, eds. Washington, DC: National Academy Press, 1994.

EXERCISES

21.1   It has been suggested that the attitudes of corporations toward environmental regulations and guidelines can be divided into three groups: unrestrained opposition, minimal compliance, and supportive compliance. One indication of evolution in these attitudes is the increasing number of corporations issuing annual or aperiodic environmental reports. Call or visit the public relations offices or environment and safety organizations of several corporations and ask for copies of these reports, if they exist. From this and/or any related information, describe and compare the environmental attitudes of those corporations.

21.2   For several years, a coalition of public interest groups has been encouraging corporations to adopt the Ceres Principles in their conduct of business. Formerly called the "Valdez Principles" to recall the petroleum spill in Alaska in 1989, the approach (below) takes a somewhat different form and content compared with that of the Business Council on Sustainable Development (Table 21.4). Contrast the two approaches from the standpoint of the corporation.

The Ceres Principles

Protection of the biosphere

Sustainable use of natural resources

Reduction and disposal of waste

Wise use of energy

Risk reduction for employees and communities

Marketing of safe products and services

Damage compensation for environments and persons adversely affected by corporate actions

Disclosure of incidents that cause environmental harm or pose health or safety hazards

Appointment of a member of the board of directors qualified to represent environmental interests, and establishment of a committee of the board to deal with environmental affairs.

Annual public self-evaluation of compliance with these principles.

21.3   Virtually all responsible firms have a written environmental policy, usually focusing on compliance issues. Far fewer have a proactive long-range policy. Using the programs discussed in this chapter as a basis, write a policy and supporting guidelines for integrating industrial ecology throughout the operations of a global, diversified manufacturing company, assuming an environmental compliance policy already exists.

21.4   You have just been appointed Chief Industrial Ecologist at a large, diversified, multinational firm that has no existing industrial ecology program. You have decided to begin by creating a self-directed design for environment coordinating team. What internal organizations should be represented on your team, and why?

CHAPTER

22 | Standards and Stimuli

22.1 INTRODUCTION

As this book is written, there is vigorous activity underway in many quarters to devise guidelines for incorporating environmental considerations into the design of industrial processes and products. It is a reasonable prediction that once there is common agreement on guidelines, the next step will be the imposition of standards and regulations. Standards are designed to ensure compliance to at least the minimum permissible degree with methods and results designed for a particular product or process. (Electrical wiring codes are examples of standards known to many.) Indeed, standards and regulations for emissions of industrial materials to the environment have been in place in many countries for two decades or more, and extension to industrial ecology principles is merely moving standards and regulations from the "end of the pipe" to the "life of the product".

What might be included in such standards, and when might they become effective? What are other external stimuli, and how compelling might they be as incentives to action? Some guidance in the direction of answers to those questions is provided by current activities of several international standards and policy-setting organizations. In this chapter, we discuss some of the most important aspects of those activities and initiatives.

22.2 INTERNATIONAL STANDARDS ORGANIZATION (ISO) ACTIVITIES

The ISO has become increasingly important in recent years as the internationalization of business has dramatically increased. Its most recent and visible set of documents has been ISO 9000, a detailed specification of actions needed to ensure that reasonable standards of quality in manufacturing have been met. An organization that qualifies for ISO registration is one that has established a quality auditing system that follows the following steps for each business area:

- Say what you do.
- Justify what you do.
- Do what you say.
- Record what you did.
- Review what you did.
- Justify or revise what you will do.

Demonstrating that a facility meets these criteria for quality at every step of its operation requires a carefully documented record of actions, and is time-consuming and expensive in its initial stages. In most cases, however, the effort expended in the qualification process is repaid many times in the product improvements that result. There are few corporations anywhere in the world who serve other than their local areas that have not engaged in major reviews and upgrades of their manufacturing processes in order to meet the standards of ISO 9000.

Given its record of generating agreement on international standards, ISO was a natural place to begin the process of generating standards for sustainable industrial development. Accordingly, 1991 saw ISO establish its Strategic Advisory Group on the Environment (SAGE), in coordination with the Geneva-based Business Council for Sustainable Development. SAGE has now been succeeded by ISO Technical Committee 207 (TC 207), with subgroups as described in the following paragraphs.

22.2.1   Life-Cycle Assessment

This subgroup is charged with reviewing existing programs to evaluate the environmental impacts of products, processes, and services over their entire life cycle. The goal is to recommend the harmonization of the various approaches on which research and development is taking place within various corporations and academic institutes.

22.2.2   Environmental Guidance for Product Standards

This subgroup is charged with preparing a guide for the integration of environmental aspects in product standards already existing or those under preparation. The approach is to generate drafts of standards that can be revised and adopted by voluntary consensus.

22.2.3   Environmental Management and Auditing

This subgroup is charged with generating a set of environmental auditing procedures, and then promoting their voluntary use by corporate management.

22.2.4   Environmental Labeling

This subgroup is charged with determining how product labeling programs can be made meaningful and consistent, and to establish a system of internationally recognized symbols and definitions. The subgroup's first task is to define terms whose meaning is unclear or ambiguous. Examples are "recyclable", "recycled", "compostable", and "biodegradable".

Regular meetings of the ISO TC 207 subgroups began in 1993 and are being carried forward with the assistance of participants from well over a dozen countries.

22.3   UNITED NATIONS ACTIVITIES

The activities of the United Nations in connection with industrial ecology have concentrated on energy efficiency, particularly within the eastern European nations. For this purpose, the United Nations Economic Commission for Europe (UNECE) has established a Working Party on Standardization Policies for Energy Efficiency Standards and Labeling Systems. Their work, begun in 1990, has thus far involved the assemblage of data on legally binding energy efficiency standards from countries and international agencies around the world. The next goal, to prepare draft energy labeling standards and standards for energy-efficient buildings and appliances, is under way.

It is likely that energy-efficiency standards for ECE countries will emerge from the activities of the Working Party within a few years. These standards will probably be voluntary, and will probably be adopted by other regional UN agencies.

22.4   INTERNATIONAL ELECTROTECHNICAL COMMISSION (IEC) ACTIVITIES

22.4.1   IEC Guidelines Document

IEC is an independent international organization that has a long record of generating standards and protocols to assist in the globalization of technology. International telecommunications technology has been a principal focus of this group. In a new initiative for IEC, an expert group produced and circulated in April 1993 a draft document titled "IEC Guide for Environmental Aspects in Product Standards". The focus of the document is on electronic equipment, and it is stated to be advisory, not prescriptive. Among the topics covered are the following:

- Life-cycle analysis, a discussion emphasizing the primitive state of LCA.
- Pollution prevention, with specific recommendations to eliminate solder cleaning and minimize the use of deionized water.
- Environmental impact assessments, with special mention of ozone-depleting substances, batteries, consumables, packaging, and emissions.

- Design for disassembly, with sections on fastening and joining and on decorative paints and finishes.

22.4.2 IEC/ISO Interrelationships

The IEC has no official relationship with ISO, and the perspective of ISO on products is considerably broader than is that of IEC. Nonetheless, the draft IEC report is likely to have substantial influence on the pace and form of ISO standards for environmental performance. Especially for the simpler IEC recommendations, such as the use of "snap technology" over screw and bolt assembly, the incorporation of IEC thinking into ISO standards within the space of a few years seems quite likely.

22.5 LABELING PROGRAMS

A number of systems designating environmentally responsible products and/or activities are being developed around the world. Some of these systems are designed by marketers on their own behalf, as shown in Fig. 22.1. The goal is to advertise the environmentally related characteristics of individual products or of the corporation itself. Many of these activities are viewed by the public as self-serving and potentially inaccurate. Although they are useful in demonstrating an environmental focus, especially if descriptive and restrained in tone, the use of first-party systems is unlikely to provide much stimulus either for customers or corporations.

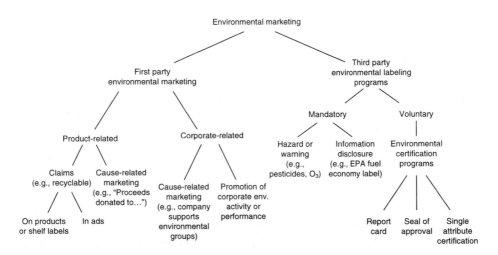

Figure 22.1 Classifications of various types of environmental marketing. (U.S. Environmental Protection Agency, Office of Pollution Prevention and Toxics, Pollution Prevention Division, *Status Report on the Use of Environmental Labels Worldwide*, EPA 742-R-9-93-001, Washington, DC, 1993.)

More visible and more stimulating are the environmental labeling programs now appearing everywhere at a great rate. Some are quasi-governmental, others strictly private. The U.S. EPA defines five types of environmental labels:

Seal of approval programs identify products or services as being less harmful to the environment than similar products or services with the same function.

Single-attribute certification programs use an independent third party to validate a particular environmental claim made by the manufacturer.

Report cards offer consumers information about a product and/or a company's environmental performance in multiple-impact categories. In this way, consumers have the option of deciding for themselves the influence of the product on the environmental impacts that they consider important.

Information disclosure labels provide specific environmentally related information about a product, and may compare that information to that for similar products.

Hazard and warning labels are mandatory warnings concerning a product's adverse environmental or health impacts.

The relationship of these types of labeling programs to each other and to the organizations that create and use them are shown in Fig. 22.1, and some characteristics of the types of labeling programs are given in Table 22.1. Note that the stimuli from these programs can be positive, negative, or neutral, and that the voluntary programs are generally those that stimulate positively.

Environmental labels demonstrate substantial creativity of design. We illustrate a selection in Fig. 22.2. Most of the environmental labels now in use, including the first eight shown in the figure, are seal of approval programs. These programs generally apply some sort of life-cycle assessment to the product and award use of the label if the

Table 22.1   Characteristics of Labeling Program Types

Type of Program	Type of Stimulus			Legal Status	
	Positive	Neutral	Negative	Voluntary	Mandatory
Seal of approval	•			•	
Single-attribute Certification	•			•	
Report card		•		•	
Information disclosure		•			•
Hazard warning			•		•

Source: Adapted from U.S. Environmental Protection Agency, Office of Pollution Prevention and Toxics, Pollution Prevention Division, *Status Report on the Use of Environmental Labels Worldwide*, EPA 742-R-9-93-001, Washington, D.C., 1993.

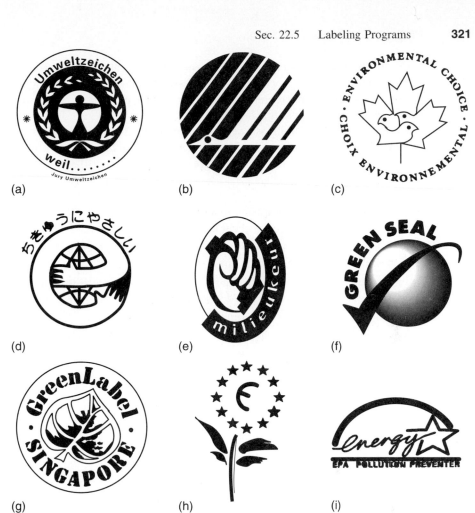

(a)   (b)   (c)

(d)   (e)   (f)

(g)   (h)   (i)

Figure 22.2   Some of the labels established to designate environmentally-responsible products. (a) German "Blauer Engel"; (b) Nordic Council "White Swan"; (c) Canadian "Environmental Choice"; (d) Japanese "Eco-mark"; (e) Dutch "Stichting Milieukeur"; (f) United States "Green Seal"; (g) Singapore "Green Label"; (h) European Community "Ecolabel"; (i) United States "Energy Star".

product meets certain specified criteria. The ninth label in Fig. 22.2 is for a single-attribute certification, the EPA Energy Star Computers program. This program permits the use of the logo on products designed so that they save energy when in use, for example, by switching to a low-power state when left idle. Computers so marked have been evaluated only for energy consumption during use, not for any environmental impacts that result from raw materials extraction, manufacturing, use, disposal, and so on.

Although procedures differ substantially for each labeling program, some idea of how these programs work can be gained by that for the European Community (EC) Eco-label. The first step is to define a product category for assessment. A category can be nominated by any interested party, and nominees are referred to "lead countries" for generating suggestions as to the suitability of the category and what the criteria should be. In step two, these draft criteria are reviewed and eventually approved by a Committee of Member States. Finally, technical boards within each of the member states evaluate applications from manufacturers of products within the category, and, consistent with the criteria, approve or disapprove the use of the Ecolabel. Every effort is made throughout the process to ensure the representation of groups from industry, environment, and consumer groups on the approving boards.

There is no such thing as a set of typical requirements for being awarded a seal of approval, but some concept of the level of detail that can be involved is demonstrated by an abbreviated list of requirements for receipt of the German Blauer Engel (Blue Angel) award for workplace computers:

- The unit must be constructed in modules and must allow the replacement of modules by the user himself without the help of special tools.
- The applicant commits himself to taking back free of charge those products that were originally delivered by him.
- The unit will minimize insoluble connections between different materials (i.e., gluing, soldering, welding, and so on).
- Coatings and composite materials will be avoided.
- Simple dismantling of appliances and modules must be possible.
- The plastics used must be recyclable under existing techniques.
- The plastic housings shall contain no more than two polymers, and they must be separable from each other.
- Batteries must be removable or replaceable by the user.

These requirements, together with a longer list not reproduced here, constitute a mini-course in design for environment.

In principle, seal of approval labels are awarded as a result of some sort of life-cycle assessment. The standards tend to be set so that a modest fraction of existing products, perhaps 10 or 20%, can successfully qualify. Because, as seen earlier in this book, LCA in its full embodiment is complex and contentious, labeling organizations tend to utilize a simplified LCA, especially as far as the impact-analysis stage is concerned. In practice, one life-cycle stage or one product characteristic (toxic content, say, or diversity and volume of packaging materials) often controls the decision, so labels seldom reflect the overall environmental impact of a product.

The plethora of labels and the diversity of criteria for their awarding make response difficult, especially for multinational corporations. Nonetheless, environmental labels are not going to disappear: they currently operate in more than 20 countries. A harmonization of these programs, or incorporation into a general international scheme, is much to be desired.

22.6   DISCUSSION

In many cases, the criteria for obtaining environmental labels for specific products are far more stringemt than existing regulations or standards. Nonetheless, it is likely that corporations will be driven by competitive pressures to satisfy the labeling criteria. The advantage of labeling systems is that they harness market forces and consumer preferences in the effort to achieve better environmental performance. At a stage of industrial ecology when life-cycle assessment methods are still relatively unfamiliar, labeling programs offer the promise of rapidly implementing LCA methodologies. However, a potential disadvantage of the labeling programs is that unless their criteria are carefully chosen, environmentally suboptimal performance might be encouraged.

Labeling programs definitely have their place in the move toward environmentally responsible products. It is clearly of little use to corporate or individual consumers to know that a product has received three labels and been denied two others, a circumstance possible today. Both the international nature of corporations and the imperfectly developed assessment situation argue, however, for programs that are carefully thought out and validated, and for those that are international in scope. Until those requirements are fulfilled, environmental labels for products are likely to contribute more to chaos than to rational consumerism.

In contrast, the existence of international standards for environmental performance of products, processes, and services, even if advisory in nature, will almost certainly stimulate the adoption of the standards by knowledgeable industrial customers. Once that process begins, as has been the case with the ISO 9000 standards, corporations have little choice but to revise their operations to meet those standards lest they lose customers.

The main issue with international standards on environmental performance is not whether they will occur, but how soon that will happen. Farsighted corporations would be well advised to take two actions: to prepare their operations so that the imposition of standards will place them in a favorable rather than unfavorable position with respect to their competitors, and to participate in the international standards-setting activities, so that the standards that emerge are a reasonable consensus of the positions and desires of all concerned.

SUGGESTED READING

Development of the Comprehensive Report on Energy Efficiency Standards and Labelling Systems, Document GE.92-31741. Geneva: United Nations Economic Commission for Europe, 1992.

Guide for Environmental Aspects in Product Standards, Draft for Comment. Geneva: International Electrotechnical Commission, March 1993.

International Standards Organization. *Sustainable Industrial Development: A Role for International Standards*, ISO Bulletin, Geneva, April 1992.

Italian Agency for Energy, Environment, and New Technologies (ENEA). *Study for the Attribution of a European Ecolabel for Refrigerators and Freezers*, Revision 2, Draft, Rome, 1993.

U.S. Environmental Protection Agency, Office of Pollution Prevention and Toxics, Pollution Prevention Division, *Status Report on the Use of Environmental Labels Worldwide*, EPA 742-R-9-93-001, Washington, DC, 1993.

———. *The Use of Life Cycle Assessment in Environmental Labeling*, EPA 742-R-9-93-003, Washington, DC, 1993.

EXERCISES

22.1 Draft a standard for environmental management and auditing for large corporations. What are the elements of the standard that are essential? What are those that are desirable but not essential? Should both types be part of the standard? Why?

22.2 Draft a standard for life-cycle assessment for large corporations. What are the elements of the standard that are essential? What are those that are desirable but not essential? Should both types be part of the standard? Why?

Satisfying Human Needs and Wants: The Future of Industrial Activity

CHAPTER 23

23.1 INTRODUCTORY NOTE

In this chapter, we depart from the pedagogical approach of the rest of the book to speculate on world society of the year 2050, and on how industrial ecology will be involved in achieving aspects of that society. Obviously, one cannot hope for an accurate prediction, because the essence of the future is its unpredictability. Rather, the purpose of the exercise is to highlight for the practitioner of industrial ecology some of the trends and emerging technologies that may shape the global economies of the future, with the hope that the discussion will not only provide a broader picture for the reader, but will inspire the creation, and critiquing, of similar scenarios.

23.2 DEMATERIALIZATION

23.2.1 Patterns of Reduced Materials Use

Dematerialization is the process by which lesser amounts of materials are used to make products that perform the same functions as their predecessors. Examples of dematerialization abound in modern society. The archtypical one is that today's palmtop computers have more capability than the "supercomputers" of 10 years ago. Another is the ability of modern stereo systems to produce sound superior to that of large, bulky systems available for sale only several years in the past. The ultimate example is the famous graph of integrated-circuit transistor packing density as a function of time (Fig. 23.1), which shows the density increasing exponentially since 1960. The most recent efforts in this direction have been to use light beams or tiny microscope tips to move

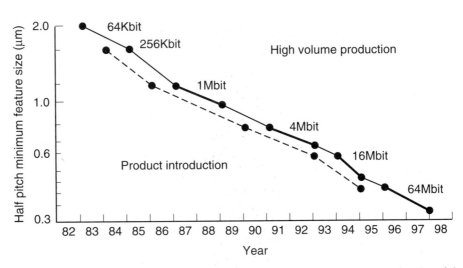

Figure 23.1   The trend in the density of transistors in integrated electronic circuits as a function of time. DRAM is dynamic random access memory. (Courtesy of C. K. N. Patel, University of California, Los Angeles.)

single atoms from place to place on a surface. Such work may eventually lead to circuits in which each electrical constituent is only a few atoms in size, with great increases in speed and much diminished power requirements.

Electronic circuits are not the only components getting smaller. Various techniques of microfabrication are being used to produce flow sensors, gear trains, and micromotors measuring less than 100 μm in diameter (Fig. 23.2). As development continues and components such as microvalves and micropumps are produced, those involved in "nanoengineering" see entire microsystems incorporating electrical, mechanical, thermal, optical, magnetic, and chemical functions operating on a single, small silicon chip. The potential applications are legion, and include the use of micromotors for security and medical applications and microrobots for assembly tasks at submillimeter size scales.

Perhaps the dominant constraint on dematerialization is that many industrial products are rather directly related to human beings and human size, and cannot be reduced arbitrarily. Personal computers, for example, have decreased substantially in size and weight over the past decade. Improvements in data storage technology have made it possible to minimize the use of floppy disk drives and, in turn, to require less use of disk drive power. Integrated circuits are lighter and smaller. Nonetheless, keyboards or notepads cannot be reduced very much in size without becoming out of scale with the human hand and thus inefficient or worthless. Thus, as long as manual interaction is required to enter information the dematerialization possibilities of the personal computer

Figure 23.2   A scanning electron micro-scope view of a magnetically driven micro-motor. The rotor diameter is 150 μm and the gears have diameters of 77, 100, and 150 μm. (Courtesy of H. Guckel, University of Wisconsin.)

are limited. A similar constraint applies to getting information from a personal computer in a usable form. Visual resolution limitations dictate the size and mode of operation of displays. Unless consumers are satisfied with some mode of oral communication, dema-terialization possibilities are again only modest.

How many other things are controlled by human sizes? It turns out that they are surprisingly many. Houses, for example, and most of the things in them, cannot change size very much without becoming unsuitable for activities like sleeping, storing food, washing clothes, and placing photographs on tables. The restriction extends to automo-biles and trains, and thus to roadways and rail lines, directional signs, and bridge trusses.

Given the preceding perspectives, what might one see as the limits to dematerial-ization? First, we can point out that materials research can produce items of similar size to their predecessors but with less use of materials. Modern automobiles, for example, are 20 to 30% lighter than those of a decade ago because of increased use of aluminum, plastics, and higher-strength steels and alloys. Dematerialization is thus achieved by changing physical properties rather than size. Another technique is to increase the life-time of a product, or at least of its major components. The increase in product cycling time requires less overall materials extraction and achieves dematerialization through a reduction in the frequency of demand. Finally, of course, are the items with which humans are not directly associated and for which size is not constrained. Not many users, for example, are aware that the cables connecting telephone switching offices are now largely made of optical fibers rather than copper wire. They carry calls and data in effortless fashion while requiring far less than 1% of the materials of the technology they replaced.

Dematerialization will cause major changes in the extractive materials industries. Those that survive will become material management full-service providers to manufacturers. In this capacity, they will continue to provide materials upon which the economy is based, but much will be recycled and refurbished, not extracted. One can also foresee that much of materials science will be taken over by these corporations, perhaps especially the large petroleum companies. The latter also can be expected to become adept at mining biomass using biological processes.

23.2.2   Substitution of Information and Intellectual Capital for Materials

The dematerialization discussion can be generalized to a trend that will become much more significant in the future: Intellectual capital and sophisticated information management will increasingly be substituted for raw material and energy inputs. In one sense, that is a simple substitution of lower-priced goods for those with higher prices, because information and intellectual resources such as computer power and information transmitting capacity are rapidly becoming cheaper whereas prices for energy and materials, particularly virgin materials, are rising. Resources that are currently undervalued by the economy and called "wastes" will be identified, tracked, and fed back into economic uses. At first, they will be recycled within facilities. Later, as increased information flow reduces transaction costs, they will be recycled within industrial sectors or among supplier–customer networks. Finally, residual streams will be designed for specialized use throughout the economy just as with primary products. A highly sophisticated and complex information infrastructure will be necessary to support such a system, and that infrastructure is rapidly being built.

23.3   NEW ENABLING TECHNOLOGIES

23.3.1   The Evolution of Energy Generation

The dominant characteristic of modern life that separates it from that of earlier eras is the ready availability of electrical energy. This energy permits us to accomplish such activities as cooking, washing, assembly of components, and maintenance of products with high efficiency and minimal effort. It is not inaccurate to say that energy is the enabling technology of modern civilization.

At present, as we have seen, the generation of energy occurs predominantly through the combustion of fossil fuels. This practice imposes a very heavy burden on the environment as a consequence of the emission of carbon dioxide, sulfur gases, particulate matter, heavy metals, and a variety of other species. In some instances, it is possible to prevent or reduce many of these problems, though at significant expense. Sulfur dioxide and particulate matter, for example, can be removed almost completely from combustion exhaust streams. This extraction solves much of the fossil fuel air pollution problems, but does require that the resulting residue that is collected must be discarded, as no useful materials recovery scheme has been devised. A more ambitious approach to cleaning combustion gases is underway in Japan, centered on technologies to separate carbon dioxide from combustion streams and bury it in the deep ocean (see Fig. 23.3). The environmental impacts of fossil fuels ideally are better avoided than mitigated, however, so storage of carbon dioxide might best be regarded as an interim strategy, though an extremely important one. If storage of emissions turns out to be implementable at reasonable cost, it may be possible to continue using the large coal resources of Earth for the next two to three centuries (i.e., until they are virtually exhausted).

A generally preferable plan to burning fossil fuels and then accomplishing engineering heroics to recover the undesirable combustion products is to avoid producing the undesirable products in the first place. This is the philosophy behind encouraging solar power, wind power, and hydropower. These three options have limitations imposed by the widely different supplies of sunshine, wind, and water, and cost is an additional issue in the case of solar energy. Nonetheless, these approaches do and will continue to make significant contributions to energy supply. Nuclear fission will doubtless also play a significant role in energy production for the next several decades, aided by improvements in fail-safe reactor design and increasing costs of alternatives. Perhaps 30 to 50 years from now, technologies such as nuclear fusion may be important, although the political acceptability of this technology remains an open question.

More likely than transformations of energy-generation facilities is a revolution in the whole area of energy provisioning. Newly developed thyristors (fast electronic switches that can operate at high voltage and high current) will be installed throughout existing power systems over the next two decades, making it much easier to transfer electricity to meet fluctuating demands and hence minimizing the number of power stations needed. In addition, a now noticeable trend is the decline of interest in large power stations with long distribution systems and a concomitant increase in small energy-generation modules (e.g., cogeneration facilities, fuel cells, batteries, photovoltaics) located nearer the customer. Electronic control of networks of these small modules can be expected to encourage energy selling and trading among small module owners. The result will be competition in energy marketing, the decentralization and deregulation of power generation, and the substitution of intelligent network technology in place of marginally needed large facilities. These structural transformations in themselves reduce power use as well as cost, to the extent that they minimize transmission losses and decrease the use of obsolescent power-generation facilities.

Perhaps the most intelligent approach of all to dealing with energy requirements is to minimize them at the design stage. A recently formed consortium is using this philosophy in designing a personal computer that only draws full electrical power when actually being used. Early estimates suggest that perhaps 80% of the power demand can be avoided with no loss of utility. This concept, an example of the trend toward substituting information management for materials and energy, is likely to be increasingly adopted as computer control of even small and inexpensive electrical products becomes routine.

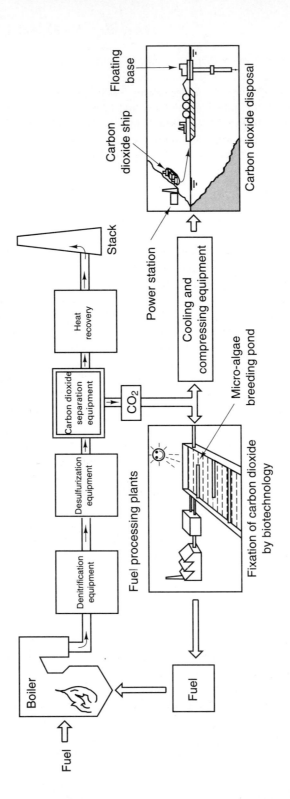

Figure 23.3   An example of a flow diagram for a system to remove carbon dioxide from the exhaust gas generated by electrical power production facilities. (Courtesy of Tokyo Electric Power Company.)

23.3.2   The New Agriculture

Agriculture and its supporting infrastructure certainly constitute an important part of humanity's interaction with the environment and thus a logical component of industrial ecology. Agriculture's importance can only increase in the next few decades, as nation-states face the requirement of feeding populations undergoing very rapid expansion. There are several characteristics of the existing global agricultural sector that differentiate it from other economic sectors. Perhaps most important is economic distortion; more than other societal activities, agriculture is marked by substantial direct and indirect subsidies. In addition, agricultural practices around the globe are highly variable: More than other sectors, agriculture reflects indigenous cultures, local climate and environmental conditions, and economic development levels. The result is significant variations in patterns of inputs and production per unit of land. Ironically, however, this variability does not extend to crop selection. Global agriculture focuses on an extraordinarily limited number of species: seven crops (wheat, rice, corn, potatoes, barley, cassava, and sorghum) provide 75% of the world's food supply. These characteristics reflect the fact that agriculturalists around the world are extremely conservative and slow to change, characteristics that impede the diffusion of new technologies or adoption of new crops. This conservatism is reinforced by extreme specialization of agricultural professionals according to commodity.

Moreover, the environmental impacts of agricultural activities are less well-controlled than in other sectors. Unlike industrial "point sources" of pollution, agricultural pollution is "nonpoint", that is, it results from many small sources over broad areas. If a factory contaminates local groundwater, for example, the plume is bounded and its source relatively easy to identify. On the other hand, groundwater contaminated by chlorinated hydrocarbon pesticides cannot be tracked, in most cases, to any specific source. Other environmental impacts are generated by soil degradation, mining of groundwater, energy consumption, and fertilizer and pesticide production, consumption, and runoff that characterize agricultural activity.

Against this background, one can see several trends (some identified by Vernon Ruttan of the University of Wisconsin) that will likely characterize the agricultural sector as it begins to evolve toward more sustainable practices:

1. In the past, most increases in food production were obtained by utilizing new agricultural land. In the future, most increases in food production will have to come from higher yields and/or different crop species.

2. The use of inexpensive, small, efficient sensors to monitor agricultural parameters and processes (soil moisture, fertilizer application, crop growth rates, insect infestation, storage of harvested products, and so on) will become increasingly feasible and widespread, and effective crop yields will improve as a result. (This is again an example of the substitution of information management for material and energy inputs.)

3. Chemically intensive and energy-intensive agricultural practices will diminish in importance, as will those that degrade and consume relatively nonrenewable

resources, especially soil and fossil water. Environmental impacts of agriculture will lead to increasingly effective regulation of the sector. As more environmental impacts are captured in prices (of fertilizer, pesticides, fuel, and so on), market distortions will tend to diminish.

4. Advances in understanding plant disease resistance and defense mechanisms and pathogen action will lead to increasingly effective strategies for growing resistant crops (plant diseases on average destroy some 12% of crops worldwide). All else equal, this will result in more useable food per unit of land. Yield, especially under conditions characterizing more sustainable agricultural practices, will also be increased. Genetic engineering advances increasingly will contribute to these trends.

5. A growing emphasis on sustainable farming practice and increasing reliance on agriculturally marginal land (frequently in arid climates) will lead to pressure to match the crop to local conditions, climate and soil, rather than modifying the environment to the crop. Crops likely to increase in importance include the grains quinoa, amaranth, triticale, millet and buckwheat, and the winged, rice, faba, and adzuki beans.

6. Products from farmed biological systems will begin to replace extractive inputs. The most obvious example is the production of fast-growing species for energy production: hemp yields 10 to 14 tons of dry fiber per hectare annually, more than five times as much as trees. Oilseeds can provide biodegradable lubricants, coatings, plastics and cosmetics, replacing petroleum-based formulations. More radically, goats, pigs, cows and other farm mammals are being genetically engineered to be walking drug factories—"bioreactors on the farm."

Over the longer term, industrial ecology will increasingly result in the substitution of biological subsystems for engineered subsystems, either on the materials side (e.g., biopolymers, biological electronics materials, bioceramics manufactured using biological, enzyme-based, processes) or on the process side (e.g., using plants or trees to recover heavy metals from aqueous residue streams). These new approaches will extend traditional "agricultural" practices into many industrial sectors, blurring heretofore clear sectoral lines, with significant potential environmental benefits.

23.3.3 Batteries as a Liberating Technology

If the wide availability of electrical power separates the twentieth century from the nineteenth, the wide availability of *portable* electrical power is the technology that will separate the twenty-first century from the twentieth. Already in the 1990s we see portable computers, stereo systems, power tools, and a variety of other objects. Several research groups are working on the design and realization of rechargeable microbatteries that could power a single computer chip, thus providing on-board memory-retention capabilities and aiding further miniaturization. As batteries become increasingly efficient, lighter in weight, and smaller in size, and as the devices they power become smaller and more energy-efficient, one can anticipate a dramatic increase in the ease, speed, and low cost of accomplishing a variety of activities.

What can one realistically anticipate as improvements in batteries over the next few decades? It seems likely that changes will be incremental rather than dramatic, at least for the most power-hungry applications. In the case of electric batteries for automobiles, researchers anticipate a doubling of vehicle range and battery life by the mid-1990s and the development of electronic units capable of recharging batteries within 15 minutes. Further doublings of performance are expected in perhaps 15 to 20 years. The former stage will produce vehicles acceptable for selected applications, and the latter will come close to making electric automobiles suitable for general use. Similar trends can be expected elsewhere. In general, where battery size and weight are not critical problems, today's batteries are perfectly adequate. Where portability is a central characteristic of batttery use, today's specialty applications will become routine in a decade or two, and many new applications of portable electricity will follow. In all instances, batteries that contain toxic heavy metals (cadmium–nickel or lead–acid batteries, for example) will be increasingly regulated and will be replaced by modern alternatives.

23.3.4 Organics Replace Metals

The replacement of metal components by organic or composite ones is one of the driving forces behind dematerialization. The organic product characteristics can be primarily structural, as with automotive body panels, or primarily electrical, as with "organic metal" conductors in high-density circuitry. It is likely that materials innovations in organics will result in increasing replacement of metals for at least two or three decades to come.

A potential limitation on the use of organic materials is the availability of feedstocks, because most organic materials are manufactured from petroleum. To the extent that petroleum resources are used for transportation or energy generation, they may be available in only limited quantities for the manufacture of products. Two possible scenarios suggest themselves. One is that better automobile batteries, increased public transportation, or other technologies will enable society to reserve petroleum for industrial use. The second is that alternative materials, such as natural gas (also in eventual short supply, however) or substitute feedstocks such as biomass may prove suitable. All things considered, it appears unlikely that organic materials will be without a feedstock of one sort or another, and reasonable to anticipate that dematerialization as a consequence of metal to organic substitution will continue into the forseeable future.

23.3.5 Cluster Processing

A factory is a system designed to produce desirable manufactured products at competitive prices. Almost everything in a factory is manufactured in stages, one activity (metal forming, say) being performed at one stage, another (welding, say) being performed at the next, and so on. As material moves from one stage to the next, it leaves behind degraded processing chemicals, it loses or gains heat, and it encounters areas of variable cleanliness. Especially in the high-technology industries where reliability is of paramount importance, substantial efforts are beginning to transform batch processes performed at least partly in open environments into continuous sequences ("clusters") of processes performed entirely in environmentally controlled spaces.

An example of this type of thinking is the vision in the electronics industry of "cluster processing", a sequential manufacturing approach in which all operations take place within a closed facility, untouched by human hands from materials input to product output. Early forms of this concept exist now in the "mini-environments" within clean-rooms; these modules are designed to function automatically and can be made much cleaner than large open spaces.

The eventual cluster processing concept as it might look for the manufacture of sil-icon integrated circuits is shown in Fig. 23.4. Rapid advances in robotics and computer control now occurring in industry are leading to factories in which a process control computer is a part of every tool. These computers evenually may make it possible to extend the cluster processing concept to assembled products as well as components; cer-tainly, today's factories have many of the necessary pieces.

The enormous challenges posed by the realization of cluster processing would be justified by the benefits that stand to be realized. Process solutions, if present at all, would be strongly minimized, because they would have to be supplied, monitored, and replenished from external sources. More likely, the feedstocks would be gases or high-purity solids that would then undergo modification by radiation or gaseous interaction to generate the final products. The desire to minimize feedstocks, so as to minimize the volume of by-products and other residual material that would need to be removed, would result in minimal materials extraction to support the process. Product yields would likely improve, and the resulting profits would provide additional corporate incentive for developing the technology. Finally, the trend toward minimization of product size would be encouraged, because hardware requirements in manufacturing would thereby be reduced to as low a level as possible.

A complete cluster processing factory is a technological dream that is doubtless a long way off. Portions of that dream are realizable today, however, and we can confi-

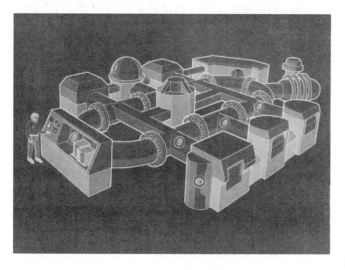

Figure 23.4   A concept for the manu-facture of electronic circuits by "cluster processing" techniques. (Courtesy of C. K. N. Patel, University of California, Los Angeles.)

dently expect that many of the products manufactured by the end of the twenty-first century's first quarter will be produced without the intervention of human hands, but with the extensive participation of human minds.

23.3.6   The Convergence of Computers and Communications Technology

Just as changes in products can produce energy efficiency and pollution reductions, so can changes in life style. These changes are coming about because of the rapid integration of powerful information processing facilities with the means to transport information rapidly from place to place, and the digitization of all forms of information: data, voice, and video. The letter on a piece of bond paper, once the only effective way of communicating, is being rapidly replaced by facsimile transmission, telephone calls, and electronic mail. As the speed, flexibility, and universality of these new technologies increase, the cost in both materials and time of transporting information from one location to another decreases very rapidly indeed.

Telecommuting is perhaps the ultimate best use of these merged technologies, because transportation plays such a significant role in many deleterious environmental impacts. At the present time, telecommuters can take advantage of telephones, fax machines, electronic mail, and modern computer links, and these devices make working at home a practical alternative for many professionals. It is estimated that, as of 1992, several million people (mostly in the United States) worked at home and telecommuted to their corporate offices, their suppliers, and their customers. Many more could and would do so but for the principal impediment: management culture, not technology. As culture changes are forced by competitive pressures and by governmental regulations such as the revisions to the U.S. Clean Air Act, management will view telecommuting across cities and, indeed, across national boundaries, as being increasingly acceptable. Before the century is out, personal television and teleconferencing will add to the convenience of working at home, a practice that will probably become widespread in the early twenty-first century.

23.4   THE ECOLOGICALLY PLANNED CITY

Technological visionaries are beginning to notice that humans tend to create unsuitable environments because the infrastructures are designed for individual activities rather than for the welfare of the society. A current example is the heating of individual homes in urban areas with coal stoves or in mountain valleys with wood stoves, an activity that is relatively salutary until a certain structural density is reached. At that point, the poor air quality resulting from the combustion products becomes undesirable for the entire community. To avoid this threshhold effect, remote combustion facilities could be used to provide a suitable level of heat, as shown in Fig. 23.5.

An even more imaginative approach to community living is to recognize that households and industries generate large quantities of materials that are taken away (such as obsolete appliances and waste paper) and import large quantities of other mate-

Figure 23.5   A vision of an ecologically engineered city in which heat is supplied by remote thermal generating facilities. (Courtesy of Tokyo Electric Power Company.)

rial (such as heating oil). Thus, one might hope to make use of discarded materials or their products to satisfy needs that currently require importing materials. One version of such schemes pictures the methane from landfills and sewage treatment plants being used as a natural gas supply for heating. Heating could also be produced by the incineration of otherwise unneeded incinerable material. Many other environmental gains are possible if engineers view communities as systems to be optimized for sustainability, just as we have advocated a systems approach to product and process design.

23.5 TRENDS FOR THE TWENTY-FIRST CENTURY

With the preceding discussion as a backdrop, it is interesting to imagine how the world will look from a technological standpoint in the next few decades. It seems likely that the combination of environmental constraints and consumer desires for functionality rather than ownership will have had profound impacts on the global economy, government policies, private corporations, and individuals. As a result of environmental constraints, for example, energy and resource conservation will have become increasingly

critical. Energy- and resource-efficient corporations will have gained substantial competitive advantages. This trend in particular may favor Japanese companies, because as
an island nation, the Japanese have already internalized a parsimonious attitude toward
energy and resources. This attitude, which at least initially was quite independent of the
environment, will serve them well in an environmentally constrained world. Americans,
on the other hand, with their "cowboy economy" and propensity for resource depletion,
may have a considerably rougher time.

Price structures, a serious impediment to industrial ecology during the 1990s, will
undergo considerable evolution in the next few decades, as more and more externalities
become captured either through market mechanisms or fees and taxes. The adjustment
of the pricing structure will be sporadic and will make business planning quite difficult.
Those corporations that fail to internalize environmental considerations into their product and process planning in the late twentieth or early twenty-first centuries will find
their costs escalating wildly and unpredictably, and will have few options when changes
have to be made rapidly.

A concomitant development will be the ascendency of materials science in many
corporations. The ability to predict environmental impacts of materials across their life
cycles and implement alternatives in response to regulatory bans and rapidly changing
costs will prove to be an important competency for any extractive or manufacturing corporation. Much of the progress in achieving sustainable manufacturing practices will
rely on new materials—superconductors, buckyball derivatives, and enzymes used in
bioprocessing factories, for example—and those corporations that stay abreast of the
"learning curve" on new materials will do well.

The crucial role of new materials will be reinforced by an explicit policy on the
part of many governments to allocate the transfer of functionality to the consumer marketplace, but to place the responsibility for the underlying product on manufacturing corporations. The trend, initiated by postconsumer product take-back legislation in Germany and Japan in the 1990s, will be extended to most product categories. Governments
will increasingly realize that environmental impacts arise predominantly from the nature
of the material stocks and flows underlying the economy rather than from the quality of
life the economy provides to consumers, and will increasingly react by imposing on
business the responsibility for materials, from extraction to rebirth to safe disposal.

On a broader front, as more and more operations of corporations became subject to
public approval processes, either formally through regulatory mechanisms or informally
through public activism, manufacturing will become a true partnership among the corporation, the community in which it exists, and the society in which it is embedded. The
idea of a corporation as responsible to only its shareholders and perhaps to its employees
or management is rapidly becoming obsolescent, although the details of balancing desirable competitive incentives against a broader social role are difficult and still evolving.
Nonetheless, in the future corporations will not quickly put profit ahead of social responsibility, so deeply will environmental concerns redefine our society.

What of various industrial sectors? Electronics manufacturing and software development will boom as the creation of intelligent resource- and energy-conserving systems
permeates the economy. Some power utilities will suffer, but many will prosper by

increasing their energy networking capabilities and becoming turn-key energy-efficiency consultants, an extention of the demand management efforts that began in the early 1990s. The transportation sector will see enormous change, best characterized by saying that customers will be offered "transparent transportation" for goods and services, where conditions and timing are specified but modes and interconnections are chosen by the service vendor. "Transparent commuting" will operate much the same way, except that many people will remain where they are and commute electronically. An economy based on intellectual capital requires an infrastructure emphasizing electronic networks, not civil engineering.

23.6 A SUMMARY OF THE VISION

The future for purposes of the practitioner of industrial ecology is essentially captured in two propositions: that it is an increasingly environmentally constrained world, and that customers will soon be buying functionality, not material. Thus, the long-term vision of industrial ecology centers both on technological development and on changes in the structure of societal demand. Industrial ecology recognizes that technology is the source of our environmental problems and that it may be the only feasible way to solve them. Technology alone cannot achieve the transformation we envision; it must work within the societal system to move closer to that goal.

SUGGESTED READING

Clark, W. C., and R. E. Munn, eds. *Sustainable Development of the Biosphere*. Cambridge, UK: Cambridge University Press, 1986.

Extended Producer Responsibility as a Strategy to Promote Cleaner Products, Proceedings from an Invitational Expert Seminar, Trolleholm Castle, ISRN LVTMDN/TMIM-92/3005-SE+126. Stockholm: Swedish Department of Environment, 1992.

Faeth, P., ed. *Agricultural Policy and Sustainability: Case Studies from India, Chile, the Philippines, and the United States*. Washington, DC: World Resources Institute, 1993.

Linden, H. R. Energy and industrial ecology, in *The Greening of Industrial Ecosystems*, B. R. Allenby and D. J. Richards, eds., pp. 38–60. Washington, DC: National Academy Press, 1994.

Office of Technology Assessment. *Biopolymers: Making Materials Nature's Way*, Report S/N 052-003-01352-5. Washington, DC: Congress, 1993.

Ruttan, V. W., ed. *Sustainable Agriculture and the Environment*. Boulder, CO: Westview Press, 1992.

Science magazine, invited papers on nanoprocessing, *254* (1991): 1312–1342.

The World Bank. *World Development Report 1992*. Oxford, UK: Oxford University Press, 1992.

Wright, K. The road to the global village, *Scientific American, 262* (3) (1990): 84–94.

EXERCISES

23.1 You are an executive in a corporation that manufactures products for sale. You anticipate a change in the structure of demand more or less as shown in Fig. 21.6. What actions do you recommend if your corporation makes (a) pencils, (b) personal computers, (c) motorcycles, (d) prefabricated houses, (e) conveyer belts, (f) pressure vessels for pharmaceutical manufacture.

23.2 Form teams of classmates and design the refrigerator of the future. What is the goal of the refrigerator design? What should the design characteristics be? Are materials innovations required to make your ideas work? If so, what are they? Is your design equally suitable for Canada, China, Chile, or Chad?

23.3 Devise several alternative scenarios for industry–government–environment conditions in the years 2000, 2010, and 2050. Include projections of customer preference, regulatory changes, and the status of the environment. What are the implications of these scenarios for corporate industrial ecologists?

A New
Industrial Revolution

Three key concepts concerning sustainable development of Earth emerge from the discussions in this book. The first, certainly not new with us, is that all industrial activity is driven by the desire of humanity for an increased standard of living. The second has to do with attempting to satisfy that driving force: that in performing its functions, industry should seek to avoid any significant perturbation of nature's cycles. The third is that the way in which industry can attempt to avoid or minimize that perturbation is by using the techniques of life-cycle analysis and design for environment in all its activities.

Two centuries ago, industry began as a small, labor-intensive, inobtrusive activity. This first industrial revolution eventually transformed the world from diffuse and labor-intensive agrarian societies to what is approaching a unified, electronically aided, urban society. This new stage has become large, obtrusive, and potentially destructive to the resources that support it. The goal of moving from this developmental stage to a state of sustainable development for the world cannot be met wholly by industrial ecology, but industrial ecology is a necessary ingredient if it is to happen.

It is worth recalling several characteristics of industrial ecology that, in one way or another, have permeated the text and that summarize the key points of the book:

- Industrial ecology is *proactive*, not *reactive*. That is, it is initiated and promoted by industrial concerns because it is in their own interest and in the interest of those surrounding systems with which they interact, not because it is imposed by one or more external factors.
- Industrial ecology is *designed-in*, not *added-on*. This characteristic recognizes that many aspects of materials flows are defined by decisions taken very early in the design process, and that optimization of industrial ecology requires every product and process designer and every manufacturing engineer to view industrial ecology

with the same intensity that is brought to bear on such issues as product quality and manufacturability.

- Industrial ecology is *flexible*, not *rigid*. Many aspects of industrial operations may need to change as new manufacturing processes become possible, new limitations arise from scientific and ecological studies, new opportunities present themselves as markets evolve, and so on.

- Industrial ecology is *encompassing*, not *insular*. In the modern international industrial world, industrial ecology calls for approaches that not only cross industrial sectors but national and cultural boundaries as well.

Practitioners of industrial ecology have the opportunity to take a step as great as was taken in the industrial revolution of the eighteenth century: to move from unconstrained use and disposal of materials to manufacturing approaches that take both products and their impacts into account in the same design and with the same degree of foresight. We believe that some segments of industry are now beginning this transformation, which can properly be called a new industrial revolution. The new revolution has the potential to move from the present resource-intensive approach to one of global sustainability. This is a bold vision, and one that will need many outstanding industrial ecologists to implement and many other segments of society to support. If it fails, the ability of the planet to support its varied life forms may fail with it. It is succeeds, it will help lead us all—trees, insects, birds, mammals, humans—to new levels of well-being.

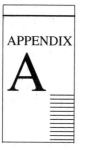

APPENDIX

A

Checklists
for Industrial
Ecology

Checklist 1 (Chapter 14)
Design Considerations in Energy Generation and Use

Product Designers

- Is the product designed with the aim of minimizing the use of energy-intensive process steps such as high heating differentials, heavy motors, extensive cooling, and so on?
- Is the product designed to utilize recycled materials wherever possible?

Process Designers

- Is the process designed with the aim of minimizing the use of energy-intensive process steps such as high heating differentials, heavy motors, extensive cooling, and so on?
- Is the process designed to optimize the use of heat exchangers and similar devices to utilize otherwise wasted heat?
- Does the process use the maximum possible amount of recycled material rather than virgin material?
- Is the process designed to utilize energy management approaches and equipment to minimize energy use?
- Is the process designed to utilize variable speed motors and other automated load controls?

Facility Engineers

- Has incandescent lighting been replaced with high-efficiency fluorescent lighting?
- Has an automatic lighting control system been installed?
- Have boilers and furnaces been checked for leaks and repaired as necessary?
- Are boilers correctly sized?
- Is the facility designed to utilize cogenerated heat and electricity from within the facility or nearby?
- Have ceilings, walls, and pipes been insulated?
- Does the industrial facility have a program to encourage good energy housekeeping?

Checklist 2 (Chapter 15)
Minimization and Design of Solid Residues in Industry

- Is manufacturing residue (mold scrap, cutting scrap, and so on) produced in connection with this assessment?
- Has manufacturing residue been minimized to the greatest extent possible?
- Have processes been designed to recycle the maximum fraction of manufacturing residue within the facility?
- Is manufacturing residue that cannot be recycled within the facility recycled outside it?
- Is biological residue produced in any manufacturing process in connection with this assessment? If so, is the residue rendered noninfectious and biologically inactive prior to disposal?
- Is radioactive residue produced in any manufacturing process in connection with this assessment? If so, is it properly handled and transported to an approved storage facility?
- Is sludge produced in any manufacturing process in connection with this assessment?
- Is excess sludge recycled rather than landfilled or incinerated?
- Has packaging material entering the facility been minimized, and designed to use the fewest possible different materials?
- Is excess packaging material recycled rather than landfilled or incinerated?

Checklist 3 (Chapter 15)
Minimization and Design of Liquid Emissions in Industry

- If trace metals are used in any manufacturing process in connection with this assessment, is their use minimized, and have substitutes been investigated?

- If nutrients are used in any manufacturing process in connection with this assessment, is their use minimized, and have substitutes been investigated?
- If solvents or oils are used in any manufacturing process in connection with this assessment, is their use minimized, and have substitutes been investigated?
- If organic species of concern are used in any manufacturing process in connection with this assessment, is their use minimized, and have substitutes been investigated?
- Are acids used in any manufacturing process in connection with this assessment?
- Have alternatives to liquid species of concern been thoroughly investigated and incorporated where possible?
- Where liquid species of concern are used, do they have the lowest possible water pollution potential?
- Have processes been designed to recycle the maximum fraction of liquid species of concern within the facility?
- Have markets and uses been identified for liquid species that leave the facility?
- Have the processes been designed to utilize the maximum amount of recycled liquid species from outside suppliers rather than virgin materials?
- Are equipment cleanouts that generate liquid or solid residues minimized to the extent possible?
- Are pumps, valves, and pipes inspected regularly to minimize leaks?

Checklist 4 (Chapter 15)
Minimization and Design of Gaseous Emissions in Industry

- Are CFCs or HCFCs used in any manufacturing process in connection with this assessment?
- Have alternatives to CFC/HCFC processes been thoroughly investigated and incorporated where possible?
- Where HCFCs are used, do they have the lowest possible ozone-depletion potential?
- Have provisions been made to eliminate all CFC/HCFC and halon use by the start of 1996?
- Are halons used in any manufacturing process in connection with this assessment?
- Are greenhouse gases used or generated in any manufacturing process in connection with this assessment?
- Are volatile organic compounds used or generated in any manufacturing process in connection with this assessment?
- If some VOCs are emitted, are the emissions designed for low reactivity?
- Are odorants used or generated in any manufacturing process in connection with this assessment?

- Are acid gases used or generated in any manufacturing process in connection with this assessment?
- Are trace metals used or generated in any manufacturing process in connection with this assessment?
- Have the processes been designed to use the minimum amounts of any of the materials mentioned above?
- Have processes been designed to capture and recycle the maximum fraction of any of the materials mentioned above from within the facility?
- Where materials of concern are feedstocks or intermediates rather than process emittants, have the processes been designed to utilize recycled quantities from outside suppliers?

Checklist 5 (Chapter 16)

Design Considerations in Materials Selection

- Are any proposed materials in restricted supply or are they likely to become so over the period during which the product manufacture is anticipated?
- Are any proposed materials toxic (includes radioactive)?
- Are any proposed materials ozone-depleting substances or global-warming substances?
- Do any proposed materials have potential disposal problems?
- If the answer to any of the above questions is yes, has materials substitution been thoroughly considered?
- If the answer to any of the above questions is yes, has materials reduction been thoroughly considered?
- Can recycled materials rather than virgin materials be specified?
- Can materials use be minimized by improved mechanical design?

Checklist 6 (Chapter 17)

Product Packaging and Transportation

- Can the product packaging be eliminated?
- Can secondary or tertiary product packaging be eliminated?
- Does the product packaging mimimize the number of different materials that are used in its manufacture?
- Does the product packaging avoid the use of toxic materials?
- Is the product packaging designed to make it easy to separate the constituent materials?
- Is the product packaging designed to minimize the packaging weight and volume?

- Is the product packaging designed for shipping in bulk as opposed to or in addition to shipping in small individual packages?
- Have efforts been made to use recyclable packaging materials?
- Have efforts been made to use refillable or reusable containers where appropriate?
- Are alternative packaging systems evaluated against one another in a structured way?
- Are products, especially hazardous products or those potentially subject to spilling or venting, transported on safe routes by trained drivers?
- Are recycling instructions clearly printed on the packaging itself?
- Are arrangements made to take back product packaging for recycling and reuse?
- Is transportation packaging design integrated with product packaging design?
- Are impacts during product installation minimized and are steps being taken to redesign products to avoid such impacts altogether?

Checklist 7 (Chapter 18)
Solid-Residue Generation During Product Use

- Is this product designed to be disposed of upon using?
- If the answer is yes, has materials use been minimized and have alternative techniques for accomplishing the same purpose been examined?
- Does use of this product require periodic disposal of solid materials such as cartridges, containers, or batteries?
- Do the consumables contain any toxic or otherwise undesirable materials?
- If toxic or undesirable materials are included, have alternative materials been thoroughly investigated and incorporated where possible?
- Have alternatives to consumables been thoroughly investigated and incorporated where possible?
- Have the processes been designed to use the minimum amounts of consumables?
- Have procedures been implemented to recycle the maximum fraction of consumables?
- Have the processes been designed to utilize recycled consumables from outside suppliers?

Checklist 8 (Chapter 18)
Liquid-Residue Generation During Product Use

- If toxic or otherwise undesirable materials are included in the product, especially if they are required for routine use or routine maintenance, have alternative materials been thoroughly investigated and incorporated where possible?
- If the product is dissipated during use, has it been designed to have minimal environmental impact?

- Does use of this product require periodic replenishment of liquid materials such as coolants or lubricants?
- Do the consumables contain any toxic or otherwise undesirable materials?
- Have alternatives to consumables been thoroughly investigated and incorporated where possible?
- Have the processes been designed to use the minimum amounts of consumables?
- Have procedures been implemented to recycle the maximum fraction of consumables?
- Have the processes been designed to utilize recycled consumables from outside suppliers?
- If the product is generally stored at the customer's site, have design provisions been taken to minimize the potential for inadvertent dissipative emissions?

Checklist 9 (Chapter 18)
Gaseous-Residue Generation During Product Use

- Does use of this product result in periodic generation of gaseous emissions such as carbon dioxide or tetraethyl lead?
- Does use of this product generate an odor (odors are signals of gaseous emissions)?
- Do the emissions contain any toxic or otherwise undesirable materials?
- If toxic or undesirable emissions result, have alternative emittants been thoroughly investigated and the use process changed where possible?
- Have the processes been designed to emit the minimum amounts of gaseous products?
- Have alternative designs not involving gaseous emissions or materials substitutions to prevent gaseous emissions been thoroughly investigated and incorporated where possible?

Checklist 10 (Chapter 18)
Energy Consumption During Product Use

- Is energy consumed during the operation of this product?
- Have enhanced insulation or other energy-conserving design features been incorporated?
- Have alternatives to energy use been thoroughly investigated and incorporated where possible?
- Have alternative-energy generation processes been thoroughly investigated and the least environmentally harmful alternative chosen?

Checklist 11 (Chapter 18)
Intentional Dissipative Emissions During Product Use

- Do dissipative emissions occur as a result of using this product?
- Are the dissipative emissions toxic or otherwise undesirable?
- Have more environmentally preferable dissipative emissions been thoroughly explored and implemented wherever possible?
- Have alternative designs not involving dissipative emissions been thoroughly explored and implemented wherever possible?

Checklist 12 (Chapter 18)
Unintentional Dissipative Emissions During Product Use

- Does this product contain any materials that have the potential to be unintentionally dissipated during use?
- Are materials that may be subject to unintentional dissipation toxic or otherwise particularly undesirable?
- Have efforts been made to eliminate the materials that may be unintentionally dissipated?
- Have efforts been made to reformulate the product so as to minimize unintentional dissipation?

Checklist 13* (Chapter 19)
Design for Maintainability

- Are subassemblies designed for ready maintainability rather than solely for disposal after malfunction?
- Are mechanical parts individually repairable or replaceable?
- Have products been designed to require a minimum of cleaning and maintenance?
- Are modules designed for ready removal? Can removal and replacement be performed by the customer if refurbished, replaced, or upgraded by the manufacturer? Are mail-back plans or other arrangements in place to encourage modular repair by customers?
- Do maintenance procedures require the use of toxic substances or other substances with adverse environmental impacts?

*This checklist is largely the conception of E. T. Morehouse, Jr., Design for maintainability, Design for Environment White Paper No. 7, Washington, DC: American Electronics Association, 1992.

Checklist 14 (Chapter 19)
Design for Recycling

- Does the product mimimize the number of different materials that are used in its manufacture?
- Does this product minimize the use of toxic materials?
- Where toxic materials are used, are they easy to identify and separate?
- Have efforts been made to avoid joining dissimilar materials together in ways difficult to reverse?
- Is the product design modular, so that obsolescence occurs with components rather than with the entire product?
- Have threaded metal inserts in plastics been avoided?
- Have efforts been made to avoid the use of plated metals?
- Where plastics are used, are they thermoplastics rather than thermosets, as far as possible?
- Are all plastic components identified by ISO markings as to their content?
- Have efforts been made to avoid painting or otherwise adulterating plastic components? Are fillers minimized or eliminated?
- Has the product been assembled with fasteners such as clips or hook-and-loop attachments rather than chemical bonds or welds?

Chemicals Identified in the 1990 U.S. Clean Air Act

CAS Number*	Chemical Name	CAS Number*	Chemical Name
79-34-5	1,1,2,2-Tetrachloroethane	91-94-1	3,3-Dichlorobenzidine
79-00-5	1,1,2-Trichloroethane	119-90-4	3,3'-Dimethoxybenzidine
57-14-7	1,1-Dimethylhydrazine	101-77-9	4,4'-Methylenedianiline
120-82-1	1,2,4-Trichlorobenzene	1011-44-4	4,4-Methylene bis(2-chloroaniline)
96-12-8	1,2-Dibromo-3-chloropropane	534-52-1	4,6-Dinitro-o-cresol, and salts
122-66-7	1,2-Diphenylhydrazine	92-67-1	4-Aminobiphenyl
106-88-7	1,2-Epoxybutane	92-93-3	4-Nitrobiphenyl
75-55-8	1,2-Propylenimine	100-02-7	4-Nitrophenol
106-99-0	1,3-Butadiene	75-07-0	Acetaldehyde
542-75-6	1,3-Dichloropropene	60-35-5	Acetamide
1120-71-4	1,3-Propane sulfone	75-05-8	Acetonitrile
106-46-7	1,4-Dichlorobenzene	98-86-2	Acetophenone
123-91-1	1,4-Dioxane	107-02-8	Acrolein
540-84-1	2,2,4-Trimethylpentane	79-06-1	Acrylamide
1746-01-6	2,3,7,8-Tetrachlorodibenzo-p-dioxin	79-10-7	Acrylic acid
95-95-4	2,4,5-Trichlorophenol	107-13-1	Acrylonitrile
88-06-2	2,4,6-Trichlorophenol	107-05-1	Allyl chloride
94-75-7	2,4-D, salts and esters	7664-41-7	Ammonia
51-28-5	2,4-Dinitrophenol	62-53-3	Aniline
121-14-2	2,4-Dinitrotoluene	7440-36-0	Antimony
95-80-7	2,4-Toluene diamine	0	Antimony compounds
584-84-9	2,4-Toluene diisocyanate	7440-38-2	Arsenic
53-96-3	2-Acetylaminofluorine	0	Arsenic compounds
532-27-4	2-Chloroacetophenone	1332-21-4	Asbestos
79-46-9	2-Nitropropane	71-43-2	Benzene
119-93-7	3,3'-Dimethylbenzidine	92-87-5	Benzidine

CAS Number*	Chemical Name	CAS Number*	Chemical Name
98-07-7	Benzotrichloride	131-11-3	Dimethyl phthalate
100-44-7	Benzyl chloride	77-78-1	Dimethylsulfate
7440-41-7	Beryllium	106-89-8	1-Chloro-2,3-epoxypropane
0	Beryllium compounds	140-88-5	Ethyl acrylate
92-52-4	Biphenyl	100-41-4	Ethylbenzene
117-81-7	bis(2-ethylhexyl) phthalate	51-79-6	Ethylcarbamate
542-88-1	bis(chloromethyl) ether	75-00-3	Chloroethane
75-25-2	Bromoform	106-93-4	Dibromoethane
7440-43-9	Cadmium	107-06-2	1,2-Dichloroethane
0	Cadmium compounds	107-21-1	Ethylene glycol
156-62-7	Calcium cyanamide	151-56-4	Aziridine
105-60-2	Caprolactum	75-21-8	Ethylene oxide
133-06-2	Captan	96-45-7	Ethylene thiourea
63-25-2	Carbaryl	75-34-3	1,1-Dichloroethane
75-15-0	Carbon disulfide	0	Fine mineral fibers
56-23-5	Carbon tetrachloride	50-00-0	Formaldehyde
463-58-1	Carbonyl sulfide	0	Glycol ethers
120-80-9	Catechol	76-44-8	Heptachlor
133-90-4	Chloramben	118-74-1	Hexachlorobenzene
57-74-9	Chlordane	87-68-3	Hexachlorobutadiene
7782-50-5	Chlorine	77-47-4	Hexachlorocyclopentadiene
79-11-8	Chloroacetic acid	67-72-1	Hexachloroethane
108-90-7	Chlorobenzene	822-06-0	Hexamethylene-1,6-diisocyanate
510-15-6	Chlorobenzilate	680-31-9	Hexamethylphosphoramide
67-66-3	Chloroform	110-54-3	Hexane
107-30-2	Chloromethyl methyl ether	302-01-2	Hydrazine
126-99-8	Chloroprene	7647-01-0	Hydrochloric acid
7440-47-3	Chromium	74-90-8	Hydrogen cyanide
0	Chromium compounds	7664-39-3	Hydrogen fluoride
7440-48-4	Cobalt	7683-06-4	Hydrogen sulfide
0	Cobalt compounds	123-31-9	Hydroquinone
0	Coke oven emissions	78-79-1	Isophorone
1319-77-3	Cresylic acid	7439-91-1	Lead
95-48-7	o-Cresol	0	Lead compounds
108-39-4	m-Cresol	58-89-9	Lindane
106-44-5	p-Cresol	108-31-6	Maleic anhydride
98-82-8	Cumene	7439-96-5	Manganese
0	Cyanide compounds	0	Manganese compounds
3457-04-4	DDE	7439-97-6	Mercury
334-88-3	Diazomethane	0	Mercury compounds
132-64-9	Dibenzofuran	108-88-3	Toluene
84-74-2	Dibutylphthalate	8001-35-2	Toxaphene
111-44-4	bis(2-Chloroethyl) ether	79-01-6	Trichloroethylene
62-73-7	Dichlorvos	121-44-8	Triethylamine
111-42-2	Diethanolamine	1582-09-8	Trifluralin
64-67-5	Diethyl sulfate	108-05-4	Vinyl acetate
601-11-7	Dimethylaminoazobenzene	593-60-4	Vinyl bromide
79-44-7	Dimethyl carbamoyl chloride	75-01-4	Vinyl chloride
68-12-2	Dimethylformamide	75-35-4	1,1-Dichloroethylene

CAS Number*	Chemical Name	CAS Number*	Chemical Name
67-56-1	Methanol	75-44-5	Phosgene
72-43-5	Methoxychlor	7803-51-2	Phosphine
74-83-9	Methyl bromide	7723-14-0	Phosphorus
74-87-3	Methyl chloride	85-44-9	Phthalic anhydride
71-55-6	1,1,1-Trichloroethane	1336-36-3	Polychlorinated biphenyls
78-93-3	2-Butanone	0	Polychlorinated organic matter
60-34-4	Methyl hydrazine	123-38-6	Propionaldehyde
74-88-4	Methyl iodide	114-26-1	Propoxur
108-10-1	Methylisobutyl ketone	78-87-5	1,2-Dichloropropane
624-83-9	Methyl isocyanate	75-56-9	Propylene oxide
80-62-6	Methyl methacrylate	91-22-5	Quinolene
1634-04-4	Methyl-*t*-butylether	106-51-4	1,6-Benzoquinone
75-09-2	Methylene chloride	0	Radionuclides (including radon)
101-68-8	Methylen diphenyl diisocyanate	7782-49-2	Selenium
121-69-7	N,N-diethylaniline	0	Selenium compounds
684-93-5	N-Nitroso-N-methylurea	100-42-5	Styrene
62-75-9	N-Nitrosodimethylamine	96-09-3	Styrene oxide
59-89-2	N-Nitrosomorpholine	127-18-4	Perchloroethylene
91-20-3	Naphthalene	7550-45-0	Titanium tetrachloride
7440-02-0	Nickel	1330-20-7	Xylene mixture
0	Nickel compounds	106-42-3	*p*-Xylene
98-95-3	Nitrobenzene	108-38-3	*m*-Xylene
56-38-2	Parathion	95-47-6	*m*-Xylene
82-68-8	Pentachloronitrobenzene	57-57-8	β-Propiolactone
87-86-5	Pentachlorophenol	90-04-0	*o*-Anisidine
108-95-2	Phenol	95-53-4	*o*-Toluidine
		106-50-3	*p*-Phenylenediamine

*The CAS number is the registry number assigned to the compound by Chemical Abstracts Service, Columbus, OH. Each compound is given a unique number by CAS, useful for computerized literature searching and for making certain of the identity of a compound.

APPENDIX
C

Materials with Disposal Restrictions

Classes of Materials

Liquids with flash point less than 60°C
Liquids with pH < 2
Liquids with pH > 12.5
Unstable or highly reactive chemicals
Radioactive materials
Ozone-depleting substances

Metals

Antimony
Arsenic (and its compounds)
Barium
Beryllium (and its alloys)
Cadmium
Chromium compounds (especially hexavalent)
Cobalt
Copper
Lead (and its compounds)
Lithium
Magnesium
Manganese
Mercury
Molybdenum

Nickel
Selenium
Silver
Tellurium (and its compounds)
Thallium
Vanadium
Zinc

Inorganic Materials

Asbestos
Carbon disulfide
Cyanides
Fluoride salts
Fluoroborates

Organic Materials

2,4-D
2,4,5-TP (Silvex)
Acetone
Benzene
Carbon tetrachloride
Chlordane
Chlorobenzene
Chloroform
Cresols
Cresylic acid
Cyclohexanone
1,4-Dichlorobenzene
1,2-Dichloroethane
1,1-Dichloroethylene
Dinitrotoluene
Endrin
2-Ethoxyethanol
Ethyl acetate
Ethylbenzene
Ethyl ether
Heptachlor
Hexachlorobenzene
Hexachlorobutadiene
Hexachloroethane
Isobutanol
Lindane
Methanol

Methyl ethyl ketone (2-Butanone)
Methyl isobutyl ketone (4-Methyl-2-pentanone)
Methylene chloride
Methoxychlor
N-butyl alcohol
Nitrobenzene
2-Nitropropane
Pentachlorophenol
Pyridine
Teflon (if heated or machined)
Tetrachloroethylene
Toluene
Toxaphene
Trichloroethylene
2,4,5-Trichlorophenol
2,4,6-Trichlorophenol
Vinyl chloride
Xylene

*The source of this list is *Guide for Environmental Aspects in Product Standards*, Draft for Comment. Geneva: International Electrotechnical Commission, March, 1993. Amended and updated by the authors.

Ozone-Depleting and Greenhouse-Warming Chemicals

Designation	Chemical Formula	Ozone-Depletion Potential[†]	Greenhouse-Warming Potential[‡]
Class I Ozone Depletion and Greenhouse Warming Species			
CFC-11	CCl_3F	1	3,400
CFC-12	CCl_2F_2	0.89	7,100
CFC-13	$CClF_3$		13,000
CFC-111	C_2FCl_5		
CFC-112	$C_2F_2Cl_4$		
CFC-113	$C_2F_3Cl_3$	0.81	4,500
CFC-114	$C_2F_4Cl_2$	0.69	7,000
CFC-115	C_2F_5Cl	0.32	7,000
CFC-211	C_3FCl_7		
CFC-212	$C_3F_2Cl_6$		
CFC-213	$C_3F_3Cl_5$		
CFC-214	$C_3F_4Cl_4$		
CFC-215	$C_3F_5Cl_3$		
CFC-216	$C_3F_6Cl_2$		
CFC-217	C_3F_7Cl		
Halon-1211	CF_2ClBr	2.2–3.5	
Halon-1301	CF_3Br	8–16	4,900
Halon-2402	$C_2F_4Br_2$	5–6.2	
Carbon tetrachloride	CCl_4	1.13	1,300
Methyl chloroform	CH_3CCl_3	0.14	
Methyl bromide	CH_3Br		
Methyl chloride	CH_3Cl		
HBFCs	Several		
Nitrous oxide	N_2O		270

Designation	Chemical Formula	Ozone-Depletion Potential[†]	Greenhouse-Warming Potential[‡]
Greenhouse-Warming (non-Ozone-Depleting) Species			
Carbon Dioxide	CO_2	0	1
Methane	CH_4	0	11
HFC-125	CHF_2CF_3		90
HFC-134a	CFH_2CF_3		1,000
HFC-152a	CH_3CHF_2		2,400
Perfluorobutane	C_4F_{10}	0	5,500
Perfluoropentane	C_5F_{12}	0	5,500
Perfluorohexane	C_6F_{14}	0	5,100
Perfluorotributylamine	$N(C_4F_9)_3$	0	4,300
Class II Ozone-Depletion and Greenhouse-Warming Species			
HCFC-21	$CHFCl_2$	0.010	
HCFC-22	CHF_2Cl	0.048	1,600
HCFC-31	CH_2ClF	0.010	
HCFC-121	C_2HFCl_4	0.004	
HCFC-122	$C_2HF_2Cl_3$	0.004	
HCFC-123	$C_2HF_3Cl_2$	0.017	90
HCFC-124	C_2HF_4Cl	0.019	440
HCFC-125	C_2HF_5	0	3400
HCFC-131	$C_2H_2FCl_3$	<0.001	
HCFC-132	$C_2H_2F_2Cl_2$	0.002	
HCFC-133	$C_2H_2F_3Cl$	0.070	
HCFC-141b	$C_2H_3FCl_2$	0.09	580
HCFC-142b	$C_2H_3F_2Cl$	0.054	1800
HCFC-221	C_3HFCl_6		
HCFC-222	$C_3HF_2Cl_5$		
HCFC-223	$C_3HF_3Cl_4$		
HCFC-224	$C_3HF_4Cl_3$		
HCFC-225	$C_3HF_5Cl_2$		
HCFC-226	C_3HF_6Cl		
HCFC-231	$C_3H_2FCl_5$		
HCFC-232	$C_3H_2F_2Cl_4$		
HCFC-233	$C_3H_2F_3Cl_3$		
HCFC-234	$C_3H_2F_4Cl_2$		
HCFC-235	$C_3H_2F_5Cl$		
HCFC-241	$C_3H_3FCl_4$		
HCFC-242	$C_3H_3F_2Cl_3$		
HCFC-243	$C_3H_3F_3Cl_2$		
HCFC-244	$C_3H_3F_4Cl$		
HCFC-251	$C_3H_4FCl_3$		

Designation	Chemical Formula	Ozone-Depletion Potential[†]	Greenhouse-Warming Potential[‡]
HCFC-252	$C_3H_4F_2Cl_2$		
HCFC-253	$C_3H_4F_3Cl$		
HCFC-261	$C_3H_5FCl_2$		
HCFC-262	$C_3H_5F_2Cl$		
HCFC-271	C_3H_6FCl		

*CFC, chlorofluorocarbon; HCFC, hydrochlorofluorocarbon; HBFC, hydrobromofluorocarbon. Absence of an entry does not imply no effect, but rather that the magnitude of the effect has not been determined. The sources of this list include *Guide for Environmental Aspects in Product Standards*, Draft for Comment. Geneva: International Electrotechnical Commission, March 1993; J. S. Nimitz and S. R. Skaggs, Estimating tropospheric lifetimes and ozone depletion potentials of one- and two-carbon hydrofluorocarbon and hydrochlorofluorocarbons, *Environmental Science and Technology, 26* (1992): 739–744; J. T. Houghton, G. J. Jenkins, and J. J. Ephraums, *Climate Change: The IPCC Scientific Assessment*, Cambridge, UK: Cambridge University Press, 1990; S. Solomon, M. Mills, L. E. Heidt, W. H. Pollock, and A. F. Tuck, On the evaluation of ozone depletion potentials, *Journal of Geophysical Research, 97* (1992): 825–842; and C. Clerbaux, R. Colin, P. C. Simon, and C. Granier, Infrared cross sections and global warming potentials for 10 alternative hydrohalocarbons, *Journal of Geophysical Research, 98* (1993): 10491–10497.

[†] The ozone depletion potential (ODP) for a specific compound i is derived by carrying out computer model calculations to determine the effect *per molecule* of that compound on the abundance of stratospheric ozone, referenced to the effect of CFC-11. Specifically,

$$ODP(i) = \frac{\displaystyle\int_z \int_\theta \int_t [\Delta O_3(z, \theta, t)]_i \cdot \cos\theta}{\displaystyle\int_z \int_\theta \int_t [\Delta O_3(z, \theta, t)]_{CFC-11} \cdot \cos\theta}$$

where z is the altitude, θ is the latitude, t is time, and ΔO_3 is the change in ozone at steady state per unit mass-emission rate.

Further information is available in S. Solomon, M. Mills, L. E. Heidt, W. H. Pollock, and A. F. Tuck, On the evaluation of ozone depletion potentials, *Journal of Geophysical Research, 97* (1992): 825–842.

[‡] The global-warming potential (GWP) for a specific compound i is derived by carrying out computer model calculations to determine the effect *per molecule* of that compound on the radiative forcing of the atmosphere referenced to the effect of carbon dioxide. Specifically,

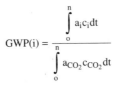

$$GWP(i) = \frac{\displaystyle\int_0^n a_i c_i \, dt}{\displaystyle\int_0^n a_{CO_2} c_{CO_2} \, dt}$$

where a_i is the instantaneous radiative forcing due to a unit increase in the concentration of the trace gas i, c_i is the concentration of trace gas i remaining t years after its release, and n is the number of years over which the calculation is performed. The values given are for a time horizon of 100 years.

Further information is available in J. T. Houghton, G. J. Jenkins, and J. J. Ephraums, *Climate Change: The IPCC Scientific Assessment*, Cambridge, UK: Cambridge University Press, 1990.

APPENDIX
E

Standards for Marking
Plastic Parts

International Standards Organization Recommendations*

1. Size, geometry, and function permitting, all plastic parts, external and internal, shall be marked.

2. Only new parts are required to be marked. The marking of parts currently being manufactured with existing tools is voluntary.

3. The ISO terminology and symbols are to be used. The capital letter alpha codes identifying the plastic species are to be set between the "larger than" and "smaller than" characters, which, according to ISO/DIS 11469, identify the marking as a plastic identification code. Marking examples of frequently used engineering plastics are:

>ABS<	acrylonitrile/butadiene/styrene
>ABS-FR<	flame retardant/ignition resistant ABS
>EP<	epoxy
>PA<	nylon (polyamide)
>PA6<	nylon 6
>PA66<	nylon 6/6
>PBT<	polybutylene terephthalate
>PC<	polycarbonate
>PE<	polyethylene
>PE-LLD<	linear low density polyethylene

*ISO/DIS 11469, Generic Identification and Marking of Plastic Parts; ISO 1043, Plastics—Symbols, Parts 1–3; SAE J1344, Marking of Plastic Parts.

>PE-LMD<	low medium density polyethylene
>PE-HD<	high density polyethylene
>PET<	polyethylene terephthalate
>PS<	polystyrene
>PS-HI<	high impact polystyrene
>PVC<	polyvinylchloride
>SAN<	styrene/acrylonitrile
>SI<	silicone

If a part consisting of a blend or mixture of polymers is to be marked, the polymer codes are to be separated by a "+" sign. For example, for an ABS/polycarbonate blend use

>ABS+PC<

4. Fillers and reinforcers are represented by two-character codes followed by a numerical value representing the percentage by weight of the filler/reinforcer. Examples of codes for some of the most frequently used fillers and reinforcing materials are

GF	glass fiber
GB	glass bead
MD	mineral powder
CF	carbon fiber

A polybutylene terephthalate/polycarbonate blend containing 30% glass fiber is to be marked in the following way:

>(PBT+PC)-GF30<

Note that the resin and the filler codes are separated by a "-" sign. Furthermore, note the proper use of brackets in case of mixtures of polymers and fillers, as indicated in the following examples:

>PC-GF20<	one polymer, one filler
>(PBT+PC)-GF30<	two or more polymers, one filler
>PC-(CF7+GF5+GB5)<	one polymer, two or more fillers
>(PPE+PS)-(GF15+MD15)<	two or more polymers, two or more fillers

5. As indicated in the figure below, the recommended height of the letters is 3 to 5 mm, although for exceptionally large or small parts the size may be adjusted. Only capital letters are to be used.

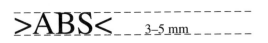

>ABS<   3–5 mm

6. On each product, at least one marking should appear on the outside of an external plastic part, although not in a location which is observed during regular use. A suitable location, for example, might be next to, or in the vicinity of, the Underwriters' Laboratory (UL) electrical safety mark. This way it will be apparent without first dismantling the product that its plastic parts have been marked for recycling.

7. Markings shall remain legible during entire part life. On molded parts, molded-in markings are preferred. Other acceptable methods are hot stamping or melt imprinting. Do not use stick-on labels to mark plastic parts.

Some corporations make one or two additions to the ISO standard. If a specific polymer blend is used, it (or a readily retrieved company code) can be placed below the ISO symbol. This information helps one locate data on polymer properties, including the concentrations of minor constituents such as plasticizers and lubricants. In the same vein, if the product is made from partially recycled material, the percentage of recycled material can be indicated above the ISO symbol. (In the second example that follows, the symbol indicates not only that 25% of the material used for the product is recycled, but that it has been recycled twice. Such information is important for the recycler, as materials properties are degraded somewhat at each recycling step.)

25%   25% × 2

>ABS + PC<   >ABS + PC<

LEXAN L-900   IBM 81-274 (2)

U.S. Recycling Markings for Plastics

Less preferable than the ISO standards, but far better than no markings at all, are the U.S. recycling marking standards for plastics. Although they make no provision for blends or mixtures of polymers or for fillers or reinforcing materials, these markings definitely aid in the recycling of the simpler plastics and have proved particularly appropriate where individual consumers are to separate the plastic prior to discarding it. The markings are as follows:

$\triangle\!\!1$ Polyethylene terephthalate

$\triangle\!\!2$ High-density polyethylene

$\triangle\!\!3$ Vinyl/Polyvinyl chloride

⚠4   Low-density polyethylene

⚠5   Polypropylene

⚠6   Polystyrene

⚠7   Blends and layered plastics

Reference Information on the Materials and Process Audit for Electronic Solder and Alternatives

This appendix contains the full set of matrices generated by the DFE analysis described in Chapter 12. Highlights of the analysis are discussed there, and details for each matrix element are presented in B. R. Allenby, *Design for Environment: Implementing Industrial Ecology*, Ph.D. dissertation, Rutgers University, 1992 (available through University Microfilms, Inc.)

Two cautionary notes are appropriate concerning the diagrams. First, in evaluating these analytical results, it should be remembered that each entry inevitably depended to some extent on the judgment of the analyst, a point also frequently true but less obvious in a more quantitative presentation. Second, the summary matrix contains the most serious impact noted in each individual matrix, regardless of which life-cycle stage it occurred in. Thus, there is no attempt to balance a serious impact in one life-cycle stage against a moderate one in another life-cycle stage, even if the latter, overall, would result in a more serious environmental impact in the minds of most people. The balancing of such impacts involves a subsequent level of value judgment, and is thus left to a separate stage of the analysis, LCA's *improvement analysis* stage.

Perhaps the most fundamental reminder to come from this analysis is that choices among options in industrial ecology generally require both data analysis and value judgments. These are two very different steps, involving very different considerations, and should be separated as cleanly as possible.

Manufacturing: Bismuth
Life stages

Manufacturing: Indium
Life stages

Manufacturing: Lead
Life stages

	Initial production	Secondary processing/ manufacturing	Packaging	Transportation	Consumer use	Reuse/ recycle	Disposal	Summary
Process compatibility								
Materials compatibility								
Component compatibility								
Performance								
Energy consumption								
Resource consumption								
Availability								
Cost	+	+						
Competitive implications								
Environment of use								

Manufacturing: Epoxy
Life stages

	Initial production	Secondary processing/ manufacturing	Packaging	Transportation	Consumer use	Reuse/ recycle	Disposal	Summary
Process compatibility								
Materials compatibility								
Component compatibility								
Performance								
Energy consumption								
Resource consumption								
Availability								
Cost								
Competitive implications					+			
Environment of use								

Social/Political: Bismuth
Life stages

	Initial production	Secondary processing/ manufacturing	Packaging	Transportation	Consumer use	Reuse/ recycle	Disposal	Summary
Regulatory status								
Legislative status								
Community status								
Community impacts								
Labor impacts	+							
Social impacts								
Significant externalities		—	—	—	—	—		

Social/Political: Indium
Life stages

	Initial production	Secondary processing/ manufacturing	Packaging	Transportation	Consumer use	Reuse/ recycle	Disposal	Summary
Regulatory status								
Legislative status								
Community status								
Community impacts								
Labor impacts	+							
Social impacts								
Significant externalities		—	—	—	—	—		

Social/Political: Lead
Life stages

	Initial production	Secondary processing/ manufacturing	Packaging	Transportation	Consumer use	Reuse/ recycle	Disposal	Summary
Regulatory status								
Legislative status								
Community status								
Community impacts								
Labor impacts								
Social impacts								
Significant externalities								

Social/Political: Epoxy
Life stages

	Initial production	Secondary processing/ manufacturing	Packaging	Transportation	Consumer use	Reuse/ recycle	Disposal	Summary
Regulatory status								
Legislative status								
Community status								
Community impacts								
Labor impacts	+							
Social impacts								
Significant externalities								

Environmental: Bismuth
Life stages

Environmental: Indium
Life stages

Environmental: Lead
Life stages

	Initial production	Secondary processing/ manufacturing	Packing	Transportation	Consumer use	Reuse/ recycle	Disposal	Summary
Local air impacts								
Water impacts								
Soil impacts								
Ocean impacts								
Atmospheric impacts								
Waste impacts								
Resource consumption								
Ancillary impacts								
Signifficant externalities								

Environmental: Epoxy
Life stages

	Initial production	Secondary processing/ manufacturing	Packaging	Transportation	Consumer use	Reuse/ recycle	Disposal	Summary
Local air impacts								
Water impacts								
Soil impacts								
Ocean impacts								
Atmospheric impacts								
Waste impacts						+		
Resource consumption								
Ancillary impacts								
Significant externalities								

Toxicity/Exposure: Bismuth
Life stages

	Initial production	Secondary processing/ manufacturing	Packaging	Transportation	Consumer use	Reuse/ recycle	Disposal	Summary
Community exposure								
Occupational exposure								
Consumer exposure								
Environmental exposure								
Mammalian acute								
Mammalian chronic								
Other acute								
Other chronic								
Bio-accumulative								

Toxicity/Exposure: Indium
Life stages

	Initial production	Secondary processing/ manufacturing	Packaging	Transportation	Consumer use	Reuse/ recycle	Disposal	Summary
Community exposure								
Occupational exposure								
Consumer exposure								
Environmental exposure								
Mammalian acute								
Mammalian chronic								
Other acute								
Other chronic								
Bio-accumulative								

Toxicity/Exposure: Lead
Life stages

Toxicity/Exposure: Epoxy
Life stages

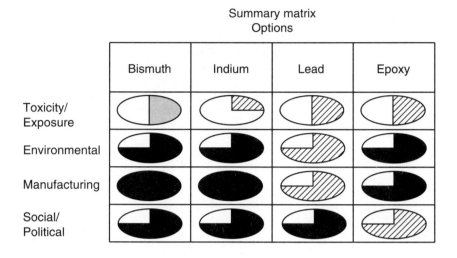

Summary matrix
Options

The Product Improvement Matrix

MATRIX ELEMENT CHECKLISTS AND EVALUATION FORMAT

The product improvement matrix is described in Chap. 20, where its relationship to the environmentally responsible product assessment matrix is explained. In this appendix, a sample of possible items appropriate to each of the matrix elements is presented. It is anticipated that different products will require different checklists and evaluations, so this appendix is presented as an example rather than as a universal formula.

Improvement Matrix Element: 1, 1
Life Stage: Resource Extraction
Environmental Concern: Materials Choice

- Are all materials the least toxic and most environmentally preferable for the function to be performed?
- Is the product designed to minimize the use of materials in restricted supply?
- Is the product designed to utilize recycled materials wherever possible?

Improvement Matrix Element: 1, 2
Life Stage: Resource Extraction
Environmental Concern: Energy Use

- Is the product designed to minimize the use of materials whose extraction is energy-intensive?

- Does the product design avoid using materials whose transport to the facility will require significant energy use?
- Does the product design avoid producing residues whose recycling will be energy-intensive?

Improvement Matrix Element: 1, 3
Life Stage: Resource Extraction
Environmental Concern: Solid Residues

- Is the product designed to minimize the use of materials whose extraction or purification involves the production of large amounts of solid residues?
- Does the product design avoid using materials whose transport to the facility will result in significant solid residues?

Improvement Matrix Element: 1, 4
Life Stage: Resource Extraction
Environmental Concern: Liquid Residues

- Is the product designed to minimize the use of materials whose extraction or purification involves the generation of large amounts of liquid residues?
- Does the product design avoid using materials whose transport to the facility will result in significant liquid residues?

Improvement Matrix Element: 1, 5
Life Stage: Resource Extraction
Environmental Concern: Gaseous Residues

- Is the product designed to minimize the use of materials whose extraction or purification involves the generation of large amounts of gaseous residues?
- Does the product design avoid using materials whose transport to the facility will result in significant gaseous residues?

Improvement Matrix Element: 2, 1
Life Stage: Product Manufacture
Environmental Concern: Materials Choice

- Is the product designed to avoid or minimize incorporating materials that are in restricted supply?
- Is the use of toxic materials avoided or minimized?
- Is the use of radioactive materials avoided or minimized?

Improvement Matrix Element: 2, 2
Life Stage: Product Manufacture
Environmental Concern: Energy Use

- Is the product designed to minimize the use of energy-intensive process steps such as high heating differentials, heavy motors, extensive cooling, and so on?
- Is the product designed to minimize the use of energy-intensive evaluation steps such as testing in a heated chamber?
- Do the processes use cogeneration, heat exchange, and other techniques for utilizing otherwise wasted energy?

Improvement Matrix Element: 2, 3
Life Stage: Product Manufacture
Environmental Concern: Solid Residues

- Have solid manufacturing residues (mold scrap, cutting scrap, etc.) been minimized and reused to the greatest extent possible?
- Has packaging material entering the facility from component suppliers been minimized, and does it use the fewest possible different materials?
- Do suppliers take back the packaging material in which their products enter the facility?

Improvement Matrix Element: 2, 4
Life Stage: Product Manufacture
Environmental Concern: Liquid Residues

- If solvents or oils are used in any manufacturing process in connection with this assessment, is their use minimized and have substitutes been investigated?
- Are liquid product residues designed for minimum toxicity and optimum reuse?
- Have the processes been designed to utilize the maximum amount of recycled liquid species from outside suppliers rather than virgin materials?

Improvement Matrix Element: 2, 5
Life Stage: Product Manufacture
Environmental Concern: Gaseous Residues

- If CFCs or HCFCs are used in any manufacturing process, have alternatives been thoroughly investigated?
- Are greenhouse gases used or generated in any manufacturing process in connection with this assessment?
- Are odorants used or generated in any manufacturing process in connection with this assessment?

Improvement Matrix Element: 3, 1
Life Stage: Product Packaging and Transport
Environmental Concern: Materials Choice

- Does the product packaging mimimize the number of different materials used?
- Does the product packaging avoid the use of toxic materials?
- Have efforts been made to use recyclable packaging materials?

Improvement Matrix Element: 3, 2
Life Stage: Product Packaging and Transport
Environmental Concern: Energy Use

- Does packaging avoid the use of materials whose extraction or processing is energy-intensive?
- Do packaging procedures avoid energy-intensive activities?
- Are product distribution plans designed to minimize energy use?

Improvement Matrix Element: 3, 3
Life Stage: Product Packaging and Transport
Environmental Concern: Solid Residues

- Has product packaging volume at all three levels (primary, secondary, and tertiary) been minimized?
- Is the product packaging designed to make it easy to separate the constituent materials?
- Are arrangements made to take back product packaging for recycling and reuse?

Improvement Matrix Element: 3, 4
Life Stage: Product Packaging and Transport
Environmental Concern: Liquid Residues

- Are refillable or reusable containers used for liquid products where appropriate?
- Does the product packaging contain any toxic or hazardous substances that might leach from it if improper disposal occurs?
- Are products, especially hazardous products or those potentially subject to spilling or venting, transported on safe routes by trained drivers?

Improvement Matrix Element: 3, 5
Life Stage: Product Packaging and Transport
Environmental Concern: Gaseous Residues

- Are product distribution plans designed to minimize gaseous emissions from transport vehicles?

- If the product contains pressurized gases, are installation procedures designed to avoid their release?
- If the packaging is recycled for its energy content (i.e., incinerated), will the incineration result in the release of any toxic materials?

Improvement Matrix Element: 4, 1
Life Stage: Product Use
Environmental Concern: Materials Choice

- If the product is designed to be disposed of upon using, has materials use been minimized and have alternative techniques for accomplishing the same purpose been examined?
- Do consumables contain any materials in restricted supply?
- Do consumables contain any toxic or otherwise undesirable materials?

Improvement Matrix Element: 4, 2
Life Stage: Product Use
Environmental Concern: Energy Use

- Has the product been designed to minimize energy use while in service?
- Have enhanced insulation or other energy-conserving design features been incorporated?
- Can the product monitor and display its energy use while in service?

Improvement Matrix Element: 4, 3
Life Stage: Product Use
Environmental Concern: Solid Residues

- Does use of this product require periodic disposal of solid materials such as cartridges, containers, or batteries?
- Have alternatives to the use of solid consumables been thoroughly investigated?
- Do intentional dissipative emissions to land occur as a result of using this product?

Improvement Matrix Element: 4, 4
Life Stage: Product Use
Environmental Concern: Liquid Residues

- Have alternatives to the use of liquid consumables been thoroughly investigated?
- Do intentional dissipative emissions to water occur as a result of using this product?
- Does this product contain any liquid materials that have the potential to be unintentionally dissipated during use?

Improvement Matrix Element: 4, 5
Life Stage: Product Use
Environmental Concern: Gaseous Residues

- Have alternatives to the use of gaseous consumables been thoroughly investigated?
- Do intentional dissipative emissions to air occur as a result of using this product?
- Does this product contain any gaseous materials that have the potential to be unintentionally dissipated during use?

Improvement Matrix Element: 5, 1
Life Stage: Recycling, Disposal
Environmental Concern: Materials Choice

- Does the product mimimize the number of different materials that are used in its manufacture?
- Does this product minimize the use of toxic materials?
- Are the different materials easy to identify and separate?

Improvement Matrix Element: 5, 2
Life Stage: Recycling, Disposal
Environmental Concern: Energy Use

- Is the product designed with the aim of minimizing the use of energy-intensive process steps in disassembly?
- Is the product designed for reuse of materials while retaining their embedded energy?
- Will transport of products for recycling be energy-intensive because of product weight or volume or the location of recycling facilities?

Improvement Matrix Element: 5, 3
Life Stage: Recycling, Disposal
Environmental Concern: Solid Residues

- Has the product been assembled with fasteners such as clips or hook-and-loop attachments rather than chemical bonds or welds?
- Have efforts been made to avoid joining dissimilar materials together in ways difficult to reverse?
- Are all plastic components identified by ISO markings as to their content?

Improvement Matrix Element: 5, 4
Life Stage: Recycling, Disposal
Environmental Concern: Liquid Residues

- Can liquids contained in the product be recovered at disassembly rather than lost?
- Does subassembly recovery and reuse generate liquid residues?
- Does materials recovery and reuse generate liquid residues?

Improvement Matrix Element: 5, 5
Life Stage: Recycling, Disposal
Environmental Concern: Gaseous Residues

- Can gases contained in the product be recovered at disassembly rather than lost?
- Does materials recovery and reuse generate gaseous residues?
- Can plastic parts be incinerated without requiring sophisticated scrubbing devices?

APPENDIX

H

The Process
Assessment Matrix

MATRIX ELEMENT CHECKLISTS AND EVALUATION FORMAT

The environmentally responsible process assesssment matrix is evaluated by checklists and protocols for each matrix element. In this appendix, a sample of possible items appropriate to each of the matrix elements is presented. Although considerable commonality of concerns and issues exists among all processes, it is anticipated that different processes will require the addition of unique considerations, so this appendix is presented as an example rather than as a universal formula.

Matrix Element: 1, 1
Life Stage: Resource Extraction
Environmental Concern: Materials Choice

- Is the process designed to avoid or minimize the use of consumable materials that are in restricted supply?
- Is the process designed to utilize recycled materials wherever possible?
- Of the potential consumable materials, are those chosen the ones whose extraction results in the lowest environmental impact?

Matrix Element: 1, 2
Life Stage: Resource Extraction
Environmental Concern: Energy Use

- Are the process consumables choices designed to minimize the use of materials whose extraction is energy-intensive?

- Does the process design avoid using consumable materials whose transport to the facility will require significant energy use?
- Does the process design avoid producing extraction residues whose recycling will be energy-intensive?

Matrix Element: 1, 3
Life Stage: Resource Extraction
Environmental Concern: Solid Residues

- Is the process designed to minimize the use of consumable materials whose extraction or purification involves the production of large amounts of solid residues?
- Does the process design avoid using consumable materials whose transport to the facility will result in significant solid residues?

Matrix Element: 1, 4
Life Stage: Resource Extraction
Environmental Concern: Liquid Residues

- Is the process designed to minimize the use of consumable materials whose extraction or purification involves the generation of large amounts of liquid residues?
- Does the process design avoid using consumable materials whose transport to the facility will result in significant liquid residues?

Matrix Element: 1, 5
Life Stage: Resource Extraction
Environmental Concern: Gaseous Residues

- Is the process designed to minimize the use of consumable materials whose extraction or purification involves the generation of large amounts of gaseous residues?
- Does the process design avoid using consumable materials whose transport to the facility will result in significant gaseous residues?

Matrix Element: 2, 1
Life Stage: Process Implementation
Environmental Concern: Materials Choice

- Is the process equipment designed to avoid or minimize incorporating materials that are in restricted supply?
- Does the process equipment minimize the number of materials used in its manufacture?
- Is the process equipment designed to utilize recycled materials wherever possible?

Matrix Element: 2, 2
Life Stage: Process Implementation
Environmental Concern: Energy Use

- Is the process equipment manufacture designed to minimize the use of energy-intensive process steps and activities such as high heating differentials, heavy motors, extensive cooling, and so on?
- Is the process equipment manufacture designed to minimize the use of energy-intensive evaluation steps such as testing in a heated chamber?
- Does the process equipment incorporate energy-efficient design elements such as variable-speed motors?

Matrix Element: 2, 3
Life Stage: Process Implementation
Environmental Concern: Solid Residues

- Has process equipment manufacture been designed so as to minimize and reuse solid manufacturing residues?
- Have solid manufacturing residues (mold scrap, cutting scrap, and so on) been minimized and reused to the greatest extent possible during process equipment manufacture?
- Have efforts been made to avoid joining dissimilar materials together during process equipment construction in ways difficult to reverse?

Matrix Element: 2, 4
Life Stage: Process Implementation
Environmental Concern: Liquid Residues

- Has process equipment manufacture been designed so as to minimize and reuse liquid manufacturing residues?
- Does process equipment maintenance require the dissipative use of solvents or lubricants?
- Does the process equipment packaging during shipment contain any toxic or hazardous substances that might leach from it if improper disposal occurs?

Matrix Element: 2, 5
Life Stage: Process Implementation
Environmental Concern: Gaseous Residues

- Has process equipment manufacture been designed so as to minimize and reuse gaseous manufacturing residues?
- If CFCs or HCFCs are likely to be used in any process equipment manufacturing process, have alternatives been thoroughly investigated?

- During eventual recycling of the process equipment, can plastic parts be incinerated for energy recovery without requiring sophisticated scrubbing devices?

Matrix Element: 3, 1
Life Stage: Process Operation
Environmental Concern: Materials Choice

- Is the use of toxic consumable materials avoided or minimized?
- Is the use of radioactive consumable materials avoided or minimized?
- Is the process designed to avoid the use of large amounts of water?

Matrix Element: 3, 2
Life Stage: Process Operation
Environmental Concern: Energy Use

- Is the process designed to minimize the use of energy-intensive process steps such as high heating differentials, heavy motors, extensive cooling, and so on?
- Is the process designed to minimize the use of energy-intensive evaluation steps such as testing in a heated chamber?
- Do the processes use cogeneration, heat exchange, and other techniques for utilizing otherwise waste energy?

Matrix Element: 3, 3
Life Stage: Process Operation
Environmental Concern: Solid Residues

- Have solid manufacturing residues (mold scrap, cutting scrap, and so on) been minimized and reused to the greatest extent possible?
- Have opportunities for sale of all solid residues as inputs into the products and processes of others been investigated, and modifications made to residues (if possible and necessary) to facilitate such transactions?
- Has packaging material entering the facility from suppliers been minimized, and does it use the fewest possible different materials?
- Do suppliers take back the packaging material in which their consumables enter the facility?

Matrix Element: 3, 4
Life Stage: Process Operation
Environmental Concern: Liquid Residues

- If solvents or oils are used in the manufacturing process, is their use minimized and have substitutes been investigated?

- Have opportunities for sale of all liquid residues as inputs into the products and processes of others been investigated, and modifications made to residues (if possible and necessary) to facilitate such transactions?
- Has the process been designed to utilize the maximum amount of recycled liquid species rather than virgin materials?

Matrix Element: 3, 5
Life Stage: Process Operation
Environmental Concern: Gaseous Residues

- If CFCs or HCFCs are used in the process, have alternatives been thoroughly investigated?
- Are greenhouse gases used or generated in the process?
- Have opportunities for sale of all gaseous residues as inputs into the products and processes of others been investigated, and modifications made to residues (if possible and necessary) to facilitate such transactions?

Matrix Element: 4, 1
Life Stage: Complementary Process Implications
Environmental Concern: Materials Choice

- Are complementary processes required to use materials that are in restricted supply?
- Are complementary processes required to use toxic materials?
- Are complementary processes required to use radioactive materials?

Matrix Element 4, 2
Life Stage: Complementary Process Implications
Environmental Concern: Energy Use

- Are complementary processes required to use energy-intensive process steps such as high heating differentials, heavy motors, extensive cooling, and so on?
- Are complementary processes required to use energy-intensive evaluation steps such as testing in a heated chamber?
- Can the primary and complementary processes share cogeneration, heat exchange, and other techniques for utilizing otherwise waste energy?

Matrix Element: 4, 3
Life Stage: Complementary Process Implications
Environmental Concern: Solid Residues

- Can solid manufacturing residues (mold scrap, cutting scrap, and so on) from complementary processes be minimized and reused?

- Does the process facilitate the use or sale of residues from complementary processes?
- Has packaging material entering the facility from complementary processes suppliers been minimized, and does it use the fewest possible different materials?

Matrix Element: 4, 4
Life Stage: Complementary Process Implications
Environmental Concern: Liquid Residues

- If solvents or oils are used in any complementary process, is their use minimized and have substitutes been investigated?
- Have opportunities for sale of all complementary process liquid residues as inputs into primary products and processes been investigated, and modifications made to residues (if possible and necessary) to facilitate such transactions?
- Is the use of recycled liquids in complementary processes precluded by primary process constraints?

Matrix Element: 4, 5
Life Stage: Complementary Process Implications
Environmental Concern: Gaseous Residues

- Are CFCs or HCFCs required in any complementary process?
- Do primary process constraints require the generation of greenhouse gases in any complementary process?
- Have opportunities for sale of all complementary process gaseous residues as inputs into products and processes been investigated, and modifications made to residues (if possible and necessary) to facilitate such transactions?

Matrix Element: 5, 1
Life Stage: Recycling, Disposal
Environmental Concern: Materials Choice

- Have consumable materials been chosen and used in light of the desired recycling/disposal option for the process (e.g., for incineration, for recycling, for refurbishment)?
- Does the process minimize the number of different materials that are used in its operation?
- Are the different material residues easy to identify and separate?

Matrix Element: 5, 2
Life Stage: Recycling, Disposal
Environmental Concern: Energy Use

- Is the process designed with the aim of minimizing the use of energy-intensive process steps in recycling consumable residues?
- Is the process designed for high-level reuse of consumable materials? (Direct reuse in a similar process is preferable to a degraded reuse.)
- Will transport of residues for recycling be energy-intensive because of weight or volume or the location of recycling facilities?

Matrix Element: 5, 3
Life Stage: Recycling, Disposal
Environmental Concern: Solid Residues

- Do solid residues from the process contain toxic materials?
- Do solid residues from the process contain radioactive materials?
- Does the process minimize the rate of generation of solid residues?

Matrix Element 5, 4
Life Stage: Recycling, Disposal
Environmental Concern: Liquid Residues

- Do liquid residues from the process contain toxic materials?
- Do liquid residues from the process contain radioactive materials?
- Is the process designed to minimize the generation of liquid residues, and to generate only liquid residues that are readily recyclable?

Matrix Element: 5, 5
Life Stage: Recycling, Disposal
Environmental Concern: Gaseous Residues

- Do gaseous residues from the process contain toxic materials?
- Do gaseous residues from the process contain radioactive materials?
- Is the process designed to minimize the generation of gaseous residues, and to generate only gaseous residues that are readily recyclable?

APPENDIX

I

The Facility Operations Matrix

MATRIX ELEMENT CHECKLISTS AND EVALUATION FORMAT

The facility operations matrix is described in Chap. 20, where its relationship to the underlying checklists and protocols is explained. In this appendix, a sample of possible items appropriate to each of the matrix elements is presented. It is anticipated that different types of facilities will require different checklists and evaluations, so this appendix is presented as an example rather than as a universal formula.

Facility Operations Matrix Element: 1, 1
Facility Activity: Site Selection and Development
Environmental Concern: Ecological Impacts

- Has the proposed site previously been used for similar activities? If not, have any such sites been surveyed for availability?
- Is necessary development activity, if any, planned to avoid disruption of existing biological communities?
- Is the biota of the site compatible with all planned process emissions, including possible exceedances?

Facility Operations Matrix Element: 1, 2
Facility Activity: Site Selection and Development
Environmental Concern: Energy Use

- Is the site such that it can be made operational with only minimal energy expenditures?

- Has the site been selected so as to avoid any energy emission impacts on existing biota?
- Does the site allow delivery and installation of construction or renovation materials with minimal use of energy?

Facility Operations Matrix Element: 1, 3
Facility Activity: Site Selection and Development
Environmental Concern: Solid Residues

- Is the site such that it can be made operational with only minimal production of solid residues?
- Have plans been made to ensure that any solid residues generated in the process of developing the site are managed so as to minimize their impacts on biota and human health?
- If any solid residues generated in the process of developing the site may be hazardous or toxic to biota or humans, have plans been made to minimize releases and exposures?

Facility Operations Matrix Element: 1, 4
Facility Activity: Site Selection and Development
Environmental Concern: Liquid Residues

- Is the site such that it can be made operational with only minimal production of liquid residues?
- Have plans been made to ensure that any liquid residues generated in the process of developing the site are managed so as to minimize their impacts on biota and human health?
- If any liquid residues generated in the process of developing the site may be hazardous or toxic to biota or humans, have plans been made to minimize releases and exposures?

Facility Operations Matrix Element: 1, 5
Facility Activity: Site Selection and Development
Environmental Concern: Gaseous Residues

- Is the site such that it can be made operational with only minimal production of gaseous residues?
- Have plans been made to ensure that any gaseous residues generated in the process of developing the site are managed so as to minimize their impacts on biota and human health?
- If any gaseous residues generated in the process of developing the site may be hazardous or toxic to biota or humans, have plans been made to minimize releases and exposures?

Facility Operations Matrix Element: 2, 1
Facility Activity: Infrastructure Interactions
Environmental Concern: Ecological Impacts

- Has the site been chosen to minimize the need for new on-site infrastructure (buildings, roads, and so on)?
- If new infrastructure must be created, are plans in place to minimize any resultant impacts on biota?
- Have provisions been made for orderly growth of infrastructure as facility operations expand in order to avoid unnecessary health or environmental impacts?

Facility Operations Matrix Element: 2, 2
Facility Activity: Infrastructure Interactions
Environmental Concern: Energy Use

- Does the existing energy infrastructure (gas pipelines, electric power cables) reduce or eliminate the need to build new systems?
- Is it possible to use heat residues from within the plant or from nearby facilities owned by others to provide heat and power?
- Is it possible to use gaseous residues from within the plant or from nearby facilities owned by others to provide heat or power?

Facility Operations Matrix Element: 2, 3
Facility Activity: Infrastructure Interactions
Environmental Concern: Solid Residues

- Is it possible to use as feedstocks solid residues from nearby facilities owned by others?
- Is it possible to use solid residues from the proposed facility as feedstocks for nearby facilities owned by others?
- Can solid residue transport and disposal operations be shared with nearby facilities owned by others?

Facility Operations Matrix Element: 2, 4
Facility Activity: Infrastructure Interactions
Environmental Concern: Liquid Residues

- Is it possible to use as feedstocks liquid residues from nearby facilities owned by others?
- Is it possible to use liquid residues from the proposed facility as feedstocks for nearby facilities owned by others?
- Can liquid residue transport and disposal operations be shared with nearby facilities owned by others?

Facility Operations Matrix Element: 2, 5
Facility Activity: Infrastructure Interactions
Environmental Concern: Gaseous Residues

- Is it possible to use gaseous residues from the proposed facility to provide heat or power for nearby facilities owned by others?
- Is it possible to use gaseous residues from the proposed facility to provide process or product feedstocks for nearby facilities owned by others?
- Is it possible to share employee transportation infrastructure with nearby facilities owned by others to minimize air pollution by private vehicles?

Facility Operations Matrix Element: 3, 1
Facility Activity: Principal Business Activity—Products
Environmental Concern: Ecological Impacts

- If the activity at this site involves extraction of virgin materials, is the extraction planned so as to minimize ecological impacts?
- Do all outputs from the site have high ratings as environmentally responsible products?
- Are products designed to use recycled materials?

Facility Operations Matrix Element: 3, 2
Facility Activity: Principal Business Activity—Products
Environmental Concern: Energy Use

- Are products designed to require minimal consumption of energy in manufacture?
- Are products designed to require minimal consumption of energy in use?
- Are products designed to require minimal consumption of energy in recycling or disposal?

Facility Operations Matrix Element: 3, 3
Facility Activity: Principal Business Activity—Products
Environmental Concern: Solid Residues

- Are products designed to generate minimal and nontoxic solid residues in manufacture?
- Are products designed to generate minimal and nontoxic solid residues in use?
- Are products designed to generate minimal and nontoxic solid residues in recycling or disposal?

Facility Operations Matrix Element: 3, 4
Facility Activity: Principal Business Activity—Products
Environmental Concern: Liquid Residues

- Are products designed to generate minimal and nontoxic liquid residues in manufacture?
- Are products designed to generate minimal and nontoxic liquid residues in use?
- Are products designed to generate minimal and nontoxic liquid residues in recycling or disposal?

Facility Operations Matrix Element: 3, 5
Facility Activity: Principal Business Activity—Products
Environmental Concern: Gaseous Residues

- Are products designed to generate minimal and nontoxic gaseous residues in manufacture?
- Are products designed to generate minimal and nontoxic gaseous residues in use?
- Are products designed to generate minimal and nontoxic gaseous residues in recycling or disposal?

Facility Operations Matrix Element: 4, 1
Facility Activity: Principal Business Activity—Processes
Environmental Concern: Ecological Impacts

- Have all process materials been optimized from a design for environment standpoint?
- Have processes been dematerialized (evaluated to ensure that they have minimum resource requirements and that no unnecessary process steps are required)?
- Do processes generate waste heat or emission of residues that have the potential to harm local or regional biological communities?

Facility Operations Matrix Element: 4, 2
Facility Activity: Principal Business Activity—Processes
Environmental Concern: Energy Use

- Have all processes been evaluated to ensure that they use as little energy as possible?
- Are processes monitored and maintained on a regular basis to ensure that they retain their energy efficiency as designed?
- Do process equipment specifications and standards require the use of energy efficient components and subassemblies?

Facility Operations Matrix Element: 4, 3
Facility Activity: Principal Business Activity—Processes
Environmental Concern: Solid Residues

- Are processes designed to generate minimal and nontoxic solid residues?
- Where solid materials are used as process inputs, have attempts been made to use recycled materials?
- Are processes designed to produce usable byproducts, rather than byproducts suitable only for disposal?

Facility Operations Matrix Element: 4, 4
Facility Activity: Principal Business Activity—Processes
Environmental Concern: Liquid Residues

- Are processes designed to generate minimal and nontoxic liquid residues?
- Where liquid materials are used as process inputs, have attempts been made to use recycled materials?
- Are pumps, valves, and pipes inspected regularly to minimize leaks?

Facility Operations Matrix Element: 4, 5
Facility Activity: Principal Business Activity—Processes
Environmental Concern: Gaseous Residues

- Are processes designed to generate minimal and nontoxic gaseous residues?
- Are processes designed to avoid the production and release of odorants?
- If VOCs are utilized in any processes, are they selected so that any releases will have minimal photochemical smog impact?

Facility Operations Matrix Element: 5, 1
Facility Activity: Facility Operations
Environmental Concern: Ecological Impacts

- Is the maximum possible portion of the facility allowed to remain in its natural state?
- Is the use of pesticides and herbicides on the property minimized?
- Is noise pollution from the site minimized?

Facility Operations Matrix Element: 5, 2
Facility Activity: Facility Operations
Environmental Concern: Energy Use

- Is the energy needed for heating and cooling the buildings minimized?
- Is the energy needed for lighting the buildings minimized?

- Is energy efficiency a consideration when buying or leasing facility equipment: copiers, computers, fan motors, and so on?

Facility Operations Matrix Element: 5, 3
Facility Activity: Facility Operations
Environmental Concern: Solid Residues

- Is the facility designed to minimize the comingling of solid waste streams?
- Are solid residues from facility operations reused or recycled to the extent possible?
- Are unusable solid residues from facility operations (including food service) disposed of in an environmentally responsible manner?

Facility Operations Matrix Element: 5, 4
Facility Activity: Facility Operations
Environmental Concern: Liquid Residues

- Is the facility designed to minimize the comingling of liquid waste streams?
- Are liquid treatment plants monitored to ensure that they operate at peak efficiency?
- Are unusable liquid residues from facility operations disposed of in an environmentally responsible manner?

Facility Operations Matrix Element: 5, 5
Facility Activity: Facility Operations
Environmental Concern: Gaseous Residues

- Is facility operations-related transportation to and from the facility minimized?
- Are furnaces, incinerators, and other combustion processes and their related air pollution control devices monitored to ensure operation at peak efficiency?
- Is employee commuting minimized by job sharing, telecommuting, and similar programs?

APPENDIX
J

Chemical Formulas Commonly Used in This Book

AsH_3	Arsine
$CBrCl_2F$	Bromodichlorofluoromethane
$CBrF_3$	Bromotrifluoromethane
$CClF_2CCl_2F$	1,2,2-Trichloro-1,1,2-trifluoroethane
CCl_2F_2	Dichlorodifluoromethane
CCl_3CH_3	1,1,1-Trichloroethane (methyl chloroform)
CCl_3F	Trichlorofluoromethane
CCl_4	Carbon tetrachloride
CFC-11 (see CCl_3F)	
CFC-12 (see CCl_2F_2)	
CFC-113 (see $CClF_2CCl_2F$)	
CH_3Br	Methyl bromide
CH_3Cl	Methyl chloride
CH_4	Methane
CO	Carbon monoxide
CO_2	Carbon dioxide
HCl	Hydrogen chloride
H_2S	Hyrogen sulfide
NH_3	Ammonia
NMHC	Nonmethane hydrocarbons (the sum of all except methane)
NO	Nitric oxide
NO_2	Nitrogen dioxide
NO_x	Oxides of nitrogen (the sum of NO and NO_2)
N_2O	Nitrous oxide
O_3	Ozone
SO_2	Sulfur dioxide

APPENDIX K

Units of Measurement in Industrial Ecology

The basic unit of energy is the *joule* ($= 1 \times 10^7$ erg). One will often see the use of the British Thermal Unit (Btu), which is 1.55×10^3 J. For very large energy use, a unit named the *quad* is common; it is shorthand notation for one quadrillion British Thermal Units. Thus, 1 quad = 1×10^{15} Btu = 1.55×10^{18} J.

The units of mass in the environmental sciences and in this book are given in the metric system. Because many of the quantities are large, the prefixes given in Table K.1 are common. Hence, we have such figures as 2 Pg = 2×10^{15} g. Where the word tonne is used, it refers to the metric ton = 1×10^6 g.

The most common way of expressing the abundance of a gas-phase atmospheric species is as a fraction of the number of molecules in a sample of air. The units in common use are *parts per million* (ppm), *parts per billion* (thousand million; ppb), and *parts per trillion* (million million; ppt), all expressed as volume fractions and therefore abbreviated ppmv, ppbv, and pptv to make it clear that one is not speaking of fractions in mass. Any of these units can be called the *volume mixing ratio* or *mole fraction*. Mass mixing ratios can be used as well (hence, ppmm, ppbm, pptm), a common example being that meteorologists use mass mixing ratios for water vapor. Because the pressure of the atmosphere changes with altitude and the partial pressures of all the gaseous constituents in a moving air parcel change in the same proportions, mixing ratios are preserved as long as mixing between air parcels can be neglected.

Particles can be mixtures of solid and liquid, so a measure based on mass replaces that based on volume, the usual units for atmospheric particles being micrograms per cubic meter ($\mu g/m^3$) or nanograms per cubic meter (ng/m^3). For particles in liquids, micrograms per cubic centimeter ($\mu g/cm^3$) is common. It is sometimes convenient to

Table K.1   Prefixes for Large and Small Numbers

Power of 10	Prefix	Symbol
+24	yotta	Y
+21	zetta	Z
+18	exa	E
+15	peta	P
+12	tera	T
+9	giga	G
+6	mega	M
+3	kilo	k
−3	milli	m
−6	micro	μ
−9	nano	n
−12	pico	p
−15	femto	f
−18	atto	a
−21	zepto	z
−24	yocto	y

compare quantities of an element or compound present in more than one phase, say, as both a gas and a particle constituent. In that case, the gas concentration in volume units is converted to mass units prior to making the comparison.

For constituents present in aqueous solution, as in seawater, the convention is to express concentration in volume units of moles per liter (designated M) or some derivative thereof [1 mole (abbreviated mol) is 6.02×10^{23} molecules]. Common concentration expressions in environmental chemistry are millimoles per liter (mM), micromoles per liter (μM), and nanomoles per liter (nM). Sometimes one is concerned with the "combining concentration" of a species rather than the absolute concentration. A combining concentration, called an *equivalent*, is that concentration that will react with 8 grams of oxygen or its equivalent. For example, 1 mole of hydrogen ions is one equivalent of H^+, but 1 mole of calcium ions is two equivalents of Ca^{2+}. Combining concentrations have typical units of equivalents, milliequivalents, or microequivalents per liter, abbreviated eq/l, meq/l, μeq/l, respectively. A third approach is to express concentration by weight, as mg/l or ppmw, for example. Concentration by weight can be converted to concentration by volume using the molecular weight as a conversion factor.

Acidity in solution is expressed in pH units, pH being defined as the negative of the logarithm of the hydrogen ion concentration in moles per liter. In aqueous solutions, pH = 7 is neutral at 25° C; lower pH values are characteristic of acidic solutions; higher values are characteristic of basic solutions.

Glossary*

Abridged life-cycle assessment (ALCA)—A simplified methodology to evaluate the environmental effects of a product or activity holistically, by analyzing the most significant environmental impacts in the life cycle of a particular product, process, or activity. The abridged life-cycle assessment consists of three complementary components, restricted inventory analysis, abridged impact assessment, and improvement analysis, together with an integrative procedure known as scoping.

Acid deposition—The deposition of acidic constituents to a surface. This occurs not only by precipitation, but also by the deposition of atmospheric particulate matter and the incorporation of soluble gases.

Acute—In toxicology, an effect that is manifested rapidly (e.g., minutes, hours, or even a few days) after exposure to a hazard; either the exposure that generates the response, or the response itself, can be called acute (compare with *Chronic*). For example, an oral poison such as cyanide would be acutely toxic.

Anthropogenic—Derived from human activities.

Aquifer—Any water-bearing rock formation or group of formations, especially one that supplies ground water, wells, or springs.

Bioaccumulation—The concentration of a substance by an organism above the levels at which that substance is present in the ambient environment. Some forms of heavy metals, and chorinated pesticides such as DDT, are bioaccumulated.

Biomagnification—The increasing concentration of a substance as it passes into higher trophic levels of a food web. Many substances that are bioaccumulated are also biomagnified.

*This glossary was compiled from various sources, including particularly: T. E. Graedel and P. J. Crutzen, *Atmospheric Change: An Earth System Perspective*, New York: W. H. Freeman, 1993; B. W. Vigon, D. A. Tolle, B. W. Cornaby, H. C Latham, C L. Harrison, T. L. Boguski, R. G. Hunt, and J. D. Sellers, *Life Cycle Assessment: Inventory Guidelines and Practices*, EPA/600/R-92/245, Washington, DC: U.S. Environmental Protection Agency, 1993.

Biosphere—That spherical shell encompassing all forms of life on Earth. The biosphere extends from the ocean depths to a few thousand meters of altitude in the atmosphere, and includes the surface of land masses. Alternatively, the life forms within that shell.

Budget—A balance sheet of all of the sources and sinks for a particular species or group of species in a single reservoir or in two or more connected reservoirs.

By-product—A useful product that is not the primary product being produced. In life-cycle analysis, by-products are treated as co-products.

Carcinogen—A material that causes cancer.

Cascade recycling—See *Open-loop recycling*

Chronic—In toxicology, an exposure or effect of an exposure that becomes manifest only after a significant amount of time—weeks, months, or even years—has passed. Many carcinogens (substances causing cancer) are chronic toxins, and low-level exposure to many heavy metals, such as lead, produces chronic, rather than acute, effects.

Climate—The temperature, humidity, precipitation, winds, radiation, and other meteorological conditions characteristic of a locality or region over an extended period of time.

Closed-loop recycling—A recycling system in which a particular mass of material is remanufactured into the same product (e.g., glass bottle into glass bottle). Also known as *Horizontal recycling*.

Co-product—A marketable by-product from a process. This includes materials that may be traditionally defined as wastes such as industrial scrap that is subsequently used as a raw material in a different manufacturing process.

Cycle—A system consisting of two or more connected *Reservoirs*, where a large part of the material of interest is transferred through the system in a cyclic manner.

Design for environment—An engineering perspective in which the environmentally related characteristics of a product, process, or facility design are optimized.

Discount rate—A rate applied to future financial returns to reflect the time value of money and inflation.

Dose-response curve—A curve plotting the known dose of a material administered to organisms against the percentage response of the test population. If the material is not directly administered, but is present in the environment surrounding the organism (e.g., water, air, sediment), the resulting curve is known as a "concentration-response curve".

Expert System—A computer-based system combining knowledge in the form of facts and rules with a reasoning strategy specifying how the facts and rules are to be used to reach conclusions. The system is designed to emulate the performance of a human expert or a group of experts in arriving at solutions to complex and not completely specified problems.

Exposure—Contact between a *Hazard* and the target of concern, which may be an organ, an individual, a population, a biological community, or some other system. The confluence of exposure and hazard gives rise to risk.

Flux—The rate of emission, absorption, or deposition of a substance from one *Reservoir* to another. Often expressed as the rate per unit area of surface.

Fossil fuel—A general term for combustible geological deposits of carbon in reduced (organic) form and of biological origin, including coal, oil, natural gas, oil shales, and tar sands.

Fugitive emissions—Emissions from valves or leaks in process equipment or material storage areas that are difficult to measure and do not flow through pollution-control devices.

Full-cost accounting—An accounting system in which environmental costs are built directly into the prices of products and services.

Global warming—The theory that elevated concentrations of certain anthropogenic atmospheric constituents are causing or will cause an increase in Earth's average temperature.

Green accounting—An informal term referring to management accounting systems that specifically delineate the environmental costs of business activities rather than including those costs in overhead accounts.

Greenhouse effect—The trapping by atmospheric gases of outgoing infrared energy emitted by Earth. Part of the radiation absorbed by the atmosphere is returned to Earth's surface, causing it to warm.

Greenhouse gas—A gas with absorption bands in the infrared portion of the spectrum. The principal greenhouse gases in Earth's atmosphere are H_2O, CO_2, O_3, CH_4, N_2O, CF_2Cl_2, and $CFCl_3$.

Hazard—A material or condition that may cause cause damage, injury, or other harm, frequently established through standardized assays performed on biological systems or organisms. The confluence of hazard and exposure create a *Risk*.

Heuristics—The intuitive rules used by human experts in arriving at decisions. In expert systems, representations of heuristics are combined with algorithms (precise definitions and procedures) to reach conclusions concerning problems whose characteristics cannot be completely specified.

Home scrap—The waste produced within a fabricating plant, such as rejected material, trimmings, and shearings. Home scrap is recirculated within the fabricating plant and does not become external waste.

Horizontal recycling—See *Closed-loop recycling*.

Impact analysis—The second stage of life-cycle assessment, in which the environmental impacts of a process, product, or facility are determined.

Improvement analysis—The third stage of life-cycle assessment, in which design for environment techniques are used in combination with the results of the first and second LCA stages to improve the environmental plan of a process, product, or facility.

Industrial ecology—An approach to the design of industrial products and processes that evaluates such activities through the dual perspectives of product competitiveness and environmental interactions.

Inventory analysis—The first stage of life-cycle assessment, in which the inputs and outputs of materials and energy are determined for a process, product, or facility.

Irreversible disassembly—Disassembly in which brute force is used to recover the bulk of the principal materials from a product, and in which no refurbishment and reuse of components or modules is possible.

LC50—The concentration of a hazard in the ambient environment that is anticipated to be lethal to 50% of a test population.

LD50—The dose of a hazard that is anticipated to be lethal to 50% of a test population. LD50 differs from LC50 in that it refers to a direct dose to the organism, as opposed to a concentration in the ambient environment (LC50); in the latter case, the actual dose received by the organism may not be known.

Leachate—The solution that is produced by the action of percolating water through a permeable solid, as in a landfill.

Life cycle—The stages of a product, process, or package's life, beginning with raw materials acquisition, continuing through processing, materials manufacture, product fabrication, and use, and concluding with any of a variety of waste-management options.

Life-cycle assessment—A concept and a methodology to evaluate the environmental effects of a product or activity holistically, by analyzing the entire life cycle of a particular material, process, product, technology, service, or activity. The life-cycle assessment consists of three complementary components—inventory analysis, impact analysis, and improvement analysis—together with an integrative procedure known as scoping.

Mutagen—A hazard that can cause inheritable changes in DNA.

Neurotoxin—A hazard that can cause damage to nerve cells or the nervous system.

New scrap—See *Prompt scrap.*

Nonpoint source—See *Source.*

NO$_x$—The sum of the common pollutant gases NO and NO$_2$.

Old scrap—See *Postconsumer solid waste.*

Open-loop recycling—A recycling system in which a product from one type of material is recycled into a different type of product (e.g., plastic bottles into fence posts). The product receiving the recycled material itself may or may not be recycled. (Also known as *Cascade recycling.*)

Overburden—The material to be removed or displaced that is overlying the ore or material to be mined.

Ozone depletion—The reduction in concentration of stratospheric ozone as a consequence of efficient chemical reactions with molecular fragments derived from anthropogenic compounds, especially CFCs and other halocarbons.

Packaging, primary—The level of packaging that is in contact with the product. For certain beverages, an example is the aluminum can.

Packaging, secondary—The second level of packaging for a product that contains one or more primary packages. An example is the plastic rings that hold several beverage cans together.

Packaging, tertiary—The third level of packaging for a product that contains one or more secondary packages. An example is the stretch wrap over the pallet used to transport packs of beverage cans.

Point source—See *Source.*

Postconsumer scrap—See *Postconsumer solid waste.*

Postconsumer solid waste—A material that has served its intended use and has become a part of the waste stream. (Also called *Old scrap* and *Postconsumer scrap.*)

Prompt scrap—Waste produced by users of semifinished products (turnings, trimmings, etc.). This scrap must generally be returned to the materials processor if it is to be recycled. (Also called *New scrap.*)

Reservoir—A receptacle defined by characteristic physical, chemical, or biological properties that are relatively uniformly distributed.

Reversible disassembly—Reverse manufacturing, in which the removal of screws, clips, and other fasteners permits refurbishment and reuse of some or all of the components and modules of a product.

Risk—The confluence of exposure and hazard; a statistical concept reflecting the probability that an undesirable outcome will result from specified conditions (such as exposure to a certain substance for a certain time at a certain concentration).

Risk assessment—An evaluation of potential consequences to humans, wildlife, or the environment caused by a process, product, or activity, and including both the likelihood and the effects of an event.

Smog—Classically, a mixture of smoke plus fog. Today the term *smog* has the more general meaning of any anthropogenic haze. Photochemical smog involves the production, in stagnant, sunlit atmospheres, of oxidants such as O_3 by the photolysis of NO_2 and other substances, generally in combination with haze-causing particles.

Solute—The substance dissolved in a *Solvent*.

Solution—A mixture in which the components are uniformly distributed on an atomic or molecular scale. Although liquid, solid, and gaseous solutions exist, common nomenclature implies the liquid phase unless otherwise specified.

Solvent—A medium, usually liquid, in which other substances can be dissolved.

Source—In environmental chemistry, the process or origin from which a substance is injected into a reservoir. *Point sources* are those where an identifiable source, such as a smokestack, can be identified. *Nonpoint sources* are those resulting from diffuse emissions over a large geographical area, such as pesticides entering a river as runoff from agricultural lands.

Stratosphere—The atmospheric shell lying just above the *troposphere* and characterized by a stable lapse rate. The temperature is approximately constant in the lower part of the stratosphere and increases from about 20 km to the top of the stratosphere at about 50 km.

Stressor—A set of conditions that may lead to an undesirable environmental impact.

Teratogen—A hazard that can cause birth defects.

Troposphere—The lowest layer of the atmosphere, ranging from the ground to the base of the stratosphere at 10–15 km altitude, depending on latitude and weather conditions. About 85% of the mass of the atmosphere is in the troposphere, where most weather features occur. Because its temperature decreases with altitude, the troposphere is dynamically unstable.

Visibility—The degree to which the atmosphere is transparent to light in the visible spectrum, or the degree to which the form, color, and texture of objects can be perceived. In the sense of visual range, visibility is the distance at which a large black object just disappears from view as a recognizable entity.

Index

DATE DUE

APR 25 1998			
MAY 0 1 2001			